SYSTEMS ANALYSIS
AND DESIGN
USING
NETWORK
TECHNIQUES

PRENTICE-HALL INTERNATIONAL SERIES
IN INDUSTRIAL AND SYSTEMS ENGINEERING

W. J. Fabrycky, and J. H. Mize, Editors

FABRYCKY, GHARE, AND TORGERSEN   *Industrial Operations Research*
GOTTFRIED AND WEISMAN   *Introduction to Optimization Theory*
MIZE, WHITE, AND BROOKS   *Operations Planning and Control*
WHITEHOUSE   *Systems Analysis and Design Using Network Techniques*

SYSTEMS ANALYSIS AND DESIGN USING NETWORK TECHNIQUES

GARY E. WHITEHOUSE

Department of Industrial Engineering
Lehigh University

PRENTICE-HALL, INC., *Englewood Cliffs, New Jersey*

Library of Congress Cataloging in Publication Data

WHITEHOUSE, GARY E
    Systems analysis and design using network techniques.

(Prentice-Hall international series in industrial
and systems engineering)
    Includes bibliographies.
    1. Network analysis (Planning)   2. System analysis.
    I. Title.
T57.85.W48   658.4′032   72–12928
ISBN 0–13–881474–0

10 9 8 7 6 5 4 3 2 1

Printed in the United States of America

PRENTICE-HALL, INC.
PRENTICE-HALL INTERNATIONAL, UNITED KINGDOM AND EIRE
PRENTICE-HALL OF CANADA, LTD. CANADA

To those who have influenced me most

My wife
My parents
Gail and Glenn
Alan Pritsker
Art Gould

CONTENTS

PREFACE

In this book I have tried to demonstrate the advantages of using network modeling techniques for the analysis of systems. Networks and network analysis are playing an increasingly important role in the description and improvement of operational systems. The ease with which systems can be modeled in network form is the fundamental reason for the significant increase in the use of networks. The network techniques discussed include PERT, CPM, decision trees, network flows, flowgraphs, and GERT. Particular emphasis is placed on the development and use of GERT: Graphical Evaluation and Review Technique.

While the material in this book lends itself to an advanced undergraduate or a graduate course in schools of business, industrial engineering, operations research, or computer science, I believe the book should be of particular interest to the practicing engineer and systems analyst. An elementary knowledge of probability theory is assumed in the development of the text. Chapter 2 presents a review of the probability theory which seems necessary for the understanding of the material presented.

Computer systems for the solution of some of the network modeling techniques are available. The computer is not necessary for the application of the techniques presented in this book, but it does help when analyzing larger systems. The programs for GERT and GERTS can be obtained by sending a magnetic tape and $25.00 (to cover handling and mailing) to:

Lehigh Computing Center   attn: Librarian
Packard Laboratory
Lehigh University
Bethlehem, Pa. 18015

Acknowledgment

I am greatly indebted to many people with whom I have been associated in the past few years. The greatest influence comes from my good friend A. Alan B. Pritsker, who was the developer of GERT and my dissertation advisor. I want to thank the countless faculty members and students from Lehigh and elsewhere who have taken a direct interest in the development of this work. My boss, Arthur Gould, gave me both moral and financial support while I was writing the draft of the book. I was supported by funds from the Alcoa Foundation during the writing of this book, and I wish to thank Arthur Doty, President of Alcoa Foundation, for his interest in this work. I would like to give particular thanks to those who have given me permission to use adaption of their work. The list includes: Roland Burgess, Salah Elmaghraby, Ron Enlow, Jim Fry, R. F. Hespos, Tom Hill, Phil Ishmael, Don McIlvain, Alan Pritsker, Lou Riccio, Carl Roth, Bill Thompson, Gary Wereberger, AIIE, APICS, TIMS, ORSA, *Technometrics*, and Addison-Wesley Publishing Co., Inc. I would also like to thank Lou Riccio, Don McIlvain, Larry Israel, Lynn Hott, John Peek, Wes Gewehr, and Jerry Goodrich for their valuable comments on earlier versions of this manuscript. I want to thank Joe Mize for his thoughtful and helpful comments on the draft of this text. Merl Miller and Virginia Huebner, my editors, have also been very helpful. I wish to thank Faith Newhall, Mickey Elkus, and Marsha Mielnik for typing portions of this book. Finally, I wish to thank my wife and kids for their understanding and support during the development of this book.

Bethlehem, Pennsylvania   GARY E. WHITEHOUSE

SYSTEMS ANALYSIS AND DESIGN USING NETWORK TECHNIQUES

INTRODUCTION—SYSTEMS MODELING AND ANALYSIS USING NETWORK TECHNIQUES

The distinguishing feature of a system is the interaction among its various components. The structure of the system is determined by this interaction. The analyst must be concerned with these interactions when analyzing a given system. Insight into the cause-and-effect relationships and into the performance of the system as a whole can be gained through this knowledge. Quantitative analysis of systems is dependent upon the structure of the system.

One means of displaying and analyzing these system interactions is by using graphic techniques. Analysts have been using graphic portrayals of physical systems for some time. However, the systems designer has considered these graphic techniques merely a different "language" that might offer clarity associated with pictorial exhibits. The early graphic displays, e.g., process flow charts, did not enhance the fundamental understanding of the structure or the dynamic behavior of the system. Recent advances in the engineering science of control systems and in probability theory have expanded the field of graphic representation with new and powerful techniques. This book presents a comprehensive study of these new graphic techniques which we will refer to as network modeling techniques.

1

1.1 Systems Modeling

Morris [2] has suggested the following steps for constructing models:

1. Identify and formulate the manager's decision in writing.
2. Identify the constants, parameters, and variables involved. Define them verbally and then introduce symbols to represent each one.
3. Select the variables that appear to be most influential so that the model may be kept as simple as possible. Distinguish between those that are controllable by the manager and those that are not.
4. State verbal relationships among the variables based upon known principles, specially gathered data, intuition, and reflection. Make assumptions or predictions concerning the behavior of the noncontrollable variables.
5. Construct the model by combining all relationships into a system of symbolic relationships.
6. Perform symbolic manipulations.
7. Derive solutions from the model.
8. Test the model by making predictions from it and checking against real-world data.
9. Revise the model as necessary.

3

In this text we will apply these steps with special emphasis on using network models to analyze the systems in question. There is a certain danger in presenting a text devoted to just one type of modeling. The modeler should not try to force-fit techniques to the problem but should try to find the most appropriate technique to solve the problem at hand.

1.2 Types of Models to be Studied

Murdick and Ross [3] have suggested an interesting classification of models.

CLASS I   FUNCTION

1. *Descriptive.* Descriptive models simply provide a "picture" of a situation and do not predict or recommend.
2. *Predictive.* Predictive models relate dependent and independent variables and permit trying out "what if" questions.
3. *Normative.* Normative models are those that provide the "best" answer to a problem. They provide recommended courses of action.

CLASS II   STRUCTURE

1. *Iconic.* Iconic models retain some of the physical characteristics of the things they represent.
2. *Analog.* Analog models are those for which there is a substitution of components or processes to provide a parallel with what is being modeled.
3. *Symbolic.* Symbolic models use symbols to describe the real world.

CLASS III   TIME REFERENCE

1. *Static.* Static models do not account for changes over time.
2. *Dynamic.* Dynamic models have time as an independent variable.

CLASS IV   UNCERTAINTY REFERENCE

1. *Deterministic.* For a specific set of input values, there is a uniquely determined output that represents the solution of a model.
2. *Probabilistic.* Probabilistic models involve probability distributions for inputs and provide a range of values of at least one of the outputs.

CLASS V   GENERALITY

1. *General.* General models are those that have applications in several functional areas.
2. *Specialized.* Specialized models are those that have application to a unique problem only.

The models discussed in this book fall into most of the classes that

Murdick and Ross discuss, but the one common category is the iconic class. The modeler using networks creates a visual representation of the system he is trying to model. For example, consider the Thief of Bagdad problem which will be analyzed in Chapter 8:

"A thief is in a dungeon that has three doors. The first leads to freedom, the second leads to a long tunnel which takes three days to traverse, and the third leads to a short tunnel which takes one day to traverse. Both tunnels lead back to the dungeon where the thief tries again to free himself. How long will it take before the thief frees himself?"

One might recognize this as a problem that could be solved using semi-Markov analysis. This is at best a difficult process. Using the methods discussed in this book we would first model the system graphically as shown in Fig. 1.1. In Fig. 1.1, node D represents the dungeon, node S represents

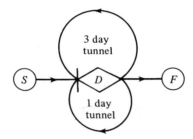

Figure 1.1. Thief of Bagdad problem.

the start of the analysis, and node F represents freedom. The large loop is the long tunnel and the small loop is the short tunnel. The iconic nature of network modeling should now be obvious. The next step in the analysis would be to enter quantitative information on our model, such as the tunnel lengths and probability of choosing each door. Next the network would be analyzed using either the techniques presented or the computer programs supplied. We will find that these procedures yield the distribution of the time the thief spends in the system.

We will now present another example to further illustrate the use of network techniques. In this example, we will use the *graphical evaluation and review technique* (GERT) to model a very simplified form of the systems equipment engineering function of the Western Electric Company. In Chapter 8 we discuss in detail the GERT approach. The steps to the solution are summarized as follows:

1. Convert a qualitative description of a system or problem to a model in stochastic network form.
2. Collect the necessary data to describe the branches of the network.
3. Determine the equivalent function or functions of the network.

4. Convert the equivalent function into the following two performance measures of the network: the probability that a specific node is realized and the moment generating function (MGF) of the time associated with an equivalent network.
5. Make inferences concerning the system under study from the information obtained in step 4.

The systems equipment engineering function of the Western Electric Company may be described as follows. Orders for equipment are received from the operating telephone companies and are distributed to power, framework, and switchboard engineering departments. Each department performs its own work and sends memoranda to the other departments requesting that certain detailed work be done. Each department then releases final specifications to the manufacturing shops and to the installers. Finally, feedback from the shops and installers may cause further work to be performed by the engineering departments which then issue appendices to the shop and installation specifications.

The first step in GERT analysis is to convert the qualitative description of the system to a stochastic network form. This has been done in Fig. 1.2, where the nodes represent logic operations and the directed branches represent the activities. The EXCLUSIVE-OR symbol used in the figure means that the realization of any branch leading into the node causes the node to be realized. However, one and only one of the branches leading into this node can be realized at a given time.

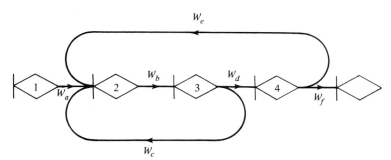

Figure 1.2. GERT model of a systems equipment engineering function.

The symbol W_i is the transmittance of the ith branch, where $W_i = p_i M_i$. The symbol p_i is the probability that the ith branch will be taken, and M_i is the moment generating function of the time required to perform the activity represented by the ith branch.

The branches of Fig. 1.2 are defined as follows:

W_a = distribution of orders to engineering departments.
W_b = engineering of specifications or appendices.
W_c = work on memoranda.

W_d = reproduction and distribution of specifications to shop and installers.
W_e = determination of errors and/or changes requiring appendices.

The nodes of Fig. 1.2 are defined as follows:

1 = receipt of orders from telephone companies.
2 = beginning of engineering work.
3 = end of engineering work and decision to send a memo or write a specification.
4 = beginning of manufacture and installation, and decision to request an
appendix or proceed with manufacture and installation.
5 = end of manufacture and installation.

The second step in GERT analysis is to collect the necessary data to describe the branches of the network. The data are shown in Table 1.1.

Table 1.1. THE BRANCHES OF FIG. 1.2

Branch	Probability	MGF of Time	Remarks
a	1.0	e^{2s}	Time is constant at 2 days
b	1.0	$e^{10(e^s-1)}$	Time is Poisson, mean = 10 days
c	0.2	$e^{5(e^s-1)}$	Time is Poisson, mean = 5 days
d	0.8	e^{2s}	Time is constant at 2 days
e	0.2	$e^{8(e^s-1)}$	Time is Poisson, mean = 8 days
f	0.8	$e^{60(e^s-1)}$	Time is Poisson, mean = 60 days

The third step is to determine the equivalent function of the network. The equivalent transmittance for the graph shown in Fig. 1.2 can be found by using Mason's rule as follows:

$$W_{1,5} = \frac{W_a W_b W_d W_f}{1 - (W_b W_c + W_b W_d W_e)}$$

Substituting the assumed values from Table 1.1 into $W_{1,5}$, we obtain:

$$W_{1,5} = \frac{0.64 e^{4S+70(e^s-1)}}{1 - (0.2 e^{15(e^s-1)} + 0.16 e^{2S+18(e^s-1)})}$$

The fourth step is to convert the equivalent function into the two performance measures of the network. These are the probability that a specific node is realized and the moment generating function of the time associated with an equivalent network. Now,

$$W_{1,5}|_{s=0} = p_{1,5} = \frac{0.64}{1 - (0.2 + 0.16)} = \frac{0.64}{0.64} = 1.0$$

which states that if node 1 has occurred, the probability of reaching node 5 is 1.0. Since node 5 is the only output of the system, this is the expected result.

The moment generating function for the time distribution to pass from node 1 to node 5 is not a trivial result, however, and may be determined as follows:

$$M_{1,5}(s) = \frac{W_{1,5}}{p_{1,5}} = W_{1,5}$$

The fifth step is to make inferences about the system using the above result. The mean of the time to pass from node 1 to node 5 may be obtained as follows:

$$\mu_{1(1,5)} = \frac{\partial}{\partial s}[M_{1,5}(s)]|_{s=0} = 83.7$$

The second moment about the origin for the distribution of time for orders to pass from node 1 to node 5 may be computed by:

$$\mu_{2(1,5)} = \frac{\partial^2}{\partial s^2}[M_{1,5}(s)]|_{s=0}$$

The variance for the distribution may then be computed if we note that:

$$\sigma_{1,5}^2 = \mu_{2(1,5)} - (\mu_{1(1,5)})^2$$

From this example of the application of GERT to the systems equipment engineering function of the Western Electric Company, it can be seen that network techniques show promise from a systems modeling standpoint.

In this book we will not dwell on the graph-theoretic aspects of the models discussed. I make an explicit distinction between graph theory and network theory. A graph defines purely structural relationships between nodes, while a network bears also the quantitative characteristics of the nodes and arcs. For those interested in graph theory, refer to the excellent text by Busacker and Saaty [1].

In this book we will consider the theory and applications of a number of network techniques. Particular emphasis will be placed on the GERT technique which was developed by Drs. Pritsker, Happ, and Whitehouse a few years ago at Arizona State University and has since received much interest from both the academic and industrial communities. We will show the development and applications of this technique along with related techniques such as flowgraph theory, PERT (project evaluation and review technique) and CPM (critical path method), project management, network flows, decision trees, and the statistical background needed to understand the material in this text.

1.3 Advantages and Disadvantages of Network Modeling Techniques

Network modeling techniques represent a valuable aid in the analysis and synthesis of systems. In a recent article Pritsker and Happ [4] discussed the advantages of network techniques:

Networks and network analyses are playing an increasingly important role in the description and improvement of operational systems. The ease with which systems can be modeled in network form is the fundamental reason for this significant increase in the use of networks. Other reasons for using networks are: (1) the need for communication mechanisms to discuss the operational system in terms of its significant features, (2) a means for specifying the data requirements for analysis of the system, and (3) a starting point for analysis and scheduling of the operational systems. The latter of these reasons was the original motive for network construction and use. The advantages that occurred outside of the analysis procedure soon justified the network approach, and further efforts toward improving and extending network analysis procedures have not kept pace with the applications of networks.

We will find that, in all the applications we will discuss, the systems could be equally represented by such techniques as differential equations, algebraic linear equations, matrices, or difference equations in a formal mathematical manner. The mathematical models lose a great deal of the insight and facility of solution available through network techniques. The mathematical models may, however, prove to be superior from a computional standpoint, especially for modeling large-scale systems.

REFERENCES

1. Busacker, R. G. and T. L. Saaty, *Finite Graphs and Networks.* New York, McGraw-Hill, Inc., 1965.
2. Morris, W., "On the Art of Modeling." *Management Science* (August, 1967).
3. Murdick, R. G. and J. E. Ross, *Information Systems for Modern Management.* Englewood Cliffs, N.J., Prentice-Hall, Inc., 1971.
4. Pritsker, A. A. B. and W. W. Happ, "GERT: Graphical Evaluation and Review Technique, Part I, Fundamentals." *Journal of Industrial Engineering*, Vol. 17, No. 5 (May, 1966).

FUNDAMENTALS OF PROBABILITY THEORY

Some familiarity with the basic notions of probability is essential for understanding the remaining chapters of this book. A rigorous development of probability theory involves logical and mathematical difficulties of high order and will not be attempted here.

The chapter is intended to serve as a brief review for those readers who have an understanding of probability theory and as a guidepost to further study of the subject for those with less preparation. Several good references are given at the end of the chapter. Drake [2] gives particularly good background for the material covered in this text.

2

2.1 Events and Probability

Our study of probability theory begins with three closely related topics: (1) the algebra of events, (2) sample space, and (3) probability measure.

2.1.1 Algebra of events. In this section we will introduce terminology which will be used throughout our discussion of probability theory.

Consider the following definitions and operations of the algebra of events:

1. *Events* are collection of points in a space.
2. The *universal event*, U, is the collection of all points in the entire space.
3. Event A' is the *complement* of event A and includes all points in the universal event which are not included in event A.
4. The *null event*, $\varnothing$, contains no points and is the complement of the universal event.
5. The *intersection* of two events A and B, designated by the notation AB, represents all points which are included in both A and B.
6. The notation $A + B$ represents the *union* of two events A and B. The union represents all points which are either in A or in B or in both.
7. Two events are *equal* if each event is included in the other event.

11

We have just viewed several concepts of the algebra of events. *Venn diagrams*, which are pictures of the events in the universal set, are often used to illustrate these concepts. For example, Fig. 2.1 illustrates a Venn diagram.

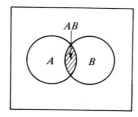

Figure 2.1. An example of a Venn diagram.

The shaded area represents AB. Any relation in the algebra of events can be visualized using Venn diagrams. In addition, every relationship in the algebra of events can be proved using the following seven axioms:

1. $A + B = B + A$
2. $A + (B + C) = (A + B) + C$
3. $A(B + C) = AB + AC$
4. $(A')' = A$
5. $(AB)' = A' + B'$
6. $AA' = \varnothing$
7. $AU = A$

It is surprisingly difficult to prove relations in the algebra of events using these axioms.

There remain two important concepts which we will use from the concepts of the algebra of events.

1. The events in a set are *mutually exclusive* if there is no point in the universal event which is included in more than one event in the set.
2. Events are said to be *collectively exhaustive* if every point in the universal event is included in at least one event in the set.

The Venn diagrams shown in Fig. 2.2 illustrate the concepts of mutually exclusive and collectively exhaustive sets of events. Events A, B, and C in Fig. 2.2(a) are mutually exclusive. Figure 2.2(b) illustrates three collectively exhaustive events, while Fig. 2.2(c) shows three events which are both collectively exhaustive and mutually exclusive.

2.1.2 Probability measure. Before we introduce the axioms of probability measure, it is necessary to introduce the concept of an *experiment*. There are usually more variables associated with the outcome of any physical experiment than the experimenter cares about. For example, in a coin

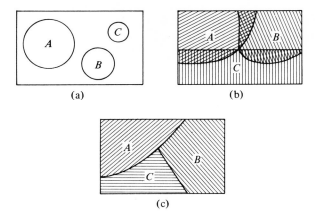

(a)   (b)

(c)

Figure 2.2. Venn diagrams illustrating mutually exclusive and collectively exhaustive sets of events.

toss we are seldom concerned with such things as the weight of the coin, the number of times the coin bounced, or the height of the flip. For most purposes all we care about in a coin flip is whether we get a head or a tail. Thus we introduce the *sample space* of an experiment. The *sample space* represents the finest grain, mutually exclusive, collectively exhaustive listing of all possible outcomes of an experiment. For the coin toss, the sample space would probably be the toss of a head or a tail.

There are many common ways of expressing sample spaces. Consider the sequential sample space for a model of three flips of a coin. T_n is defined as a tail on the nth toss and H_n as a head on the nth toss. This leads to the sequential space picture in Fig. 2.3. In this figure the experiment is pictured as moving from left to right. Each point located at the end of a branch represents the event corresponding to the intersection of all events encountered in moving from left to right in the tree. For example, $H_1 T_2 T_3$ represents a head on the first toss, a tail on the second toss, and a tail on the final toss. The eight sample points and labels constitute the sample space for our three flips of a coin. Note that they represent a mutually exclusive, collectively exhaustive set of events.

If an experiment has numerical outcomes, a coordinate system can be a useful representation of a sample space. Figure 2.4 illustrates the sample space for an experiment involving two tosses of a die.

To complete the description of the experiment we must assign *probabilities* to the events in the sample space. The *probability* of an event represents the relative likelihood that the experiment will yield the event. By combining three new axioms with the seven axioms for the algebra of events, one achieves a system for associating probabilities with events in a sample space. This set

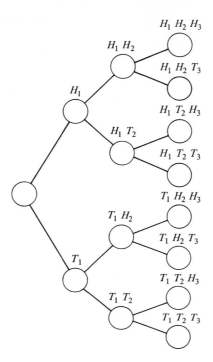

Figure 2.3. Sequential sample space for three flips of a coin.

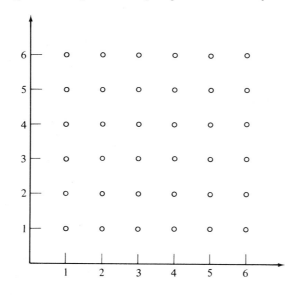

Figure 2.4. Coordinate system sample space representation of two tosses of a die.

of ten axioms will also yield a system for computing the probabilities of more complex events. If we define $P(A)$ to be the probability of event A, then the new axioms become:

1. For any event A, $P(A) \geq 0$
2. $P(U) = 1$
3. If $AB = \varnothing$, then $P(A + B) = P(A) + P(B)$

Using our axioms and Venn diagrams we can show that for the space shown in Fig. 2.1:

$$P(A + B) = P(A) + P(B) - P(AB)$$

or $\quad P(A + B) = 1 - P(A'B')$

or $\quad P(A + B) = P(AB') + P(A'B) + P(AB)$

etc.

Viewing Fig. 2.1 we might wish to consider the situation that results when a single experiment is performed and it yields the result that the outcome has attribute B. Knowing this result we see that the experimental observation is either in (AB) or $(A'B)$. We might be interested in the probability that A occurs given that B has been observed. This type of probability measure is known as *conditional probability* measure and is designated as $P(A \mid B)$.

From our diagram we can see that

$$P(A \mid B) = \frac{P(AB)}{P(B)}$$

Using this concept we observe that $P(AB)$ also equals $P(A \mid B)P(B)$ and $P(B \mid A)P(A)$.

It may happen that the conditional probability of A given B is the same as the original probability of A. If this is the case, the event A is said to be *independent* of B and we have the expression:

$$P(AB) = P(A)P(B) \qquad \text{if } A \text{ is independent of } B$$

If A is independent of B, then it can be shown that B is independent of A.

The results just stated can easily be generalized to the case of n events $A_1, A_2, \ldots, A_n$ as follows:

1. If $A_1, A_2, \ldots, A_n$ are mutually exclusive in pairs, then

$$P(A_1 + A_2 + \cdots + A_n) = P(A_1) + P(A_2) + P(A_3) + \cdots + P(A_n)$$

2. For any n events $A_1, A_2, A_3, \ldots, A_n$

$$P(A_1 A_2 A_3 \cdots A_n) = P(A_1)P(A_2 \mid A_1)P(A_3 \mid A_1 A_2) \cdots P(A_n \mid A_1 A_2 \cdots A_{n-1})$$

3. If $A_1, A_2, A_3, \ldots, A_n$ are independent, then

$$P(A_1 A_2 A_3 \cdots A_n) = P(A_1)P(A_2) \cdots P(A_n)$$

Consider the following simple example to illustrate the points just discussed.

EXAMPLE:

An urn contains twenty balls, of which five are black and fifteen are red. Two balls are drawn at random from the urn. What is the probability that
 (a) both are black?
 (b) both are red?
 (c) one is red and the other black?

Solution:

(a) Let A be the event "the first ball drawn is black" and B be the event "the second ball drawn is black." Then:

$$P(AB) = P(A)P(B \mid A) = (\tfrac{5}{20})(\tfrac{4}{19}) = \tfrac{1}{19}$$

(b) By similar reasoning, the probability of both balls being red is $(\tfrac{15}{20})(\tfrac{14}{19}) = \tfrac{21}{38}$.
(c) Let

$$E_1 = \text{event "both are red"}$$
$$E_2 = \text{event "both are black"}$$
$$E_3 = \text{event "one is red and one is black"}$$

Since E_1, E_2, and E_3 are mutually exclusive and collectively exhaustive, then:

$$P(E_1 + E_2 + E_3) = 1$$

and $$P(E_1 + E_2 + E_3) = P(E_1) + P(E_2) + P(E_3)$$

Thus $$P(E_1) + P(E_2) + P(E_3) = 1$$

and $$P(E_3) = 1 - P(E_1) - P(E_2)$$
$$= 1 - \tfrac{1}{19} - \tfrac{21}{38} = \tfrac{15}{38}$$

To obtain this result in another way, define two mutually exclusive events:

$$A = \text{event "first is red and second is black"}$$
$$B = \text{event "first is black and second is red"}$$

Then:

$$P(A + B) = P(A) + P(B)$$
$$= (\tfrac{15}{20})(\tfrac{5}{19}) + (\tfrac{5}{20})(\tfrac{15}{19}) = \tfrac{15}{38}$$

2.2 Random Variables

Often there are reasons to associate one or more numbers with each possible outcome of an experiment. For example cost, height, or weight might be measured. In this section we will investigate methods for studying experiments whose outcomes may be described numerically.

2.2.1 Discrete random variables.
A *random variable* is a set of numbers $x_1, x_2, \ldots, x_n$, one for each state, so that the result of a trial is not only the state E_i but also a number of interest x_i. For example, if a trial consisted of flipping ten coins, a random variable could be the number of heads showing. Another example could be the toss of two dice where the random variable could be the sum of points showing on both dice.

For the dice game the random variable could be described by a *frequency function* which is a complete listing of the random variables together with their probabilities of occurrence. The frequency function is as follows:

x	$P(x)$
2	$\frac{1}{36}$
3	$\frac{2}{36}$
4	$\frac{3}{36}$
5	$\frac{4}{36}$
6	$\frac{5}{36}$
7	$\frac{6}{36}$
8	$\frac{5}{36}$
9	$\frac{4}{36}$
10	$\frac{3}{36}$
11	$\frac{2}{36}$
12	$\frac{1}{36}$
	$\frac{36}{36}$

This frequency function is derived from the fact that there are 36 different outcomes from tossing two dice. Four of these throws (4, 1), (3, 2), (2, 3), and (1, 4) sum to 5, so the probability of a 5 is $\frac{4}{36}$. Note that the sum of all the probabilities is one. Many random variables can be defined for the same sample space. For example, the random variable in the dice experiment might be the product of the dice thrown or the smallest number thrown on the two dice.

The frequency function is often characterized by two measures: its mean and variance. These values do not completely describe a frequency function but are commonly used and are useful in many applications.

If x_i $(i = 1, 2, \ldots, n)$ represents all the possible values that a random variable can assume and if $P(x_i)$ represents the probability of the random variable, then the mean (μ) and variance (σ^2) of a distribution are defined as follows:

$$\mu = \sum_{i=1}^{n} x_i P(x_i)$$

$$\sigma^2 = \sum_{i=1}^{n} (x_i - \mu)^2 P(x_i) = \sum_{i=1}^{n} x_i^2 P(x_i) - (\mu)^2$$

The mean is a measure of the value expected from an experiment, while the variance is a measure of the dispersion about this expected value.

Another important concept is that of the *cumulative frequency function* for a random variable. If the range of possible values for x is arranged in increasing order, then the cumulative distribution is defined, for any number y, as the probability of the event $(x \le y)$. If we use $G(y)$ to designate the cumulative frequency function, we observe that:

$$G(y) = 0 \qquad \text{for} \quad y < x_1$$

$$G(y) = \sum_{j=1}^{i} P(x_j) \quad \text{for} \quad x_i \le y < x_i + 1$$

$$G(y) = 1 \qquad \text{for} \quad y \ge x_n$$

There are a number of discrete frequency functions which we will use in this text. These are listed below with a few of their properties.

BINOMIAL DISTRIBUTION

$$P(k) = \frac{(N - k)!k!}{N!} p^k (1 - p)^{N-k} \quad \text{for} \quad k = 0, 1, \ldots, N$$

$$0 < p < 1$$

$$\mu = Np$$
$$\sigma^2 = Np(1 - p)$$

POISSON DISTRIBUTION

$$P(k) = \frac{\lambda^k e^{-\lambda}}{k!} \quad \text{for} \quad k = 0, 1, \ldots, \infty \quad \text{and} \quad \lambda > 0$$

$$\mu = \lambda$$
$$\sigma^2 = \lambda$$

UNIFORM DISTRIBUTION

$$P(k) = \frac{1}{N + 1} \quad \text{for} \quad k = 0, \ldots, N$$

$$\mu = \frac{N}{2}$$

$$\sigma^2 = \frac{N^2}{12} + \frac{N}{6}$$

GEOMETRIC DISTRIBUTION

$$P(k) = p(1-p)^{x-1} \quad \text{for} \quad x = 0, 1, \ldots, \infty$$
$$0 < p < 1$$

$$\mu = \frac{p}{1-p}$$

$$\sigma^2 = \frac{p}{1-p^2}$$

2.2.2 Continuous distributions. In the preceding section we discussed random variables whose possible values are discrete. In this section random variables which are defined over a continuum of values will be discussed. Continuous random variables are defined on sample spaces with a non-denumerable infinity of points. Although the theoretical development for the continuous distribution space is considerably different from that in the discrete case, the results are comparable. As in the discrete case the cumulative frequency function is defined as the probability of the random variable, x, being less than or equal to y. We call this probability $P(x \leq y)$. Since $P(x \leq y)$ is defined for each value y on the range of x, it is actually a function of y which will be called $F(y)$.

If a and b are two values in the range of x with $a < b$, $P(a < x \leq b)$ can be expressed in terms of the cumulative frequency function as follows:

$$P(x \leq b) = P(x \leq a) + P(a < x \leq b)$$

thus $\qquad P(a < x \leq b) = F(b) - F(a)$

In this section we will be concerned only with cumulative distribution functions $F(a)$ that possess derivatives $f(a)$ at all points on the range of x. In such cases $f(a) = F'(a)$ and we have

$$F(a) = \int_{-\infty}^{a} f(x)\,dx \quad \text{and} \quad P(a < x \leq b) = \int_{a}^{b} f(x)\,dx$$

$f(x)$ is defined as the probability density function of x and has the following properties:

1. $\int_{-\infty}^{\infty} f(x)\,dx = 1$
2. $P(x = a) = 0$

3. $f(x)\,dx$ is the probability that the random variable takes on a value between x and $x + dx$.

As in the discrete case, the probability distribution of the continuous random variable is often defined in terms of its mean and variance.

$$\mu = \int_{-\infty}^{\infty} x f(x)\,dx$$

$$\sigma^2 = \int_{-\infty}^{\infty} (x - \mu)^2 f(x)\,dx = \int_{-\infty}^{\infty} x^2 f(x)\,dx - (\mu)^2$$

The following are examples of some of the continuous probability distributions we will be exposed to in this text.

NORMAL DISTRIBUTION (Fig. 2.5)

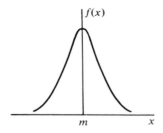

Figure 2.5. Normal distribution.

$$f(x) = \frac{1}{\sigma\sqrt{2\pi}} e^{-(x-m)^2/2\sigma^2} \quad \text{for} \quad -\infty \leq x \leq \infty$$

$$\mu = m$$

$$\sigma^2 = \sigma^2$$

RECTANGULAR DISTRIBUTION (Fig. 2.6)

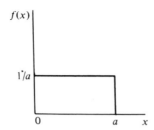

Figure 2.6. Rectangular distribution.

$$f(x) = \frac{1}{a} \quad \text{for} \quad 0 \leq x \leq a$$

$$\mu = \frac{a}{2}$$

$$\sigma^2 = \frac{a^2}{12}$$

EXPONENTIAL DISTRIBUTION (Fig. 2.7)

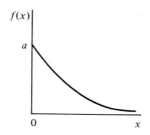

Figure 2.7. Exponential distribution.

$$f(x) = ae^{-ax} \quad \text{for} \quad 0 \le x \le \infty$$
$$a > 0$$

$$\mu = \frac{1}{a}$$

$$\sigma^2 = \frac{1}{a^2}$$

BETA DISTRIBUTION (Fig. 2.8)

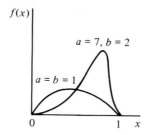

Figure 2.8. Beta distribution.

$$f(x) = \frac{(a + b + 1)!}{a!b!}x^a(1 - x)^b \quad \text{for} \quad 0 \le x \le 1$$
$$a > -1$$
$$b > -1$$

$$\mu = \frac{a + 1}{a + b + 2}$$

$$\sigma^2 = \left(\frac{a + 2}{a + b + 3}\right)\left(\frac{a + 1}{a + b + 2}\right)$$

2.2.3 Expectation and moments. In the previous section the mean, μ, and variance, σ^2, were introduced as measures to characterize probability distributions. These measures are actually moments of the distribution. The mean is often referred to as the *expected value* and designated as $E(x)$ as stated previously.

$$E(x) = \mu = \begin{cases} \sum_{i=1}^{n} x_i P(x_i) & \text{if } x \text{ is a discrete random variable} \\ \int_{-\infty}^{\infty} x f(x)\, dx & \text{if } x \text{ is a continuous random variable} \end{cases}$$

For the discrete case it is observed that $E(x)$ is just the sum of the products of the possible values of the random variable x and their respective probabilities. Assume that a variable is distributed such that

$$P(x) = \frac{x}{55} \quad \text{for} \quad x = 1, 2, \ldots, 10$$
$$= 0 \quad \text{otherwise}$$

then

$$E(x) = \mu = 1(\tfrac{1}{55}) + 2(\tfrac{2}{55}) + \cdots + 10(\tfrac{10}{55})$$
$$= \tfrac{385}{55}$$

For the continuous case the expected value can also be easily obtained. If x has an exponential distribution with parameter a, its expected value is found as follows:

$$E(x) = \mu = \int_{-\infty}^{\infty} x f(x)\, dx = \int_{0}^{\infty} x a e^{-ax}\, dx = \frac{1}{a}$$

It is not necessary to confine our discussion of expectation to the random variable x. It is also possible to define the expectation of a function of x, say, $g(x)$. The expectation of $g(x)$ can be defined as:

$$E(g(x)) = \begin{cases} \sum_{i=1}^{n} g(x_i) P(x_i) & \text{if } x \text{ is a discrete random variable} \\ \int_{-\infty}^{\infty} g(x) f(x)\, dx & \text{if } x \text{ is a continuous random variable} \end{cases}$$

If $g(x) = x^j$, where j is a positive integer, then the expectation of x^j is called the jth moment about the origin of the random variable x and is given by:

$$E(x^j) = \begin{cases} \sum_{i=1}^{n} x_i^j P(x_i) & \text{if } x \text{ is a discrete random variable} \\ \int_{-\infty}^{\infty} x^j f(x)\, dx & \text{if } x \text{ is a continuous random variable} \end{cases}$$

If $g(x) = (x - E(x))^j = (x - \mu)^j$, where j is a positive integer, then the expectation of $(x - \mu)^j$ is called the *j*th moment about the mean of the random variable $\dot{x}$ and is given by:

$$E(x - E(x))^j = E(x - \mu)^j = \begin{cases} \sum_{i=1}^{n} (x_i - \mu)^j P(x_i) & \text{if } x \text{ is a discrete random variable} \\ \int_{-\infty}^{\infty} (x - \mu)^j f(x) \, dx & \text{if } x \text{ is a continuous random variable} \end{cases}$$

When $j = 2$, we have $E(x - \mu)^2$, which is the variance discussed in the previous section. The square root of the variance, σ, is called the standard deviation of the random variable x. Hill [4] has shown that it is possible to describe probability distribution in terms of its mean, variance, and higher moments.

2.3 Transforms

In this section a brief introduction to transform theory will be presented. The importance of this material will become apparent in later chapters.

2.3.1 The moment generating function. Let $f(x)$ be any probability distribution function; then its moment generating function, $M(s)$, is defined as:

$$M(s) = E(e^{sx}) = \int_{-\infty}^{\infty} e^{sx} f(x) \, dx$$

The moment generating function is derived in the same manner as any other expected value. For example, consider the exponential distribution:

$$f(x) = ae^{-ax} \qquad x \geq 0$$

$$M(s) = \int_{-\infty}^{\infty} e^{sx} f(x) \, dx = \int_{0}^{\infty} ae^{sx} e^{-ax} \, dx = \left(1 - \frac{s}{a}\right)^{-1}$$

Table 2.1 gives some moment generating functions for common distributions.

The moment generating function derives its name from the fact that the moments of a distribution are easily obtainable from $M(s)$. Consider the *n*th derivative with respect to s of $M(s)$.

$$M(s) = \int_{-\infty}^{\infty} e^{sx} f(x) \, dx$$

$$\frac{d^n M(s)}{ds^n} = \int_{-\infty}^{\infty} (x)^n e^{sx} f(x) \, dx$$

Table 2.1. PROBABILITY LAWS AND THEIR MOMENT GENERATING FUNCTIONS.

Distribution	Probability Law	$M(s)$
Binomial	$P(x) = \binom{n}{x} p^x q^{n-x}, \quad x = 0, 1, \ldots, n$ $q = 1 - p$ $0, \quad$ otherwise	$(pe^s + q)^n$
Exponential	$f(x) = ae^{-ax}, \quad x \geq 0$ $0, \quad x < 0$	$\left(1 - \dfrac{s}{a}\right)^{-1}$
Gamma	$f(x) = \dfrac{a}{\Gamma(b)} (ax)^{b-1} e^{ax}, \quad x \geq 0$ $0, \quad x < 0$	$\left(1 - \dfrac{s}{a}\right)^{-b}$
Geometric	$P(x) = pq^{x-1}, \quad x = 1, 2, 3, \ldots$ $0, \quad$ otherwise	$\dfrac{pe^s}{1 - qe^s}$
Negative binomial	$P(x) = \binom{x + r - 1}{x} p^r q^x, \quad x = 1, 2, \ldots$ $0, \quad$ otherwise	$\left(\dfrac{p}{1 - qe^s}\right)^k$
Normal	$f(x) = (1/\sigma\sqrt{2\pi}) e^{-(1/2)((x-\mu)/\sigma)^2},$ $-\infty < x < \infty$	$e^{s\mu + (1/2)s^2\sigma^2}$
Poisson	$P(x) = \dfrac{e^{-\lambda}\lambda^x}{x!}, \quad x = 0, 1, 2, 3, \ldots$ $0, \quad$ otherwise	$e^{\lambda(e^s - 1)}$
Uniform	$f(x) = \dfrac{1}{(b - a)}, \quad a < x < b$ $0, \quad$ otherwise	$\dfrac{e^{sa} - e^{sb}}{(b - a)s}$
Constant	$P(x) = 1, \quad x = t$ $0, \quad$ otherwise	e^{ts}

The right side of the last equation, when evaluated at $s = 0$, is recognized as $E(x^n)$. Thus, once we obtain the moment generating function of a probability distribution function, we can find all the moments by repeated differentiation rather than by performing other integrations.

From the preceding discussion the following useful results can be summarized:

$$M(s)|_{s=0} = 1$$

$$E(x) = \mu = \left.\frac{dM(s)}{ds}\right|_{s=0}$$

$$E(x^2) = \left.\frac{d^2 M(s)}{ds^2}\right|_{s=0}$$

$$\sigma^2 = E((x - \mu)^2) = E(x^2) - [E(x)]^2$$
$$= \left.\frac{d^2 M(s)}{ds^2} - \left[\frac{dM(s)}{ds}\right]^2\right|_{s=0}$$

Of course, when certain moments of a probability distribution function do not exist, the corresponding derivatives of $M(s)$ will be infinite when evaluated at $s = 0$.

Now let's consider using the moment generating function to calculate the moments of the exponential distribution.

$$M(s) = \left(1 - \frac{s}{a}\right)^{-1}$$

$$E(x) = \frac{dM(s)}{ds}\bigg|_{s=0} = \frac{1}{a}\left(1 - \frac{s}{a}\right)^{-2}\bigg|_{s=0} = \frac{1}{a}$$

$$E(x^2) = \frac{d^2M(s)}{ds^2}\bigg|_{s=1} = \frac{2}{a^2}\left(1 - \frac{s}{a}\right)^{-3}\bigg|_{s=0} = \frac{2}{a^2}$$

$$\sigma^2 = E(x^2) - (E(x))^2 = \frac{1}{a^2}$$

Another important property of moment generating functions is that the transform of the sum of two independent random variables is equal to the product of the transforms of the individual variables; that is, for $w = x + y$, then:

$$M_w(s) = M_{x+y}(s) = M_x(s)M_y(s)$$

For example, assume that x is normally distributed with $\mu = 6$ and $\sigma^2 = 6$, and y is normally distributed with $\mu = 10$ and $\sigma^2 = 16$.

$$M_w(s) = M_{x+y}(s) = M_x(s)M_y(s) = (e^{6s+3\sigma^2})(e^{10s+8\sigma^2}) = e^{16s+11\sigma^2}$$

Thus w is normally distributed with $\mu = 16$ and $\sigma^2 = 22$.

2.3.2 The z transform. Consider the discrete probability distribution:

$$P(x) = \tfrac{1}{3}, \qquad x = 1, 2, 3$$
$$= 0, \qquad \text{otherwise}$$

The moment generating function for this distribution would be:

$$M(s) = E(e^{sx}) = \tfrac{1}{3}e^{s} + \tfrac{1}{3}e^{2s} + \tfrac{1}{3}e^{3s}$$

Although the moment generating function is defined for the probability distribution function of any random variable, it is convenient to define one additional type of transform for certain types of distributions. If $P(x)$ is the distribution for a discrete random variable which can take on only non-negative integer experimental values ($x = 1, 2, \ldots$), we define the z transform of $P(x)$ to be $M(z)$ given by:

$$M(z) = E(z^x) = \sum_{x=0}^{\infty} z^x P(x)$$

Thus $M(z)$ for the distribution mentioned at the beginning of this section becomes:

$$M(z) = \tfrac{1}{3}z + \tfrac{1}{3}z^2 + \tfrac{1}{3}z^3$$

From the definition of $M(z)$

$$M(z) = P(0) + zP(1) + z^2 P(2) + \cdots$$

we see that it is possible to determine individual probabilities from the expression:

$$P(x) = \frac{1}{x!}\left[\frac{d^x}{dz^x}M(z)\right]\bigg|_{z=0}, \qquad x = 1, 2, 3, \ldots$$

The moments for a probability distribution can also be obtained from its z transform as follows:

$$M(z) = \sum_{x=0}^{\infty} z^x P(x)$$

$$\frac{dM(z)}{dz}\bigg|_{z=1} = \sum_{x=0}^{\infty} x z^{x-1} P(x)\bigg|_{z=1} = E(x)$$

$$\frac{d^2 M(z)}{dz^2}\bigg|_{z=1} = \sum_{x=0}^{\infty} x(x-1) z^{x-2} P(x)\bigg|_{z=1} = E(x^2) - E(x)$$

Thus:

$$E(x) = \frac{dM(z)}{dz}\bigg|_{z=1}$$

$$E(x^2) = \frac{d^2 M(z)}{dz^2}\bigg|_{z=1} + \frac{dM(z)}{dz}\bigg|_{z=1}$$

$$\sigma^2 = \frac{d^2 M(z)}{dz^2}\bigg|_{z=1} + \frac{dM(z)}{dz}\bigg|_{z=1} - \left[\frac{dM(z)}{dz}\bigg|_{z=1}\right]^2$$

Let us consider applying the z-transform analysis to the geometric distribution.

$$P(x) = p(1-p)^{x-1}, \qquad x = 1, 2, \ldots \quad \text{and} \quad 0 < p < 1$$

$$M(z) = E(z^x) = \sum_{x=0}^{\infty} z^x P(x) = \sum_{x=1}^{\infty} p z^x (1-p)^{x-1}$$

$$= \frac{zp}{1 - z(1-p)}$$

$$1 + a + a^2 + \cdots + a^k = \frac{1 - a^{k+1}}{1 - a} \quad \text{for} \quad |a| < 1$$

$$E(x) = \frac{dM(z)}{dz}\bigg|_{z=1} = \frac{1}{p}$$

$$E(x^2) = \frac{d^2M(z)}{dz^2}\bigg|_{z=1} + \frac{dM(z)}{dz}\bigg|_{z=1} = \frac{2-p}{p^2}$$

$$\sigma^2 = E(x^2) - (E(x))^2 = \frac{1-p}{p^2}$$

2.4 The Markovian Property

Consider a system which may be described at any time as being in one of a set of mutually exclusive, collectively exhaustive states $S_1, S_2, \ldots, S_m$. According to a set of probabilistic rules, the system may at certain times undergo changes of state.

Let $S_i(n)$ be the event that the system is in state S_i immediately after the nth transition. The probability of this event may be written as $P(S_i(n))$. Each transition in the process just discussed may be described by transition probabilities of the form:

$$P(S_j(n)\,|\,S_a(n-1)S_b(n-2)\ldots)$$
$$1 \le j, a, b, c \ldots < m, \qquad n = 1, 2, 3$$

This expression specifies that the probabilities associated with each trial are conditional on the entire past history of the process.

There is a very important subclass of problems which exhibits the following property:

$$P(S_j(n)\,|\,S_a(n-1)S_b(n-2)S_c(n-3)\ldots)$$
$$= P(S_j(n)\,|\,S_a(n-1)) \qquad \text{for all } n, j, a, b, c, \ldots$$

In other words, the state of the system on the nth transition is conditioned only on the state of the system on the $(n-1)$th trial. This property is known as the *Markovian property*. In this text we will not consider processes for which the conditional transition probabilities $P(S_j(n)\,|\,S_i(n-1))$ depend on the transition number. Thus we may define the state transition probabilities for a Markov process to be $p_{ij} = P(S_j(n)\,|\,S_i(n-1))$, where $1 \le i, j \le m$; p_{ij} independent of n. p_{ij} is the conditional probability that the system will be in state S_j immediately after the next transition, given that the present state of the system is S_i. The quantity p_{ij} has the properties:

$$0 \le p_{ij} \le 1 \quad \text{for} \quad i, j = 1, 2, \ldots, m$$
$$\sum_j p_{ij} = 1 \quad \text{for} \quad i = 1, 2, \ldots, m$$

There are three types of Markov processes which have received much attention in the literature: (1) discrete, (2) continuous, and (3) semi-Markov processes. For the discrete parameter model, the time to move from one state to another is a constant, usually equal to one. For the continuous parameter Markov process, the transition time between states is distributed exponentially. In the semi-Markov process, the time between transitions is permitted to have any distribution and to depend not only upon the current state of the process, but also upon the state to which it is going. It is apparent that the discrete and continuous Markov processes are special cases of the semi-Markov process.

EXERCISES

1. Use Venn diagrams and the algebra of events to prove or disprove the relations:
    (a) $(A + B + C)' = A' + B'A' + A'B'C'$
    (b) $(A'B')' = AB + AB' + A'B$
    (c) $A + B + C = A + BA' + CB'A'$

2. If $P(A) = 0.4$, $P(B') = 0.7$, and $P(A + B) = 0.6$, find:
    (a) $P(B)$
    (b) $P(AB)$
    (c) $P(A \mid B)$
    (d) $P(A \mid B')$

3. Consider events A, B, and C with $P(A) > P(B) > P(C) > 0$. If A and B are mutually exclusive and collectively exhaustive, events A and C are independent. Can C and B be mutually exclusive?

4. Consider the density function:

$$f(x) = cx(1 - x), \quad \text{for} \quad 0 \le x \le 1$$
$$0, \quad \text{elsewhere}$$

    Find c, $F(y)$, μ, σ^2, and $P(x \le 0.5)$.

5. Find μ and σ^2 for the following discrete probability distributions:
    (a) $P(1) = 0.5$, $P(4) = 0.25$, $P(5) = 0.25$
    (b) Binomial distribution
    (c) $P(x) = \frac{1}{10}$, $x = 11, 12, \ldots, 20$

6. Find μ and σ^2 for the following continuous probability distributions:
    (a) $f(x) = \frac{1}{100}$, $100 \le x \le 200$
    (b) Exponential distribution

7. Given:

$$f(x) = k(1 - x^3) \quad \text{for} \quad 0 \le x \le 1$$

    Find k, $F(y)$, $E(3x^2 - 2x)$, μ, σ^2, σ, and $M(s)$.

8. Consider the following discrete probability distribution:

$$P(x = x_1) = 0.6$$

$$P(x = 2) \quad = 0.3$$
$$P(x = x_2) = 0.1$$

Find x_1 and x_2 such that $\mu = 1$ and $\sigma^2 = 10$.

9. Given:

$$M_x(s) = \left(1 - \frac{s}{6}\right)^{-1}$$

$$M_y(s) = \left(1 - \frac{s}{10}\right)^{-1}$$

Find $M(s)$, μ, and σ^2 of
(a) $w = x + y$
(b) $w = x + 2y$

10. Given that:

$$M(s) = \frac{k}{2 - s}$$

Find μ, σ^2, and $E(x^3)$.

11. Given:

$$M(z) = A(1 - 3z)^3$$

Find μ, $E(x^3)$, and $p(x = 2)$.

12. Find $M(z)$ and $M(s)$ for:
(a) $P(x = 4) \quad = 0.6$
$\quad\;\; P(x = 5) \quad = 0.2$
$\quad\;\; P(x = 10) = 0.2$
(b) Poisson distribution

REFERENCES

1. Cramer, H., *The Elements of Probability Theory and Some of Its Applications.* New York, John Wiley & Sons, Inc., 1955.

2. Drake, A. W., *Fundamentals of Applied Probability Theory.* New York, McGraw-Hill, Inc., 1967.

3. Feller, W., *An Introduction to Probability Theory and Its Applications*, Vol. 1, 2nd ed. New York, John Wiley & Sons, Inc., 1957.

4. Hill, T. W., "On Determining a Distribution Function Known Only by Its Moments and/or Moment Generating Function." Unpublished PhD. Dissertation, Arizona State University, 1969.

5. Meyer, P. L., *Introductory Probability and Statistical Applications.* Reading, Mass., Addison Wesley Publishing Co., Inc., 1965.

6. Parzen, E., *Modern Probability Theory and Its Applications.* New York, John Wiley & Sons, Inc., 1960.

ACTIVITY NETWORKS: PERT AND CPM

During the last few years, business and industry have found it both convenient and practical to do much of their work in segments or projects. *Projects have come to occupy such an important position in industry that much time and effort have been spent developing project management techniques. The basis of most of the more successful project management techniques is the* activity network *or* project network *model.*

Two major variations of this model are now being applied successfully to increasing numbers of complex engineering projects of all types. The first is called PERT—project evaluation and review technique. The second is CPM —critical path method.

PERT was originally developed as a joint effort by Booz, Allen, and Hamilton and the Navy's Special Project Office. Its basic feature is the estimation of time spans. Each project job-duration estimate is made at a single facility, cost, or investment level. A computational procedure then generates an estimated overall project duration and derives a measure of the certainty of meeting this estimate.

CPM was developed by J. E. Kelley, Jr. and Morgan Walker [1]. Unlike PERT, this method considers duration estimates over a range of facility or

3

cost levels and, as a result, provides a range of project durations with an associated range of project costs. CPM computations establish the absolute minimum cost of attaining any feasible project duration. The basic activity-network procedure consists of three phases: the planning phase, the scheduling phase, and the control phase. These phases will be discussed in this chapter.

3.1 Arrow Diagrams

The heart of critical path methods is a network portrayal of the plan for carrying out the program. Such a network shows the precedence relationships of the elements of the program leading to the program's completion and is called an *arrow diagram*. The idea of using the arrow diagram is not new; however, it wasn't until 1958, when PERT and CPM were available, that the technique became popular. The arrow diagram is an outgrowth of the familiar Gantt chart.

3.1.1 Drawing arrow diagrams. The development of the arrow diagram occurs in the planning phase of the project. Costs, resources, and times are neglected at this point. We ask three questions about each element of the project:

1. What immediately precedes this element?
2. What can be done concurrently?
3. What immediately follows this job?

With these questions answered we are prepared to draw an arrow diagram of a project. The diagram has two elements:

1. *Arrow* or *arc* represents the activities of the project under consideration.
2. *Node* or *event* represents the intersection of activities.

The node rule for an arrow diagram is that no activity can start from an event until all activities entering this event are complete.

Consider a project for which the analysis has resulted in the definition of five activities: *A*, *B*, *C*, *D*, and *E*. Furthermore, assume the following precedence relationships between these activities:

1. *A* and *B* can be started simultaneously.
2. Activity *C* can be started only upon the completion of *A*.
3. Activity *D* can be started when *B* is completed.
4. Both *C* and *D* must be complete before *E* may be started.

In the analysis, estimated durations of these activities have not been specified. Consequently, the lengths of the arrows used in the diagram representing these activities do not correspond to any specific lengths of time. Only the relative positions of the arrows have any significance. The five-activity project for which the precedence relationships have been defined may therefore be shown as an arrow diagram in Fig. 3.1(a).

Suppose the precedence relationships for this example were changed slightly to read:

1. *A* and *B* can be started simultaneously.
2. Activity *C* can be started only upon the completion of *A*.
3. Activity *D* can be started *when A and B are completed.*
4. Both *C* and *D* must be complete before *E* may be started.

The arrow diagram as shown in Fig. 3.1(a) is no longer appropriate for the new problem because it violates the restriction that activity *D* be dependent upon activity *A*. We might consider redrawing the network as shown in Fig. 3.1(b), but this is also incorrect. This diagram implies that activity *C* is dependent upon activity *B*, which is not the case. To display this relationship, it will be necessary to introduce another type of activity called a *dummy activity*. A dummy activity, represented by a broken-line arrow, has a duration of zero. If we introduce a dummy as shown in Fig. 3.1(c), the new pre-

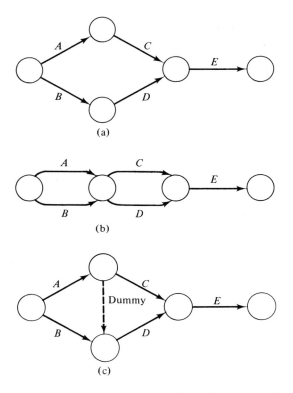

(a)

(b)

(c)

Figure 3.1. An example of the use of dummies in an activity network.

cedence relationship can be modeled. The arrow diagram says that activity *D* cannot start until both the dummy activity and activity *B* are completed. But the dummy must wait for activity *A* to be completed, so activity *D* is dependent on activities *A* and *B*.

In real life we must determine the interrelationships among the activities in the project we are modeling. The following example illustrates how you proceed to establish your own interrelationships.

EXAMPLE:

Consider an example from everyday life—a grease job and oil change for your car. Make a list of all the jobs involved, arranging them in more or less chronological order. Your list might look something like this:

Hoist car
Remove drain plug, drain oil
Grease underside fittings

Inspect tires and exhaust system
Check differential and transmission levels
Replace drain plug
Lower car
Refill crankcase
Grease upper fittings
Oil generator and distributor
Check radiator and battery

Assume there are two service station attendants to work on your car. Now prepare a diagram of the work, using arrows to represent the jobs you have listed. It is logical to hoist the car first so that the oil can be drained at the outset. Therefore, start the arrow diagram with a job arrow marked "hoist car."

The next job on the list is "remove drain plug and drain oil." What other job(s) must be completed before this job can start? Obviously, "hoist car" is the only job that must be done before the oil can be drained.

What job(s) can be done while this is being done? There are three such jobs—"grease underside fittings," "inspect tires and exhaust system," and "check differential and transmission." It is logical to expect that one attendant will drain the oil and inspect tires and exhaust *while* the oil is draining. The other attendant will grease the underside fittings and check differential and transmission while the first is draining the oil and inspecting.

What job(s) cannot start until this job ("remove drain plug and drain oil") is done? All of the remaining jobs must wait until the oil is drained. Obviously, the next job is to replace the drain plug and then lower the car. This done, one of the attendants can refill the crankcase while the other greases the upper fittings, oils the generator and distributor, and checks the radiator and battery. Add these job arrows in their logical positions. Our arrow diagram will then look like Fig. 3.2.

We have been dealing with simple arrow diagrams showing the plans for very simple operations. However, complicated operations are diagramed in exactly the same way, by adding arrows step by step. With a little practice, diagraming can be done very quickly. All that is needed is a sheet of paper, a pencil, a practical knowledge of the work to be done, and a detailed analysis of the sequence in which the different jobs are to be carried out.

A word of caution: It is estimated that seventy percent of the work required in the implementation of a successful project-network endeavor is spent in the planning phase. Since both the scheduling and control phases, and the overall success of the project, are solely dependent upon the work done in the planning phase, it is extremely important that the arrow diagram be representative of the procedures and activities that will actually be used on the project. If this is not the case, then the value of project management is negated.

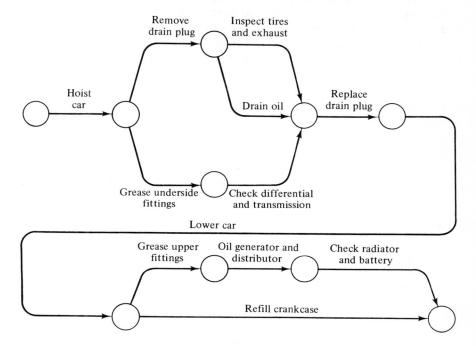

Figure 3.2. Arrow diagram for servicing a car.

In summary, the following rules can be stated for establishing an arrow diagram:

1. Each activity is represented by one and only one arrow.
2. The length of the arrow has no meaning with respect to the importance or duration of an activity.
3. The arrow direction merely indicates the general progression of time; it has no vectorial significance.
4. The intersection of activities is called an event and is represented by a circle. All activities begin and end in events. Events are identified by numbers within the event circles.
5. Only the relative positions of activities with common events have significance. Arrows originating at an event indicate activities that can only begin after all activities terminating at the event have been completed.
6. If one event takes precedence over another but there is no activity relating them, a dummy activity is used. Dummy activities are represented by a broken-line arrow and have no duration or cost.

3.1.2 The addition of time to our model. Now that we are *experts* on drawing arrow diagrams, let us add time to the diagram. Adding the time dimension will allow scheduling of the various elements of the project.

Estimating the time required for each task is usually done by one or more people thoroughly familiar with the task. This estimate is the total elapsed time required from the start to the finish of a task and is called the *time duration*. It is the time required to accomplish the task most economically. Later adjustments can be made to lengthen or shorten this time duration as required. Some systems have extensive procedures for arriving at time durations based on several estimates involving statistics and probability. At present we assume one estimate per task. To illustrate the inclusion of time into an arrow diagram, consider Fig. 3.3(a). The numbers on the arrows represent the time estimates to complete that activity. For example, task *F* is estimated to take eight time units.

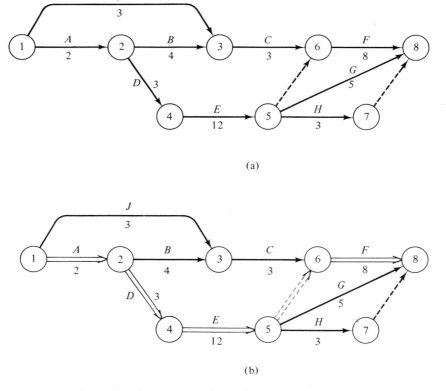

(a)

(b)

Figure 3.3. An example of a project activity diagram.

In the arrow diagram, Fig. 3.3(a), the project starts with tasks *A* and *J*. At the completion of task *A*, both tasks *B* and *D* may start. Task *C* follows *B* and *J*, and task *E* follows *D*. There is no relationship between tasks *C* and *E* except that they both follow *A*. Tasks *G* and *H* can start at the completion of task *E*. Task *F* can start upon the completion of task *C* and task *E*. Since it is not necessary for tasks *G* and *H* to follow *C*, they are not connected to task *C*, and the relationship of tasks *E* and *F* is shown by the use of a dummy. The project is finished upon completion of tasks *F*, *G*, and *H*. Since tasks *G* and *H* would have the same numbers of their start and finish nodes, a dummy was used to connect task *H* to the finish. This allowed task *H* to have different node numbers from task *G*. This unique identification is required by some computer packages developed to analyze activity networks.

By inspection of Fig. 3.3(a), we can see that route or path *A*, *B*, *C*, *F* would require 17 days, path *A*, *D*, *E*, *G* would require 22 days, path *A*, *D*, *E*, *H* would require 20 days, and path *A*, *D*, *E*, *F* would require 25 days. Therefore, path *A*, *D*, *E*, *F* is the *critical path* as shown in Fig. 3.3(b). If any shortening of the project is to be accomplished, it must first be along this path *A*, *D*, *E*, *F*. Shortening of tasks *B*, *C*, *G*, or *H* by using overtime, double-shifting, or more men, machines, or materials would not shorten the completion time of the project. Shortening of the tasks on the critical path (task *A*, *D*, *E*, and *F*) would shorten the project time, and it is these tasks which must receive special attention. Shortening of tasks along the critical path can eventually cause other paths to become critical. For example, if task *F* is shortened to less than five days, task *G* becomes critical since the critical path would then be *A*, *D*, *E*, *G*.

In the example shown in Fig. 3.3(b), all of the relationships that have been mentioned are apparent by inspection and mental arithmetic. What happens, however, when the tasks increase in number from 9 to 90 or to 900? Exactly the same procedure is followed up to and through the creation of the arrow diagram and the list of node numbers, durations, and task descriptions. A computer is usually used to do the work of analyzing the project and creating the schedule data.

We will now present an algorithm for determining the critical path and other scheduling information in a mechanical fashion. The technique can either be applied by hand or programmed for the computer.

The network is first solved from source to sink by what are often referred to as *forward-pass rules*. These rules can be stated as follows:

1. The earliest event time for the source event is assumed to be zero.
2. Each activity is assumed to start as soon as the event at which it starts is realized. The earliest finish time for an activity is equal to its earliest start time plus its duration.

3. The earliest event time is equal to the longest of the earliest finish times of the activities merging on the event. Return to step 2.

Next the network is analyzed in similar manner working backwards from the sink. The rules are as follows and are often called *backward-pass rules*:

1. The earliest event time for the sink is equated to its latest allowable event time.
2. The latest allowable finish time for an activity is equated to its successor event's latest allowable time. The latest start time for an activity is its latest finish time minus its duration.
3. The latest allowable time for an event is the smallest of the latest start times for the activities emanating from the event. Return to step 2.

This algorithm can be stated in equation form using the following nomenclature:

t = single estimate of mean activity duration time
T_E = earliest event occurrence time
T_L = latest allowable event occurrence time
ES = earliest (activity) start time
EF = earliest (activity) finish time
LS = latest allowable (activity) start time
LF = latest allowable (activity) finish time

FORWARD-PASS RULES

1. $T_E = 0$ (for the source node)
2. $ES = T_E$ (for the predecessor event)
   $EF = ES + t$
3. $T_E = \max (EF_1, EF_2, \ldots, EF_n)$ for an event with n activities merging on it
   Return to step 2.

BACKWARD-PASS RULES

1. $T_L = T_E$ (for the sink node)
2. $LF = T_L$ (for successor event)
   $LS = LF - t$
3. $T_L = \min (LS_1, LS_2, \ldots, LS_n)$ for an event with n activities originating from it
   Return to step 2.

If you are applying this algorithm, you can either apply it directly on the arrow diagram or solve it in a tabular form. Figure 3.4 shows these rules as applied to the arrow diagram introduced in Fig. 3.3. These results are also tabulated in Table 3.1.

Figure 3.4 and Table 3.1 are derived as follows. Referring to Fig. 3.3, we see that tasks *A* and *J* are started immediately at zero time. They have time durations of 2 days and 3 days, respectively; therefore, their earliest starting times are zero and their earliest finish times are 2 days and 3 days, respectively. At the finish of task *A*, tasks *B* and *D* may start. Since task *A* has an earliest finish of 2 days, tasks *B* and *D* have an earliest start date of 2 days. Adding their time durations to their earliest start times, we find that they have earliest finish times of 6 days and 5 days, respectively. Task *C* can start at the finish of both tasks *B* and *J* and cannot start until both are finished; therefore, it can start on the 6th day. Its earliest finish would be 9 days. Continuing through the project, we can see that task *F* has an earliest finish of 25 days. This is the earliest the project can be completed under the conditions set forth in the arrow diagram and the time duration estimated. Reversing the procedure, we can calculate the latest start and latest finish for each of the tasks. The latest finish of the last task in the project is the same as the earliest finish of this task. Subtracting the time duration from this date gives the latest start of that task.

Table 3.1. CRITICAL PATH CALCULATION

Nodes		*t*	Description of Tasks	ES	EF	LS	LF	TF	FF
1	2	2	Task *A*	0	2	0	2	*	0
1	3	3	Task *J*	0	3	11	14	11	3
2	3	4	Task *B*	2	6	10	14	8	0
2	4	3	Task *D*	2	5	2	5	*	0
3	6	3	Task *C*	6	9	14	17	8	8
4	5	12	Task *E*	5	17	5	17	*	0
5	6	0	Dummy	17	17	17	17	*	0
5	7	3	Task *H*	17	20	22	25	5	0
5	8	5	Task *G*	17	22	20	25	3	3
6	8	8	Task *F*	17	25	17	25	*	0
7	8	0	Dummy	20	20	25	25	5	5

(*—Critical Path)   (Project completion 25)

The next two items of the scheduling procedure concern what is called *float*. Float is a measure of allowable delay or leeway. There are two types of floats normally considered, *total float* and *free float*. Total float (*TF*) is the time any given task may be delayed before it will affect the project completion time. We compute it by subtracting the earliest finish from the latest finish. Free float (*FF*) is the time any task may be delayed before it will affect the earliest starting date of any of the tasks immediately following. We compute it by subtracting the earliest finish time from the earliest start date of the very next task. These quantities are reported in Table 3.1 and Fig. 3.4. We notice that those items on the *critical path* can be identified by the fact that their total floats equal zero.

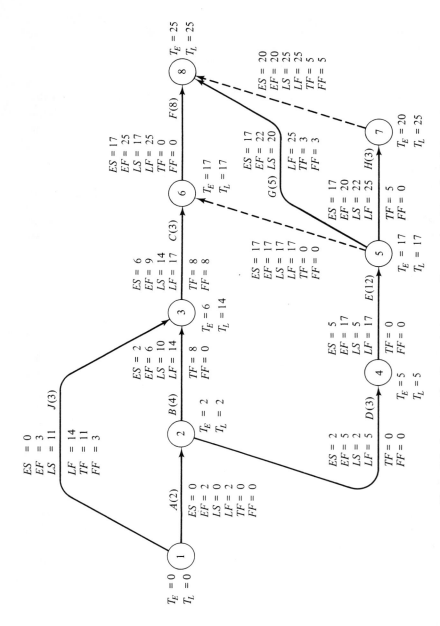

Figure 3.4. An example of a project arrow diagram showing calculations.

40

3.2 PERT: Program Evaluation and Review Technique

3.2.1 Introducing uncertainty into the activity times. PERT introduces uncertainty into the time estimates for activities and hence in the project duration. It is therefore well suited for situations where there is either insufficient data to predict activity durations or where the project activities involve research and development.

The first step in a PERT analysis is to create an arrow diagram as described in Section 3.1. PERT uses an activity duration called the *expected mean time* (t_e) together with an associated measure of uncertainty of this activity duration. The uncertainty may be expressed as either the standard deviation (σ_e) or the variance (V_e) of the duration. These two quantities are related by the relationship $V_e = \sigma_e^2$.

PERT uses three time estimates for each activity:

a, the optimistic time (which should have only a very low probability of occurring);

m, the most likely time (or the mode of the beta distribution);

b, the pessimistic time (which also should have only a very low probability of occurring).

PERT then assumes that these three estimates can be used to describe a beta distribution for each activity duration (see Fig. 3.5). Then t_e, the expected time for the activity, can be computed:

$$t_e = \frac{a + 4m + b}{6}$$

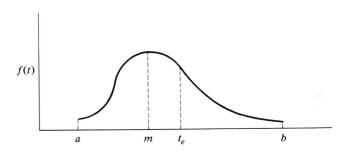

Figure 3.5. PERT beta distribution.

The standard deviation, σ_e, and variance, V_e, for each activity can also be computed:

$$\sigma_{t_e} = \frac{b-a}{6}$$

$$V_{t_e} = \left(\frac{b-a}{6}\right)^2$$

The logic behind these calculations will be discussed in Section 3.2.3. For example, for the PERT network shown in Fig. 3.6, the activity going from event 3 to event 5 has time estimates of $a = 0$, $m = 6$, $b = 18$. The parameters for this activity would be calculated as follows:

$$t_e = \frac{0 + 4(6) + 18}{6} = 7$$

$$\sigma_{t_e} = \frac{18-0}{6} = 3$$

$$V_{t_e} = \left(\frac{18-0}{6}\right)^2 = 9$$

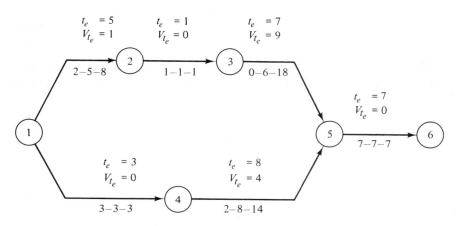

Figure 3.6. Sample PERT network.

The analysis of the arrow diagram proceeds in a fashion similar to that described in Section 3.1.2 with particular emphasis on determining the variance of the T_E and T_L of the events.

To permit the finding of the variance of the earliest ($\sigma_{T_E}^2$) and latest ($\sigma_{T_L}^2$) times, we assume that the elapsed times for individual activities are statistically independent. For only one path to an event, the earliest time equals the elapsed times to the event. Therefore the expected value and the variance of the earliest time equal the sum of the expected values and the sum of the variances, respectively, to the event. For example, there exists only one path to event 3 in Fig. 3.6, that being from 1–2–3. The expected value of the earliest time (μ_{T_E}) is $5 + 1 = 6$, and the expected value of the variance of the earliest time is $1 + 0 = 1$. For the case of one path, the expected latest time (μ_{T_L}) is calculated in a similar manner.

Now consider the case where there exists more than one path to an event. For event 5, two possible paths exist: 1–2–3–5 and 1–4–5. The earliest time of event 5 equals the greater of the total elapsed times along these paths. However, it is difficult to find the exact expected value and variance for the maximum time. Therefore, the largest total elapsed time is said to always occur on the path with the largest expected total elapsed time. The expected total elapsed times to event 5 are 13 and 11. Path 1–2–3–5 has the larger value (13), so it is assumed that the actual total elapsed time is greater on this path. With this assumption, an approximation for the expected value of the total of the elapsed time is 13. The variance of T_E and T_L are calculated under the same assumption, i.e., the path with the largest expected time will always dominate other paths. Therefore, for our example, an approximation for the variance for the T_E for node 5 equals 10 $(1 + 0 + 9)$.

Slack for the PERT-type network is defined as the expected latest time for an event (μ_{T_L}) minus the expected earliest time for an event (μ_{T_E}). Slack is similar to the float defined in Section 3.1.2, except that it is defined with respect to events instead of activities.

The calculations for a PERT network are usually summarized as shown in Table 3.2 which describes Fig. 3.6.

Table 3.2. PERT CALCULATIONS

Event No.	Earliest Time (T_E) Expected Value μ_{T_E}	Variance $\sigma_{T_E}^2$	Latest Time (T_L) Expected Value μ_{T_L}	Variance $\sigma_{T_L}^2$	Slack
1	0	0	0	10	0
2	5	1	5	9	0
3	6	1	6	9	0
4	3	0	5	4	2
5	13	10	13	0	0
6	20	10	20	0	0

To find the critical path, examine the slack column of Table 3.2. Where there is a zero slack, an event is in a critical path. The critical path for the given example is shown to be 1–2–3–5–6. This path is the path with the longest expected time through the network, and any increase in any of the activity on the critical path will yield a corresponding increase in the total project time.

3.2.2 Uses of the probabilistic information.

Once the expected earliest time of an event (μ_{T_E}) and its standard deviation (σ_{T_E}) have been determined, it is possible to use probability theory to calculate the chances of meeting a specific scheduled time (T_s) for a node. Based on the central limit theorem, the earliest completion for an event is assumed to have a normal probability distribution with a mean of μ_{T_E} and a standard deviation of σ_{T_E}.

Suppose m independent tasks are to be performed in order (one might think of these as the m tasks which lie on the critical path of a network), and let $t_1, t_2, \ldots, t_m$ be the actual duration of each of these tasks. Note that these are random variables with true means $\mu_1, \mu_2, \ldots, \mu_m$ and true variances $\sigma_1^2, \sigma_2^2, \ldots, \sigma_m^2$, and these actual times are unknown until the specific tasks are performed. Now define $T = t_1 + t_2 + \cdots + t_m$ and note that T is also a random variable and thus has a distribution. The central limit theorem states that, if m is large and the variables are independent, the distribution of T is approximately normal with mean T and variance V_T given by:

$$T = \mu_1 + \mu_2 + \cdots + \mu_m,$$

and

$$V_T = \sigma_1^2 + \sigma_2^2 + \cdots + \sigma_m^2$$

That is, the mean of the sum is the sum of the means; the variance of the sum is the sum of the variances; and the distribution of the activity times will be normal regardless of the shape of the distribution of actual activity performance times.

To calculate the probability of meeting a scheduled time T_s for a particular event, it is necessary to visualize a normal distribution centered at μ_{T_E} as illustrated in Fig. 3.7. The probability of meeting the desired scheduled

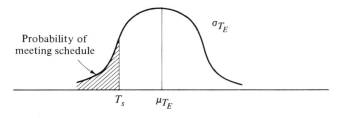

Figure 3.7. Calculation of the probability of meeting schedule.

time T_s is obtained by finding the area under the normal curve to the left of T_s.

The familiar Z statistic of the normal distribution can be calculated if we realize that $T_s = \mu_{T_E} + Z\sigma_{T_E}$. Therefore:

$$Z = \frac{T_s - \mu_{T_E}}{\sigma_{T_E}}$$

The Z value can be converted to a probability by means of Table 3.3 or a more accurate table available in all probability and statistics books.

Table 3.3. APPROXIMATE VALUES OF THE STANDARD NORMAL FUNCTION

Z	Cumulative Probability	Z	Cumulative Probability
-2.0	0.02	$+0.1$	0.54
-1.5	0.07	$+0.2$	0.58
-1.3	0.10	$+0.3$	0.62
-1.0	0.16	$+0.4$	0.66
-0.9	0.18	$+0.5$	0.69
-0.8	0.21	$+0.6$	0.73
-0.7	0.24	$+0.7$	0.76
-0.6	0.27	$+0.8$	0.79
-0.5	0.31	$+0.9$	0.82
-0.4	0.34	$+1.0$	0.84
-0.3	0.38	$+1.3$	0.90
-0.2	0.42	$+1.5$	0.93
-0.1	0.46	$+2.0$	0.98
0	0.50		

For the PERT network described in Fig. 3.6, let us find the probability of completing the project within 18 time units. This is the same as asking what is the probability of realizing event 6 before time 18? From Table 3.2 we find $\mu_{T_E} = 20$ and $V_{T_E} = 10$. Realizing that $\sigma_{T_E} = \sqrt{V_{T_E}} = \sqrt{10}$, we can calculate Z:

$$Z = \frac{18 - 20}{\sqrt{10}} = -0.64$$

Therefore, we can see from Table 3.3 that there is a probability of about 0.26 that the project will be finished in 18 time units or less. This approach for finding the probability of meeting a schedule can be applied to any event in the network, not just to the last event.

Another use for the probabilistic information in PERT is in finding the probability that a particular event will have slack. Since slack $= T_L - T_E$ and both T_L and T_E are random variables, then slack must be a random variable. Slack is defined as the difference of two independent normal distributions. It can be shown that under this definition slack is also normally distributed with a mean equal to $\mu_{T_L} - \mu_{T_E}$ and a variance equal to $\sigma_{T_L}^2 + \sigma_{T_E}^2$. The probability of having slack will equal the area to the right of zero in Fig. 3.8. Figure 3.8 pictures the distribution of slack at a particular event.

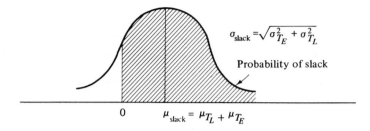

Figure 3.8. Calculation of the probability of slack.

The Z value associated with the zero point is that

$$0 = \mu_{\text{slack}} + Z\sigma_{\text{slack}}$$

and

$$Z = \frac{-\mu_{\text{slack}}}{\sigma_{\text{slack}}} = \frac{-(\mu_{T_L} - \mu_{T_E})}{\sqrt{\sigma_{T_L}^2 + \sigma_{T_E}^2}}.$$

The Z value can be converted to a probability using Table 3.3, which represents the area to the left of a particular Z value. Since we are seeking the area to the right, we take the quantity found in Table 3.3 and subtract it from one (i.e., the complement of Table 3.3).

For Fig. 3.6, what is the probability that event 4 will have slack? From Table 3.2 we find that:

$$\mu_{T_E} = 3$$
$$\sigma_{T_E}^2 = 0$$
$$\mu_{T_L} = 5$$
$$\sigma_{T_L}^2 = 4$$

Therefore:

$$Z = \frac{-2}{\sqrt{0 + 4}} = -1$$

From Table 3.3, we find a probability of 0.16. Taking its complement, we find that the probability of slack at event 4 is 0.84.

3.3. A critical study of the assumptions used in the analysis and solution of a PERT network.* PERT has been successfully used in both industry and government for some time. However, the methods and assumptions used in the development of PERT are still subject to constant discussion [9, 11, 12, 13, 19, 21, 22]. This section will discuss the errors and inconsistencies that have been found in the PERT assumptions. This section is intended not to downgrade PERT but simply to inform the reader of its weaknesses so that he can have fuller understanding of the technique. Most of the material and almost all the data presented in this section are drawn from an excellent article by MacCrimmon and Ryavec [9].

PERT is essentially a network analysis. Between each two points in the network are given three time estimates for completion. The two extremes are the optimistic and pessimistic estimates. The third is the mode of the time distribution. This time distribution is estimated by the beta distribution and defined by the function:

$$f(t) = k \cdot (t - a)^{\alpha} \cdot (b - t)^{\beta}, \qquad a \leq t \leq b, \qquad a, b \geq 0$$

where a and b are the pessimistic and optimistic times, respectively. The mean of the distribution is assumed to be $(a + 4m + b)/6$, and the standard deviation is given by $(b - a)/6$. It should be pointed out that the true distribution of the times is unknown, so all of the above equations are just approximations. There are, however, some properties of the distribution that are known. The first is that there are two nonnegative abscissa intercepts. This must be true because there cannot be negative times. The second hypothesis is that the distribution is continuous. This is not necessarily true, but a continuous function will approximate a discrete distribution closely enough to make this a sound assumption. The third assumption is that the distribution must be unimodal. The fact that there can be only one most likely time bears this out. Because the beta distribution possesses all of these qualities, it seems reasonable that it can be used to approximate the true distribution of activity time.

The first group of assumptions and errors that should be considered are those dealing with the specific activity times. The most important of these is the assumption that the beta distribution approximates the true distribution over the range from the optimistic time estimate to the pessimistic estimate. As was previously pointed out, the beta distribution does have three characteristics of the time estimates, but there are also some inconsistencies. Along with this, it is also necessary to include the corollary assumptions about the mean and the standard deviation.

---

*The material in this section is based upon an adaptation of Reference 9 and is used with the permission of ORSA.

From the equation of the beta distribution, we see that there are four variables (a, b, α, β). The variables a and b are the limits of the distribution. The problem is that there are four unknowns but only three equations to solve. To remedy this, the originators of PERT introduced the equation for the standard deviation: $(b - a)/6$. The question is whether this is the best relationship that can be developed. Let us assume that there are two independent random variables that are defined over the interval (a, b). Statistically the standard deviation would be $\sqrt{2}(b - a)/6$; however, the sum is defined over the region $(2a, 2b)$. In this case, the PERT assumption of $(b - a)/6$ predicts the incorrect result of $(2b - 2a)/6$.

Before we look at some ways to correct this problem, the error associated with these assumptions will be found. The first error is that related to the beta distribution assumption. To calculate the maximum error, we hypothesize two possible extreme distributions: (1) the times are almost a straight, uniform distribution and (2) the times are almost a pure delta distribution. Using these two extremes over the interval $(0, 1)$, the following worst errors are:

Worst Error in the Mean

$$\max\left\{\left[\frac{(4m + 1)}{6} - \frac{1}{2}\right], \left[\frac{(4m + 1)}{6} - m\right]\right\} = \frac{1}{3}(2m - 1)$$

Worst Error in the Standard Deviation

$$\max\left\{\left[\sqrt{\frac{1}{12}} - \frac{1}{6}\right], \left[0 - \frac{1}{6}\right]\right\} = \frac{1}{6}$$

These values are calculated on the basis of the following parameter values:

Distribution	Mean	Standard Deviation
Beta	$\frac{1}{6}(4m + 1)$	$\frac{1}{6}$
Uniform	$\frac{1}{2}$	$\sqrt{\frac{1}{12}}$
Delta	m	0

We can also calculate the errors that are introduced by the assumption that the standard deviation is $\frac{1}{6}(b - a)$ and that the mean is as above. For the following conditions, the results are as shown:

Interval $(0, 1)$
Actual mode $\alpha/(\alpha + \beta)$
Actual mean $(\alpha + 1)/(\alpha + \beta + 2)$
Actual standard deviation $\sqrt{(\alpha + 1)(\beta + 1)/(\alpha + \beta + 2)^2(\alpha + \beta + 3)}$

Worst Error in the Mean

$$\left| \frac{1}{6}(4m + 1) - \frac{m(\alpha + 1)}{(\alpha + 2m)} \right|$$

Worst Error in the Standard Deviation

$$\left| \frac{1}{6} - \sqrt{\frac{m^2(\alpha + 1)(\alpha - \alpha m + m)}{(\alpha + 2m)^2(\alpha + 3m)}} \right|$$

The actual percentage values associated with these two errors are 33% for the mean and 17% for the standard deviation. If the restrictions of $1 \leq \alpha \leq \beta$ and $|\frac{1}{2} - m| \leq \frac{1}{6}$ are placed on the equations, the errors drop to 4% for the mean and 7% for the standard deviation.

The presentation of these errors and inconsistencies in the PERT assumptions does not do anything to improve the results obtained from a PERT analysis. Therefore, we must look for some changes or modifications to help correct the inconsistencies.

One possible improvement in estimating the mean and variance is to adopt the mean as the third time estimate instead of the mode. For this derivation, let $x_1 = a$, $x_2 = $ mean, $x_3 = b$, $\alpha = p - 1$, $\beta = q - 1$. Using these values, we get the following estimates of the mean and standard deviation:

$$\mu = \frac{(pb + qa)}{(p + q)}$$

$$\sigma = \sqrt{\frac{(b - a)^2 pq}{(p + q)^2(p + q + 1)}}$$

This change in time estimate changes the mean and standard deviation values in two ways. First, it greatly improves the chance of statistical accuracy by removing the statistical inconsistencies inherent in the previous values. This method also allows for a greater skew variation of the distribution.

Another possible alternative is to use an entirely different probability distribution such as the gamma distribution which is defined as follows:

$$f(t) = \frac{\delta^\lambda t^{\lambda - 1} \gamma^{-\delta t}}{\Gamma(\lambda)}, \qquad 0 < t \leq \infty, \qquad \gamma, \lambda > 0$$

$$= 0 \qquad\qquad\qquad\qquad\qquad \text{otherwise}$$

The main advantage in using the gamma distribution is that the parameters can be evaluated by only two time estimates instead of the three that the beta distribution requires. These estimates can be intermediate points which are easier to establish than precise end points of an unknown distribution.

The last alternative deals with this subject of just what time estimates have to be. If we change the meaning of *a* and *b* to some percentage point instead of an absolute extreme, the results should be more accurate. When an extreme is estimated, it is often necessary to extrapolate to get values, but if we were to estimate the 0.025 and 0.975 points of the distribution, it may be easier to get accurate results.

Thus far all of the attention has been focused on the activity times in a PERT network, but this is only half of the problem. There is a significant problem with the analysis of the network as a whole.

The method used to find the project completion time distribution is to find the path that has an expected value not less than that of all the other paths. After this is done, PERT assumes that the project duration is normally distributed with a mean equal to the expected value of the largest path and a standard deviation equal to the standard deviation of this same path. Using these results, the PERT calculated mean is never greater than the true mean, and the standard deviation is usually greater than the actual standard deviation.

Different types of networks will be analyzed separately and in combination. It will be assumed that the activity time distribution is known, so that all the error is from the network assumptions and configurations.

For a single straight-line path that is clearly longer than any other path, and if the other paths have no effect on its completion time distribution, the PERT assumptions make no error in finding the project mean and standard deviation. For the case of two parallel paths, PERT will take the sum of the means and standard deviations of the longest path. The problem occurs when two paths have mean lengths very close to the same value. The second path will affect the completion time distribution. Numerically, for two paths each having a mean of $\frac{1}{2}$, the mean of the maximum time distribution is actually 0.63. For three paths which have no cross-connection, the error goes up and the maximum time distribution mean is 0.69. Thus, as the number of paths increases, the error of the PERT assumptions will also increase. One factor which will decrease this tendency is cross-connecting paths. To demonstrate this, consider the following three networks. The first is two parallel paths with the same mean lengths. It has an error of about 13%. The second situation is three paths all with the same mean length. This will result in an error of over 18%. The last set of paths is one in which two parallel paths are cross-connected with all paths still having the same expected length. For this case the error is just over 15%. This shows that cross-connection can lower the potential error and also that the number of paths has a direct bearing on the extent of the error. Generalizing the information on parallel paths, if there are two paths, the closer the mean length of the paths, the greater the error in both the mean and standard deviation. Some data on this show that, for two paths whose lengths are equal, the mean error is about -17% and the standard deviation error is $+39\%$. For a ratio of one to two the figures are -0.5% and $+4\%$. For a ratio of one to four the figures are both 0.0%. (There is an inconsistency in results where a 1:1 ratio produced 17% in one case and 13% in the other. This occurred because the lengths were not the same. The longer the path, the lower the error.) When series and parallel paths are combined, the error comes from the parallel sections only; so the magnitude of the error depends on the relative lengths of the two types.

3.3 CPM: Critical Path Method

3.3.1 Introducing cost data into the activity network.   CPM was developed within the construction industry where previous experience in similar work can be used to predict time durations and cost within a tight range. It is difficult to give a universally acceptable description of CPM because many companies have developed their own versions of a cost model based on activity network. This section adopts the philosophy of the system developed by the General Electric Company [4, 6, 14]. Like PERT, CPM consists of three phases—the planning phase, the scheduling phase, and the control phase. The initial step of the planning phase is the preparation of the detailed analysis of the project. This is usually accomplished by the use of the arrow diagram. The second step in the planning phase is the assignment of estimated durations and costs to each activity on the arrow diagram. Although there are many methods currently used to determine time and cost estimates, a procedure utilizing two time/cost estimates will be illustrated in this section. These two estimates are called the "normal" and "crash" time/cost estimates. Based upon the relationship between time and money (i.e., the shorter the time the higher the cost, and the longer the time the smaller the cost), these two estimates can be defined as follows:

1. Normal time and cost—the duration to complete the project requiring the least amount of money.
2. Crash time and cost—the minimum possible time to complete the activity and the associated increased cost.

After the assignment of time/cost estimates to all activities on the arrow diagram, the planning phase of CPM is complete.

Since each activity is assigned a duration range and related cost, obviously there is a range of durations available for the project depending upon the individual time selections for each activity. Each one of these various project durations also presents a different project cost. In the scheduling phase, the mathematics of CPM is used to develop these various project durations from which the optimum schedule will be selected. In each instance only the lowest possible cost for each different project duration is produced. The following problem is presented to illustrate what is accomplished in the scheduling phase of CPM.

CPM PROBLEM

Using the basic network model illustrated in Fig. 3.9 and assigning time/cost estimates to each activity as shown in Table 3.4, we are ready to begin the scheduling processes of CPM.

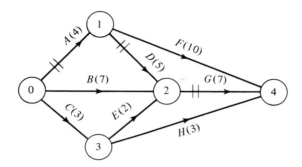

Figure 3.9. All normal CPM network.

Table 3.4. COST TABLE

	Normal		Crash	
Activities	*Days*	*Dollars*	*Days*	*Dollars*
A	4	$100	3	$200
B	7	280	5	520
C	3	50	2	100
D	5	200	3	360
E	2	160	2	160
F	10	230	8	350
G	7	200	5	480
H	2	100	1	200
	Total	$1,320	Total	$2,370

From our definition of the critical path, the longest project duration using normal time estimates would be 16 days along path *A–D–G*. To shorten the project's duration, either activity *A*, *D*, or *G* will have to be expedited. To ensure that this acceleration will be done at the lowest possible cost, we must calculate one further figure for each activity which we will call the *activity cost slope*. Using activity *B* as an example and assuming a linear relationship, we can present the cost curve for activity *B* graphically, as shown in Fig. 3.10.

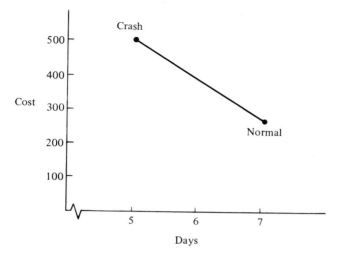

Figure 3.10. Time cost curve for activity *B*.

The cost slope of this curve is computed by the formula:

$$\frac{\text{crash cost} - \text{normal cost}}{\text{normal time} - \text{crash time}}$$

For activity *B*, the cost slope is:

$$\frac{\$520 - \$280}{7 \text{ days} - 5 \text{ days}} = \frac{\$240}{2 \text{ days}} = \$120 \text{ per day}$$

If we perform this calculation for each activity, an additional column can be added to our cost table as shown in Table 3.5.

Table 3.5. COST TABLE

	Normal		Crash		Cost
Activities	*Days*	*Dollars*	*Days*	*Dollars*	*Slope*
A	4	$100	3	$200	$100
B	7	280	5	520	120
C	3	50	2	100	50
D	5	200	3	360	80
E	2	160	2	160	—
F	10	230	8	350	60
G	7	200	5	480	140
H	2	100	1	200	100
	Total	$1320	Total	$2370	

With this information we can now develop various project durations using the logic of CPM:

16-day schedule—normal duration for this project—cost $1320.

15-day schedule—the least expensive way to gain one day would be to reduce activity *D* for an additional cost of $80, project cost $1400. If the other activities on the critical path were reduced, activity *A* or *G* would have been more costly.

14-day schedule—another day can also be gained by expediting *D* for an additional cost of $80, project cost $1480. In the 14-day schedule, it is apparent that there are three critical paths, activities *A–F*, *A–D–G*, and *B–G*. Further cuts in activity *D* will not help since the project's duration is also determined by other activities.

13-day schedule—the cheapest way to attain a 13-day duration is to cut *A* and *G* each one day for an additional cost of $240. However, in doing this we can extend *D* by one day (allocation of one day of float), gaining $80. The net result is an additional cost of $160, project cost $1640.

12-day schedule—activites *F* and *G* can be accelerated for an additional $200, project cost $1840.

11-day schedule—to cut down to 11 days, an activity on each of our three critical paths must be expedited. The cheapest combination of reductions would be *B*, *D*, and *F* for an additional cost of $260 and a project cost of $2100.

The project cannot be performed in less than 11 days; therefore, we have determined six different project durations and costs from which we must choose a schedule.

Selection of a particular schedule from among the many alternatives depends mainly upon your project objectives. If time is the primary concern, a schedule between 11 and 16 days can be selected directly. If the completion date calls for a project duration of less than 11 days, it will be necessary to reconsider your arrow diagram to see if any efficiencies can be gained.

If total project cost is the deciding factor, more information is required than the direct project investment already calculated. Indirect expenses, lost revenue, penalties, and similar costs must be added to the direct costs. Total indirect expenses increase as the project duration increases. The sum of the direct and indirect expenses gives a U-shaped total project cost curve. The optimum schedule for implementing the project is the schedule corresponding to the minimum point on this curve. The relationship among direct, indirect, and total project costs for our example is shown in Table 3.6 and graphically in Fig. 3.11. From the table, the optimum schedule is 14 days at a cost of $2880.

A very flat total cost curve is not uncommon in practical applications. This flatness indicates that, for only a nominal additional investment, the

Table 3.6. TOTAL COST TABLE

Days	16	15	14	13	12	11
Direct costs	1320	1400	1480	1640	1840	2100
Indirect costs	1600	1500	1400	1300	1200	1100
Total cost	2920	2900	2880	2940	3040	3200

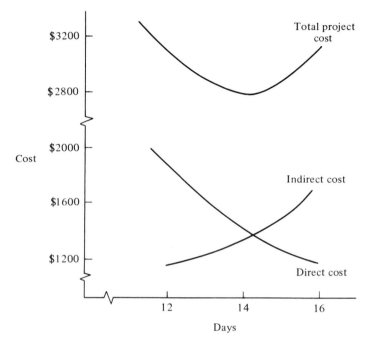

Figure 3.11. Total cost analysis for selecting the "optimum" project duration.

project duration can be reduced. In our case we can reduce the project duration to 13 days at a cost of $60 ($2940 — $2880) or to 12 days at a cost of $160.

3.3.2 The development of activity cost curves. Since CPM is so dependent upon the development of time/cost curves for the activities of the project network, we will now discuss the subject further.

As can be seen in the following example, there is a direct relation between the time and cost for any activity. This relationship takes into account the men, resources, and method used and the efficiency achieved. For example, consider a precision milling operation that takes 56 man-hours of work and by the nature of the work only one person can work on it at a time. Then a number of possibilities exist for scheduling the work. For example:

1. One man can work for seven 8-hour-day shifts over a period of seven days.
2. Two men work on two shifts and finish in four calendar days—four day shifts, three second shifts.
3. Three men work on three shifts and finish in three calendar days—three day shifts, two second shifts, two third shifts.

Obviously, conditions 2 and 3 are more costly than condition 1 because of the overtime premium costs. The cost curve covering this situation is shown in Fig. 3.12. This example shows that a direct time/cost relationship exists

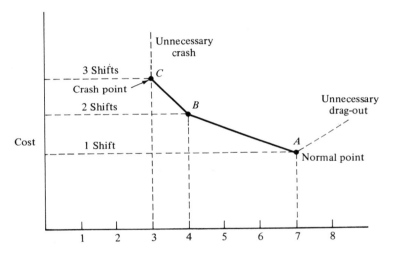

Figure 3.12. An example of the construction of an activity cost curve.

for any job. The requirement of the critical path method is not to produce such a curve (sufficient data seldom exist), but to estimate two important points on the curve. These are the "crash" and "normal" points where:

1. For the "normal" point *normal cost* is given as the minimum job cost, and the associated minimum time is defined as *normal time*.
2. For the "crash" point *crash time* is the minimum possible time, with *crash cost* being the associated minimum cost.

CPM approximates nonlinear cost curves by a linear function between the normal and crash durations by assuming a constant slope for the entire activity. This is usually a reasonable approximation of a nonlinear cost curve.

However, if greater precision is desired (and if cost data permits), a more precise approximation of the nonlinear function is possible.

To illustrate, consider an activity, *A*, with normal and crash points as shown in Fig. 3.13. Normally, the cost curve for activity *A* would be assumed

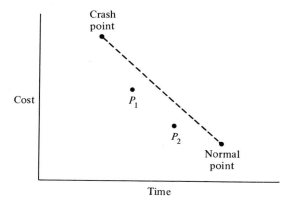

Figure 3.13. Linear approximation of a cost curve.

as a straight line between the normal and crash points as illustrated. However, if it is known that the actual cost curve for activity *A* passes through other points (P_1 and P_2) not on the straight line from normal to crash point, then it may be desirable to analyze the project using the known costs for activity *A* rather than the assumed costs.

The method for handling these closer approximations begins by breaking the original activity *A* into the number of pseudoactivities required to represent the closer approximation to the cost curve as shown in Fig. 3.14.

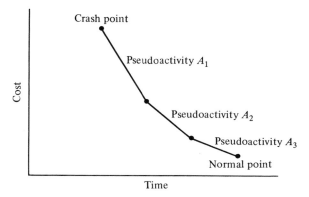

Figure 3.14. Approximate cost curve.

To determine the time and cost characteristics of the pseudoactivities, the cost curve in Fig. 3.14 is assumed to have the following coordinates:

COORDINATES

Point	Time	Cost
Crash point	d	$\$_2 + b_3 = \$_3 = \$_C$
Point P_1	$d + a_1 = d_1$	$\$_1 + b_2 = \$_2$
Point P_2	$d_1 + a_2 = d_2$	$\$_N + b_1 = \$_1$
Normal point	$d_2 + a_3 = d_3 = D$	$\$_N$

The cost curve is redrawn in Fig. 3.15 to show these coordinates.

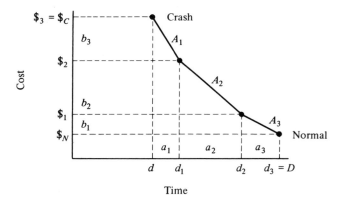

Figure 3.15. Approximate cost curve.

If the output for pseudoactivities A_1, A_2, and A_3 are to be related to the original activity A so that the cost of A equals the cost of $A_1 + A_2 + A_3$ and the duration of A equals the duration of $A_1 + A_2 + A_3$, and if the effect on the minimum cost/time solution is to remain unchanged, the parameters for pseudoactivities must be as follows:

Pseudoactivity	Normal		Crash	
	Time	Cost	Time	Cost
A_1	d_1	0	d	b_3
A_2	a_2	0	0	b_2
A_3	a_3	$\$_N$	0	$\$_1$

We see that the sums of the normal times and costs equal the normal time and cost for the original activity A, and the sums of the crash times and costs equal the crash time and cost for the original activity A. The same procedure is used regardless of the number of pseudoactivities used.

To further illustrate the cost approximating technique, consider the following activity A with a normal cost of $300, a crash cost of $600, a normal duration of 200 units, and a crash duration of 100 units of time. Consider further that, at a duration of 125 units, the cost is $400 and, at the duration of 150 units, the cost is $350. This example is shown in Fig. 3.16.

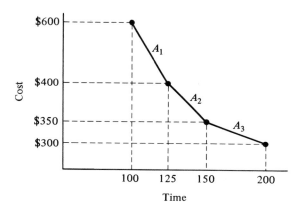

Figure 3.16. Example of a cost curve approximation.

Following the rules given before, we would describe the pseudoactivities A_1, A_2, and A_3 as follows:

	Normal		Crash	
Pseudoactivity (i, j)	Time	Cost	Time	Cost
$A_1(10, 100)$	125	0	100	200
$A_2(100, 101)$	25	0	0	50
$A_3(101, 11)$	50	300	0	350

In the arrow diagram, activities treated in this manner should be indicated as shown in Fig. 3.17.

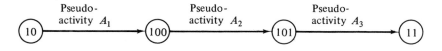

Figure 3.17. Network modification for a cost curve approximation.

Most of the existing computer packages for the solution of the CPM network will also allow the modeling of discontinuous cost curves in addition to the continuous curve.

3.3.3 Some thoughts on the mathematical optimization of the CPM system.

As discussed previously, the basic project representation used in CPM is the activity-oriented project arrow diagram. For notational convenience, a letter i can be associated with each node or event. Then each activity is uniquely defined by a node pair (i, j).

A time/cost curve such as the one shown in Fig. 3.18 is assumed to

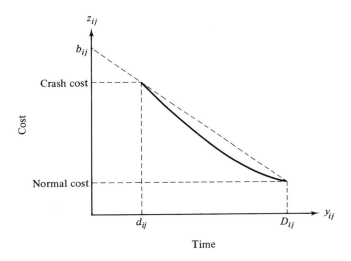

Figure 3.18. Typical activity time/cost curve.

exist for each activity. It will be assumed that all such curves are nondecreasing, convex, and can be approximated by a straight line of the form:

$$z_{ij} = a_{ij}y_{ij} + b_{ij}; \qquad a_{ij} \le 0, \qquad b_{ij} \ge 0 \qquad (3\text{-}1)$$

where y_{ij} is the duration of activity (i, j). Furthermore, the crash duration d_{ij} and the normal duration D_{ij} obey the relationship:

$$0 \le d_{ij} \le y_{ij} \le D_{ij} \qquad (3\text{-}2)$$

The overall project direct cost is found simply by summing z_{ij} for all $(i, j) \in P$ for the particular set of scheduled y_{ij} of interest. Thus:

$$\text{Project direct cost} = \sum_{(i,j) \in P} (a_{ij}y_{ij} + b_{ij}) \qquad (3\text{-}3)$$

The problem to be considered here may be stated as follows: Minimize Eq. (3-3) subject to Eq. (3-2) and also to

$$y_{ij} \leq t_j - t_i \tag{3-4}$$

and

$$t_0 = 0, \qquad t_n = \lambda \tag{3-5}$$

where t_i, t_j are the earliest occurrence times for the ith and jth events, and t_n is the earliest occurrence time of the last (nth) event of a project containing $n + 1$ events. Since t_0, the occurrence time of the zeroth event, is defined as zero, the overall project duration is λ which is considered to be a parameter. Thus, the problem is to find a set of y_{ij} (a schedule) and a resulting *minimum* project direct cost for each possible value of λ in the interval in which λ is defined. The process of generating such schedules is called *project expediting*.

We will now briefly review some of the methods used to solve this mathematical formulation of the CPM system. A more complete discussion is given in References 2, 4, and 10.

BRUTE FORCE [1]

The initial attempt to expedite a project under the stated restraints resulted in the *brute force method*. This method consists of determining all the critical paths through the network, choosing that activity or combinations of activities with the minimum negative cost slope, and then expediting that particular activity or combination until it reaches its crash limit or until a new critical path appears. The discussion in Section 3.3.1 is an application of a form of a brute force method.

The major disadvantage of the brute force method is that it does not provide a formal means for determining combinations of y_{ij} increments involving both shortening and lengthening certain activities. Thus it will not necessarily lead to optimum schedules, although the schedules so obtained tend toward the optimum.

Although several modifications of the brute force approach have been tried in attempts to allow for possible increases in y_{ij}'s, they were not successful due to the highly combinatorial nature of the problem when attacked in this manner.

LINEAR PROGRAMMING APPROACH

Since we are dealing with an optimization problem, linear programming readily comes to mind as the appropriate tool.

With the project duration being a parameter, instead of formulating the problem as a simple linear program, a *parametric* linear program formulation would be advantageous.

The major difficulty encountered in formulating a parametric linear program is the introduction of the parameter λ. It appears that this can be accomplished if we sum the y_{ij}'s along each path from the initial to the terminal event and restrict these sums to be less than or equal to λ (where λ is the total project duration).

The parametric programming formulation thus becomes:

$$\text{Minimize} \quad \sum_{i,j \in P} z_{ij} = \sum_{i,j \in P} (a_{ij} y_{ij} + b_{ij})$$

subject to:

$$y_{ij} \leq D_{ij}$$
$$y_{ij} \geq d_{ij}$$ for all activities

$$\sum_{\text{path } 1} y_{ij} \leq \lambda$$

$$\vdots$$

$$\sum_{\text{path } i} y_{ij} \leq \lambda$$ for all paths

$$\vdots$$

For the example discussed in Section 3.3.1 the formulation would appear as follows:

$$\text{Minimize } Z = (500 - 100y_{01}) + (1120 - 120y_{02}) + (200 - 50y_{03})$$
$$+ (600 - 80y_{12}) + (160 - 0y_{32}) + (830 - 60y_{14})$$
$$+ (1180 - 140y_{24}) + (300 - 100y_{34})$$

subject to:

$$y_{01} \leq 4 \qquad\qquad y_{01} \geq 3$$
$$y_{02} \leq 7 \qquad\qquad y_{02} \geq 5$$
$$y_{03} \leq 3 \qquad\qquad y_{03} \geq 2$$
$$y_{12} \leq 5 \qquad\qquad y_{12} \geq 3$$
$$y_{32} \leq 2 \qquad\qquad y_{32} \geq 2$$
$$y_{14} \leq 10 \qquad\qquad y_{14} \geq 8$$
$$y_{24} \leq 7 \qquad\qquad y_{24} \geq 5$$
$$y_{34} \leq 2 \qquad\qquad y_{34} \geq 1$$
$$y_{01} + y_{14} \leq \lambda$$
$$y_{01} + y_{12} + y_{24} \leq \lambda$$

$$y_{02} + y_{24} \leq \lambda$$
$$y_{03} + y_{32} + y_{24} \leq \lambda$$
$$y_{03} + y_{34} \leq \lambda$$

There are standard parametric programming algorithms that can be applied to this formulation. There are, however, some major disadvantages to the parametric linear programming formulation. It requires that all paths in P from the initial to final event be identified. For large projects it is obvious that the total number of paths is very large and it becomes difficult to identify all the paths. A second disadvantage is that, computationally, the formulation is not very efficient. The coefficient matrix of the linear programming formulation consists mostly of zeros so that much of the computation deals with zero elements which contain no useful information concerning the project. In short, information is too sparsely distributed by the above parametric linear programming formulation.

THE FLOW APPROACH

The flow approach to the project expediting problem may be stated briefly. Having formulated a *primal* linear program, we can derive a *dual* program having the characteristics of a flow problem. The flow problem is then solved by efficient flow algorithms rather than by the highly inefficient (in this particular application) standard linear programming methods. In Chapter 5 we consider in detail the flow approach to solving the CPM model. Thus, discussion of this approach will be deferred until that time.

3.4 Project Control

All the methods described so far have been for planning and scheduling, and they are employed before the project begins. The final phase of project management, the control phase, is needed to provide the project manager with the information for managing his project. This monitoring is achieved by a comparison of the actual status of the project against the CPM or PERT schedule. The outgrowth of this comparison is a revised strategy. The monitoring phase is usually performed using a computer, although limited success can be achieved by using hand calculations. By periodically feeding the computer such data as (1) completed activities, (2) changes in the durations of activities, and (3) changes in the detailed structure of the network, the project manager obtains from the computer a monitor report illustrating the current status of the project. The monitor indicates those activities which are ahead of or behind schedule. Using new and improved data, the computer predicts when the project will be completed. It identifies critical activities and those about to become critical. The manager thus can see how the total project is progressing and can determine where maximum effort should be placed.

In the next chapter we will consider applications of the methods presented in this chapter. A case problem will be presented to illustrate project management using critical path methods. A number of extensions to classical PERT and CPM will also be considered.

EXERCISES

1. Make an arrow diagram for the following:
   (a) O = start of job
         K = end of job
         O precedes A and B
         A precedes C and M
         B precedes E, F, G
         C precedes I, H
         E precedes D
         M precedes D
         D, F also precede I, H
         H, G precede J
         I, J precede K
   (b) Same as above, except:
         D now only precedes I and
         F only precedes H, while
         C still precedes both.

2. Make an arrow diagram for the following:
   (a) A must precede B and C
       B must precede D and E
       C must precede E
       D and E must precede F
   (b) A must precede B, C, and D
       B, C, D must precede E
   (c) A and B must precede C
       A must precede D

3. Draw the activity arrow diagram for the following:
       O is the start of the job
       O precedes A and B
       A precedes G and H
       B precedes E, F, and C
       C, E, and G precede M
       F precedes G

H precedes *I*

M and *I* precede *J*

J is the end of the job

4. Make an arrow diagram for the following problem of servicing a car. Assume that you must remove the air filter before oiling the generator and distributor, and consider the following activities:
    1. Hoist car
    2. Remove drain plug
    3. Grease underside fittings
    4. Inspect tires and exhaust system
    5. Check differential and transmission levels
    6. Replace drain plug
    7. Lower car
    8. Refill crankcase
    9. Check oil level
    10. Grease upper fittings
    11. Open hood
    12. Oil generator and distributor
    13. Check radiator and battery
    14. Drain oil
    15. Wipe excess grease off upper fittings
    16. Remove and clean air filter
    17. Replace air filter
    18. Close hood

5. Considering Problem 1, construct a table showing the earliest start, latest start, earliest finish, latest finish, total float, and free float for each activity. The time assignments are as follows:

$$O = 2, \quad A = 6, \quad B = 1, \quad E = 4, \quad M = 9$$
$$C = 7, \quad D = 3, \quad F = 12, \quad H = 6, \quad G = 10$$
$$I = 3, \quad J = 5, \quad K = 3$$

6. Comment briefly on the following statement: "In considering the expected duration of a project, λ, as calculated by PERT, the estimate tends to be less than the actual expected duration of the project."

7. Draw a cost curve considering the following information:
    (a) Forty man-hours of work are required.
    (b) Miss *A* is normally assigned.
    (c) Misses *B* and *C* can be made available if necessary.
    (d) First shift operation costs $2.00/hour.
        Second shift operation costs $2.20/hour.
        Third shift operation costs $2.42/hour.
    (e) Typists are assigned for eight-hour periods.
    (f) One typewriter is available.
        Can rent additional typewriters at $5.00 per 24-hour period, including delivery and pickup.

(g) No girl works overtime (more than 8 hours in any 24-hour period).

(h) There is no additional expense for second and third shift supervision.

8. For the arrow diagram in Fig. 3.19, construct a table showing the earliest start, latest start, earliest finish, latest finish, total float, and free float for each activity. Also, indicate the critical path of the diagram.

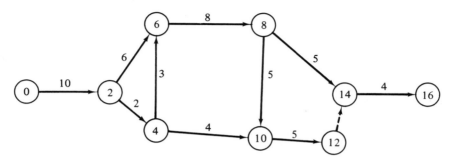

Figure 3.19.

9. Given the following information and the arrow diagram in Fig. 3.20:

Activity	D	d	Normal Cost	Crash Cost
A	0	0	0	0
B	15	10	7000	8500
C	6	4	2000	3000
D	2	2	500	500
E	4	4	3500	3500
F	5	3	4000	4400
G	7	5	3000	4000
H	6	6	1000	1000
I	0	0	0	0
J	4	4	3000	3000

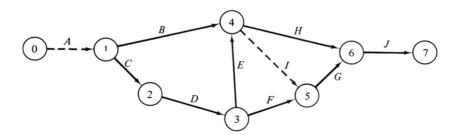

Figure 3.20.

(a) Construct a table showing earliest start and finish, latest start and finish, cost slope, and the two types of floats.

(b) What is the critical path of this project? What will the project cost?

(c) What is the shortest possible time in which the project can be completed? What will the critical elements be in this case?

(d) Derive *minimum cost schedules* for all days between the all-normal and the all-crash schedules?

10. Given the activity diagram in Fig. 3.21 and the following data:

Activity	D	d	Normal Cost	Crash Cost
A	10	7	700	1000
B	12	9	800	1400
C	7	7	1000	1000
D	13	10	1200	1500
E	3	3	500	500
F	7	6	600	900
G	0	0	0	0
H	2	1	1000	5000
I	7	5	1300	1600
J	4	4	2000	2000

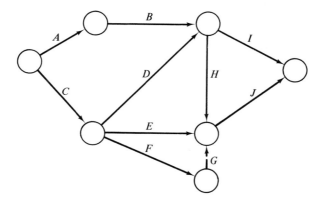

Figure 3.21.

(a) Considering normal time, calculate the total and free floats for activities *B*, *D*, and *E*.

(b) If you wish to crash your schedule to 26 days, what would the optimum cost of the project be? Show your work along with the scheduled duration of each activity for the 26-day schedule.

(c) If activity *B* were crashed to 11 days, what would you estimate its cost to be? If the actual cost for 11 days was $900, how would you refine your diagram and cost figures to show this change? How would this affect your optimum cost answer to part (b)?

(d) If activity *D* must be completed on day 19 because of government inspection, write any additional restraints and dummies needed to assure completion *on* that day.

11. Assume that a cost curve was represented by the following pseudoactivities:

	Normal		Crash	
	Cost	Time	Cost	Time
A_1	0	13	$100	10
A_2	0	2	100	0
A_3	0	2	75	0
A_4	$300	2	350	0

Plot the cost curve. Would you expect the common crashing procedure to be effective for this set of data? Why?

12. Given the cost diagram in Fig. 3.22, how would you approximate it? Explain why you would expect your answer to work.

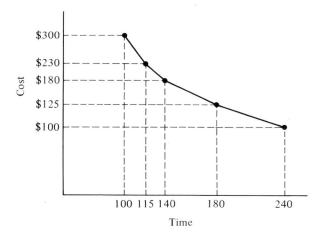

Figure 3.22.

13. Updating problem: The original CPM arrow diagram appeared as shown in Fig. 3.23. After fifteen days:

0–2	Complete
2–6	Complete
2–10	Complete

2–4   Complete

4–16   Five more days left

6–8   Ten days remaining

8–18   Re-estimated to four days—not done

All remaining jobs not started

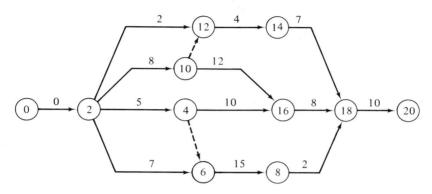

Figure 3.23.

Draw a new diagram and evaluate the position of the company with respect to its original scheduled completion.

14. Discuss differences between CPM and PERT. Discuss the statistical development of (a) probability of slack and (b) probability of meeting the schedule.

15. (a) Complete the following table for PERT diagram shown in Fig. 3.24:

Event No.	T_E. Mean	Var.	T_L. Mean	Var.	Slack	Probability of Slack
1						
2						
3						
4						
5						
6						
7						

(b) What is the critical path?

(c) If the scheduled completion time is 28 days, what is the likelihood of meeting the schedule?

(d) If event 3 is scheduled for day 3, what is the probability of meeting the schedule?

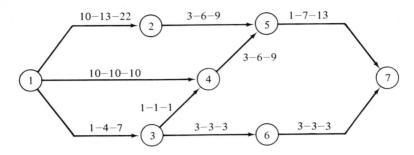

Figure 3.24.

16. Write a computer program to evaluate an activity network.

17. Modify the activity network algorithm presented in this chapter so that it finds not only the longest path through a network, but also the second and third longest paths?

REFERENCES

1. Antill, J. M. and R. W. Woodhead, *Critical Path Methods in Construction Practice*. New York, John Wiley & Sons, Inc., 1966.

2. Battersby, A., *Network Analysis for Planning and Scheduling*. London, MacMillan and Co., Ltd., 1967.

3. Clark, Charles E., "The PERT Model for the Distribution of an Activity Time." *Operations Research*, Vol. 10, No. 3 (May–June, 1962).

4. Clingen, C. T., *Methods of Expediting Project Schedules in View of Time and Cost*. Report T. I. SRGICD15, General Electric Computer Department, Phoenix, Arizona, 1962.

5. Everts, Harry F., *Introduction to PERT*. Boston, Allyn and Bacon, Inc., 1964.

6. *G.E. 225 Application, Critical Path Method Program*. Bulletin CPB198B, General Electric Computer Department, Phoenix, Arizona, 1962.

7. Kelley, J. E., Jr., "Critical Path Planning and Scheduling: Mathematical Basis." *Operations Research*, Vol. 9, No. 3 (May–June 1962).

8. Levin, R. and C. Kirkpatrick, *Planning and Control with PERT/CPM*. New York, McGraw-Hill, Inc., 1966.

9. MacCrimmon, K. R. and C. A. Ryavec, "An Analytical Study of the PERT Assumptions." *Operations Research*, Vol. 12, No. 1 (1964), pp. 16–37.

10. Moder, Joseph J. and Cecil R. Phillips, *Project Management with CPM and PERT*. New York, Reinhold Publishing Corporation, 1964.

11. Murray, J. E., "Consideration of PERT Assumptions." *IEEE Transactions on Engineering Management*, Vol. EM-10, No. 3 (1963), pp. 94–99.

12. VanSlyke, Richard M., "Monte Carlo Methods and the PERT Problem." *Operations Research*, Vol. 11, No. 5, (1963), pp. 839–860.

13. Welsh, D. J. A., "Errors Introduced by a PERT Assumption." *Operations Research*, Vol. 13, No. 1 (1965), pp. 141–143.

14. Zalokar, F. J., *The Critical Path Method, A Presentation and Evaluation.* Report of the Real Estate and Construction Operation of General Electric, Schenectady, New York, 1964.

PROJECT MANAGEMENT—IMPLEMENTATION OF ACTIVITY NETWORK STUDIES

In Chapter 3 we introduced the concept of activity networks with emphasis on PERT and CPM. In this chapter we will investigate some applications of the activity network approach. Particular emphasis will be placed on a case study which will be used to demonstrate the applicability of activity network analysis in the industrial environment.

Many people have proposed modifications to PERT and CPM which they feel represent improvement over the standard systems. We will briefly consider some of these modifications in this chapter.

4

4.1 Applications of Project Management Techniques

Most of the best-known applications of project management network analysis have been in the fields of construction, large-scale maintenance, and weapons development. Our successful moon landings were managed using project management techniques. Kelley [9] reports the following list of applications:

- All types of construction and maintenance.
- Retooling programs for high-volume production.
- Low-volume production scheduling.
- Scientific missile countdown procedures.
- Budget planning.
- Mobilization, strategic, and tactical planning.
- New product launching.
- Assembly and testing of electronic systems.
- Installation, programming, and debugging of computer systems.

Battersby [3] cited such specific applications as:

- The Sun Maid Raisin Growers, by timing construction of a plant to the growing season, saved an estimated $1,000,000.

- The Catalytic Construction Company reduced the average duration of 47 construction works by 22 percent and the expediting costs by 18 percent.
- The producers of the Broadway play "Morgana" used network analysis to control the staging of the play.
- Plans for the economic growth of underdeveloped countries have been hypothesized using PERT.
- Surgical procedures such as open-heart surgery and kidney transplant have been scheduled by networks.
- Mrs. T. Russell has modeled the preparation of Sweet and Sour Pork for a banquet application.

We can conclude that network analysis can be applied to a wide range of problems yielding the following advantages:

FOR MANAGEMENT

1. More realistic time and cost estimates.
2. Improved program control through early detection of uncompleted events and isolation of problem areas through critical path analysis.
3. Good simulation of alternative plans and schedules.

FOR OPERATING PERSONNEL

1. Participation in PERT implementation provides for better visualization of the individual tasks.
2. Interim schedule objectives are more meaningful.
3. Clearer delineation of task and decision responsibilities.
4. Improved communication among operating personnel.

4.2 Case Problem—Launching a New Product

To illustrate the use of project management techniques, we will now consider a case study example. Assume that you have been appointed project manager to direct the launching of a new product. We will assume that the product is a consumer item such as an air conditioner or vacuum cleaner.

4.2.1 Development of the activity network. Your company has had no experience with similar products; so detailed market studies will be performed. A list of the steps involved in the launching of this product is developed next. Our analysis might lead to the following activities broken down into departmental responsibility.

MARKET ANALYSIS DEPARTMENT ACTIVITIES

1. *Develop product planning specifications.* This is where the CPM program is begun, where the tasks are defined and assigned, and where the monitorships are established.
2. *Conduct market research.* This activity determines the demand for the product.

3. *Develop price demand schedule.* This activity determines the various grades, models, and prices of the product and relates them to the potential market.
4. *Conduct profit-and-loss analysis.* From the work of the market analysis section, a profit-and-loss estimate is made, and it is determined whether to continue on with product development and what the characteristics of the product will be.
5. *Conduct product appraisal.* This activity determines whether or not the design presented by engineering meets requirements of cost analysis and permits a profitable undertaking. It uses information on costs from engineering with known direct and indirect costs of all types and grades of product under consideration in making its determination.
6. *Determine price.* This activity determines the price from cost data and market research so that it will be profitable and competitive.

ENGINEERING DEPARTMENT ACTIVITIES

1. *Conduct engineering research.* This is the activity where the preliminary design is made and where necessary information is gathered for making decisions regarding make-or-buy components.
2. *Conduct patent search.* This activity determines whether or not the product is patentable or infringes on the other patents. It also determines if a need to license patent rights for manufacture exists.
3. *Prepare cost estimates.* This engineering function estimates production costs and material costs.
4. *Develop laboratory model.* In conjunction with engineering design, this activity makes a model of the product. The model is used for testing and as an example for production of the final product. The product is made by production people from drawings from the design section, and it is modified as tests and cost figures become available and as improved design drawings become available.
5. *Design final product.* This activity uses information from the laboratory-model testing and from market research activities.
6. *Issue drawings and specifications.* This engineering activity issues necessary drawings and specifications for manufacturing processes.
7. *Determine manufacturing methods.* This activity establishes efficient manufacturing methods and decides what equipment is needed; it determines job requirements at each station in the manufacturing process.
8. *Prepare service literature.* This activity prepares literature concerning servicing of the product for service people, outlets, and customers.
9. *Train service organization.* This activity trains manufacturing outlets' service people in the proper methods of servicing the product.
10. *Test.* This engineering function is the reponsibility of the quality-control section.

SALES AND ADVERTISING

1. *Train sales force.* This activity instructs the firm's sales force concerning its new product so that the sales force can obtain new sales outlets and orders for the new product.

2. *Prepare advertising.* This activity prepares an advertising campaign stressing the new product's outstanding features.
3. *Establish distribution outlets.* This activity is accomplished by the sales force and takes initial orders from dealers.
4. *Release advertising.* This activity contracts for advertising display and will be coordinated with dealer and product availability.
5. *Design and procedure packaging.* This activity develops and procures attractive and secure packaging.

PURCHASING DEPARTMENT ACTIVITIES

1. *Procure raw materials.* This activity covers the purchase of raw materials for timely delivery to the production process.
2. *Procure "buy" items.* This activity covers purchases of items such as screws, tools, and product components used in the production process.

PRODUCTION DEPARTMENT ACTIVITIES

1. *Train production personnel.* This activity sets up the production line and trains personnel in their assigned jobs.
2. *Manufacture "make" items.* This is the production line.
3. *Assemble.* This activity assembles make items and buy items into finished product.
4. *Box, pack, and ship.* This activity represents the last step before the product is available to the consumer.

We can now consider the interrelationships among these activities and develop an activity network. The activity network is shown in Fig. 4.1 and was developed using the following logic.

The market analysis department is responsible for the product planning specifications, market research, and price demand schedule activities. Since they are all interrelated and concerned with consumer preferences and the associated markets, they are placed first. From these specifications of the market, the creative engineering research and development activities are carried on in sequence. At this time distribution outlets are also established for the product.

Product appraisal consists of a review of the preceding activities, resulting in final product specifications which are transmitted to the final design group. In parallel with this activity, cost estimates are made and combined with the distribution cost data to determine the total cost of producing and selling the product. In the latter portion of the final design phase, a profit-and-loss analysis is accomplished to establish a market price for the product. This price analysis utilizes the cost estimates and distribution information to calculate and "engineer in" a fair profit for the company.

Work on the project can be stopped at any phase of development; however, the most logical point would be at event 22. Beyond this point tooling, material, and machines will be committed to the production of

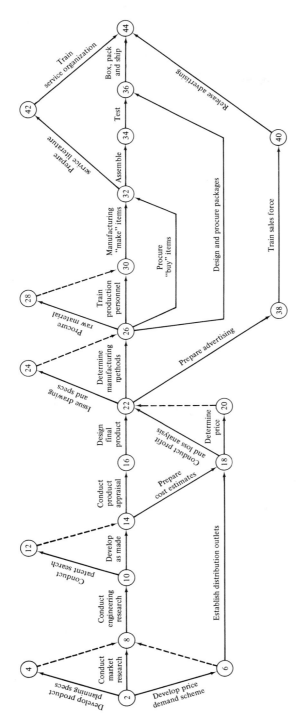

Figure 4.1. Activity network for "launching a new product."

77

the product. We will consider this as a "point of no return" in committing resources to the production of the product.

If the product meets the profit criteria, it is then committed to the production phase. At this point production personnel determine manufacturing methods including make-or-buy decisions. After the make-or-buy decisions are made, drawings are released for bid or use in the production department. Advertising material for the product is started at this time.

At the same time that the purchased items are procured, the packaging is designed and procured. Production personnel are then trained, raw materials procured, and the manufacturing process begun. The service literature is prepared only after both the make items and buy items are on hand. This will make the preparation of product description and photographic copy much simpler to perform. After this literature has been prepared and made available, the service organization is trained.

While the assembly, testing, and packing phases are under way, the sales force is trained using the actual assembly-line-produced items for demonstrations. After sales personnel are trained, but just prior to shipment, the advertising is released. With the shipment of the product, the CPM network analysis is complete.

The interrelationships of our activities have been determined and we have developed an activity network describing the project. Our next step is to determine the time and cost data associated with each activity. The normal and crash information is shown in Table 4.1. We now have enough information to use an available computer package. Using the normal time for each activity, the network would yield a project duration of 52 weeks. The results of the analysis are shown in Fig. 4.2. The earliest and latest times for each node are shown on the figure along with the critical path.

4.2.2 Determination of the optimum project length.

When consumer products are marketed, timing is extremely important. The longer it takes to get a product on the market, the lower the anticipated profits from the endeavor. We know, however, that it will cost us money (crashing cost) to expedite the project. Thus, there is a cost tradeoff. We can also save on indirect project costs if we decide to expedite the project. Historically, the indirect costs for a project of this nature could be calculated from the following formula:

$$\text{Indirect costs} = \$600,000 + \$10,000 \times (\text{project duration})$$

The marketing department's preliminary estimates of the market potential of this product are shown in Table 4.2.

The procedure for determining the optimum project duration is as follows:

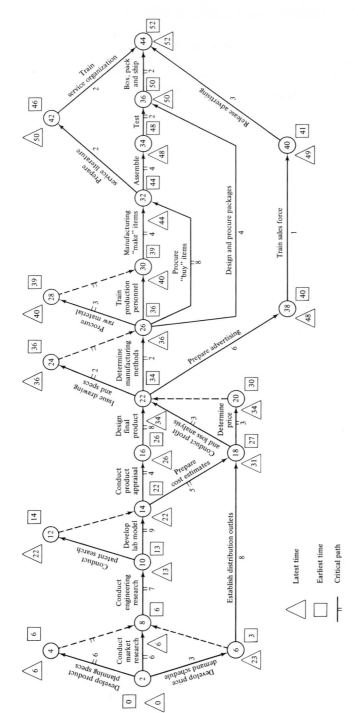

Figure 4.2. All normal schedule for "launching a new product."

Latest time ◁
Earliest time ▢
Critical path ⫫

79

1. Develop the computer input for this project. The computer will determine the direct costs and schedules for various project durations.
2. Determine the total project costs for each feasible schedule calculated by the computer by adding indirect costs (computed initially by the formula $600,000 + $10,000 \times$ Project Duration) to the direct cost.
3. Determine the profit for each schedule by subtracting total costs as determined in step 2 from the anticipated market values provided.
4. Select the schedule that will yield the maximum profit.

Table 4.1. TIME AND COST INFORMATION FOR LAUNCHING A PRODUCT

Activity Description	Normal Duration	Normal Cost	Crash Duration	Crash Cost
Conduct market research	6	30,000	2	125,000
Develop price demand schedules	3	6,250	1	27,400
Develop product planning specifications	6	28,120	4	48,500
Conduct engineering research	7	33,750	4	159,590
Conduct patent search	1	10,000	1	10,000
Prepare cost estimates	5	9,380	2	26,420
Develop laboratory model	9	51,250	5	158,760
Conduct product appraisal	4	15,650	2	51,950
Conduct profit-and-loss analysis	3	5,630	1	24,570
Design final product	8	40,620	5	151,870
Train sales force	1	5,000	1	5,000
Prepare advertising	6	18,750	3	47,500
Issue drawings and specifications	2	3,120	1	7,100
Determine price	3	5,130	1	16,420
Establish distribution outlets	8	56,250	5	132,500
Release advertising	3	4,380	3	4,380
Determine manufacturing methods	2	5,630	1	14,200
Procure raw material	3	3,750	2	9,250
Procure "buy" items	8	11,880	5	28,880
Prepare service literature	2	5,000	1	12,000
Design and procure packaging	4	10,500	1	73,700
Train production personnel	1	9,370	1	9,370
Manufacture "make" items	4	68,750	1	237,380
Assemble	4	49,380	3	95,750
Train service organization	2	11,500	1	27,500
Test	2	12,500	2	12,500
Box, pack, and ship	2	5,000	1	11,230
		$511,540		$1,528,900

Table 4.3 shows the results of applying our procedure to this project. Since our objective is to select the project duration with the maximum profit, we select the 46-week schedule. The optimum schedule is shown in Fig. 4.3. The profit curve is fairly flat, so the company might select a somewhat different schedule based on intangible considerations.

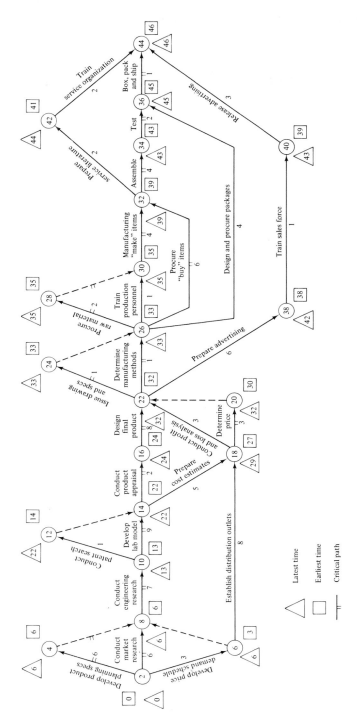

Figure 4.3. Optimum profit schedule for "launching a new product."

81

Table 4.2. MARKET POTENTIAL OF THE NEW PRODUCT

If Shipped by Week	Total Anticipated Market
30	$2,750,000
31	2,695,000
32	2,650,000
33	2,610,000
34	2,575,000
35	2,540,000
36	2,510,000
37	2,485,000
38	2,460,000
39	2,440,000
40	2,422,000
41	2,406,000
42	2,390,000
43	2,375,000
44	2,360,000
45	2,341,000
46	2,335,000
47	2,325,000
48	2,315,000
49	2,310,000
50	2,305,000
51	2,300,000
52	2,295,000

4.2.3 Comparison of minimum duration schedules. The activity networks shown in Figs. 4.4 and 4.5 illustrate the extent of cost reduction possible through the use of the CPM scheduling technique. A minimum duration schedule can be found if we assume that all activities are scheduled at their crash time and cost. This schedule is shown in Fig. 4.4 and takes 32 weeks. The computer calculated an optimum 32-week schedule which is shown in Fig. 4.5.

Project duration for the optimum schedule, as determined by the CPM scheduling technique, is equal in duration to the all-crash schedule. Also, indirect costs are assumed to be the same for both schedules. Direct costs, however, vary considerably. Direct cost for the all-crash schedule amounts to $1,528,900, whereas direct cost for the optimum 32-week schedule as determined through use of the CPM scheduling technique is only $1,114,860. In this case the use of the CPM scheduling technique indicates a schedule which, if followed, would result in a savings of $414,040 in direct cost, for a 27.1% saving over the all-crash direct cost. This is an outstanding example of the savings that are possible through use of the CPM scheduling technique.

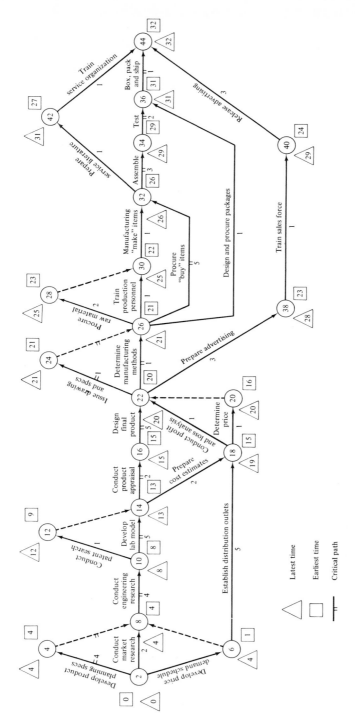

Figure 4.4. All crash schedule for "launching a new product."

Table 4.3. TOTAL PROFIT ANALYSIS FOR THE PRODUCT-LAUNCHING

Duration	Direct Cost	Indirect Cost	Anticipated Market	Profit
52	516,540	1,120,000	2,295,000	658,460
51	522,200	1,110,000	2,300,000	667,800
50	528,430	1,100,000	2,305,000	676,570
49	539,600	1,090,000	2,310,000	680,400
48	552,150	1,080,000	2,315,000	682,850
47	570,300	1,070,000	2,325,000	684,700
46	588,450	1,060,000	2,335,000	686,550*
45	615,327	1,050,000	2,341,000	675,673
44	642,205	1,040,000	2,360,000	677,795
43	669,082	1,030,000	2,375,000	675,918
42	695,960	1,020,000	2,390,000	674,040
41	729,900	1,010,000	2,406,000	666,100
40	763,840	1,000,000	2,422,000	658,160
39	805,920	990,000	2,440,000	654,080
38	838,000	980,000	2,460,000	642,000
37	879,947	970,000	2,485,000	635,053
36	921,893	960,000	2,510,000	628,107
35	963,840	950,000	2,540,000	626,160
34	1,006,610	940,000	2,575,000	628,390
33	1,052,980	930,000	2,610,000	627,020
32	1,114,860	920,000	2,650,000	615,140

*Optimum schedule.

4.2.4 Project monitoring—a project review after 20 weeks. We find that after 20 weeks all work is progressing according to schedule; however, a revised "Anticipated Market Forecast" has been issued and the indirect labor cost estimates have changed. To date, $800,000 ($600,000 + $200,000) has been spent, but it is estimated that for the remainder of the project indirect labor costs will be at the rate of $7500/week. Table 4.4 shows the revised market forecast. Should we continue under the previous schedule or should we change? If we change, what would our new schedule be?

A procedure for analyzing this problem follows:

1. Examine the selected optimum schedule (46 weeks) to determine which activities either have been completed, are in process, or have not yet been started.
2. Prepare updating computer input assigning "zero time" to those activities which have been completed and "time remaining" to those activities still in process.
3. Make a computer run to determine new direct costs based on the revised input data.

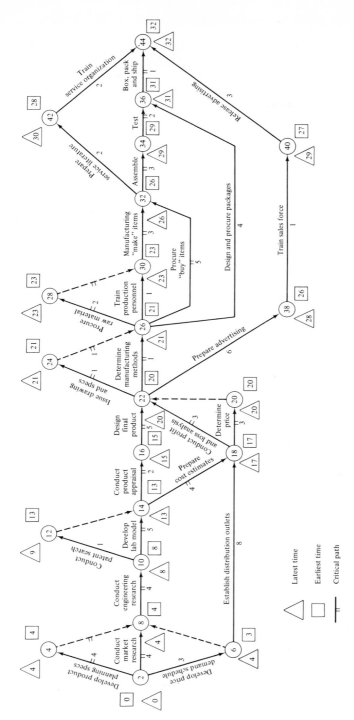

Figure 4.5. 32-day optimum schedule for "launching a new schedule."

Legend:

△ Latest time

□ Earliest time

╫ Critical path

85

4. Compute total costs for each schedule printed out by the computer by adding the revised indirect costs to the direct costs.
5. Determine the profit for each schedule by subtracting total costs as determined in step 4 from the revised market values provided.
6. Compare the anticipated profit that results from staying on the existing schedule with that anticipated from changing to a new schedule as indicated in step 5.

Table 4.4. REVISED MARKET SURVEY

Weeks After 20th Week	Anticipated Sales
10	2,800,000
11	2,775,000
12	2,750,000
13	2,725,000
14	2,700,000
15	2,680,000
16	2,660,000
17	2,640,000
18	2,600,000
19	2,560,000
20	2,530,000
21	2,500,000
22	2,450,000
23	2,410,000
24	2,375,000
25	2,350,000
26	2,340,000
27	2,330,000
28	2,320,000
29	2,310,000
30	2,305,000
31	2,300,000
32	2,290,000

Table 4.5 shows the results of applying our procedure to the modified network. Since our objective is to maximize profit, the 48-week schedule appears to be the most attractive. The anticipated profit would be $757,850 for the 48-week schedule compared to the $756,550 if we continue with the 46-week schedule. The savings are not great if we relax the schedule to 48 weeks; but if we base our decisions exclusively upon maximizing expected profit, we would adopt the schedule shown in Fig. 4.6. We observe that the minimum duration schedule is no longer 32 weeks but is now 41 weeks.

In summary, the CPM cost optimization procedure can have definite advantages when all aspects of this problem are considered. The task of estimating the duration and cost of the various activities must be carefully made. If time and cost estimates are poor, the whole CPM concept suffers;

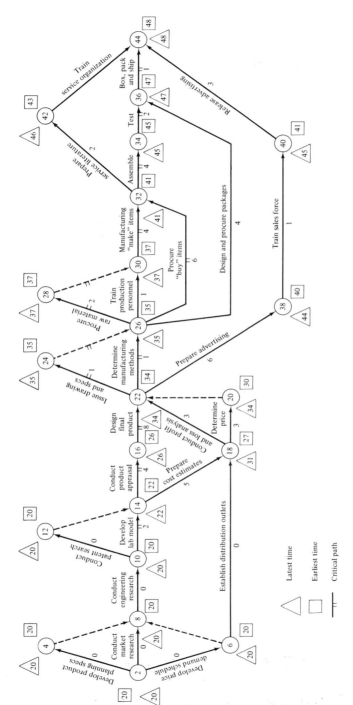

Figure 4.6. Revised schedule for "launching a new product."

87

Table 4.5. REVISED PROFIT ANALYSIS FOR LAUNCHING A PRODUCT

Project Duration	Direct Cost	Indirect Cost	Anticipated Market	Profit
52	516,540	1,040,000	2,290,000	733,460
51	522,200	1,032,500	2,300,000	745,300
50	528,430	1,025,000	2,305,000	751,570
49	539,600	1,017,500	2,310,000	752,900
48	552,150	1,010,000	2,320,000	757,850
47	570,300	1,002,500	2,330,000	757,200
46	588,450	995,000	2,340,000	756,550
45	625,530	987,500	2,350,000	736,970
44	662,610	980,000	2,375,000	732,390
43	705,380	972,500	2,410,000	732,120
42	751,750	965,000	2,450,000	733,250
41	813,630	957,500	2,500,000	728,870

however, if these elements are correctly estimated, the CPM scheduling technique can be a very powerful management tool. Several of the more important advantages have been mentioned previously. These are cost optimization for selected project durations, optimization of the all-crash schedule, and optimization of an updated schedule.

4.3 Use of the Computer in Project Network Analysis*

The use of the computer has been alluded to throughout Chapters 3 and 4. Authors [2, 3, 11] have discussed and compared the use of computers and hand calculations. They state such rules of thumb as: If the network exceeds 20–25 activities, the computer should be used. It is my feeling that, with the abundance of computer packages, every effort should be made to take advantage of them. The computer, however, can only perform the calculations of scheduling and control. The human must supply the arrow diagram, time and cost estimates, and decisions. The computer is important in that it relieves the manager of many laborious calculations; but the manager will make or break the system.

The following is a condensation of the fifteen points of comparison of PERT and CPM computer packages suggested by Phillips [15]. It is presented to give the reader a feel for the features available in addition to the standard PERT and CPM calculations.

---

*The material in this section was adapted from material presented in Reference 15 and is used with the permission of the American Institute of Industrial Engineers.

1. *Capacity.* The capacity of the programs is usually expressed in terms of the number of activities permitted and varies from a few hundred to 75,000. For longer programs it is sometimes necessary to divide the network into two or more parts.
2. *Event numbering.* Many CPM programs and a few PERT programs require that events be numbered in ascending order. This inhibits the flexibility of the network and causes event-numbering bookkeeping problems.
3. *Multiple initial and terminal events.* CPM programs usually require that there be only one initial event and one terminal event. Most PERT programs permit multiple initial and terminal events.
4. *Scheduled dates.* PERT programs will usually accept scheduled dates assigned to the terminal events and will make backward passes from these dates to establish times when events should be started. CPM gives only elapsed time.
5. *Calendar dates.* PERT was developed for research and for the sake of management. One criterion was the use of calendar dates for the input and output phases. This gives the starting and finishing times in terms of specified dates.
6. *Output sorts.* This is the way in which the critical path output data are arranged or sorted. There are several methods, such as by the total slack, successor event number, expected date, or the latest allowed date.
7. *Graphical output.* Some programs will plot graphs, but very few have the capacity to generate network diagrams directly from the computer.
8. *Network condensation.* Routines have been developed recently which condense large networks into smaller ones. The result of this is a network listing the key events only.
9. *Updating facility.* Revision may occur in a number of ways, and the programs vary somewhat in the manner of making these revisions.
10. *Error detection.* Some programs will not detect loops, while others not only detect the loop, but will report the event numbers in the loop. Other errors which are not so easy to find are such things as nonunique activity, improper time estimates, and excessive terminal ends.
11. *Activity versus event orientation.* Critical path programs may be either *event-oriented* or *activity-oriented.* This means that the input and output data are associated with either activities or events. However, the distinction between the two is not a substantive one with respect to computational practice.
12. *Statistical analysis.* Until a few years ago, PERT and CPM were clearly distinguished from each other in that PERT was associated with a three-estimate probabilistic approach while CPM used a single-estimate deterministic approach. Lately the two terms have been used loosely and applied to both in actual practice, and they are no longer mutually exclusive.
13. *Cost optimization.* The early CPM and at least one PERT program assumed that the time/cost relationship for each activity was an inverse one. In reality, though, they may be both inverse and direct.

14. *Cost control.* These features deal with summarization of budgeted and expended cost by time periods.
15. *Resource allocation.* Project managers must be concerned with the proper utilization of men, equipment, and facilities. Some programs have been written which, to some extent, attempt to schedule available resources to minimize overload and delays.

An example of a computer program to perform critical path calculations is presented in Reference 13.

4.4 Other Project Management Techniques

Many techniques closely related to PERT and CPM have been developed in recent years. In this section we will consider a few representative examples of these techniques.

4.4.1 Line-of-balance technique. With an ever increasing amount of project-type business in modern industry, the problem of monitoring the progress of the project with respect to completion requirements is becoming most severe. Coordination of demand upon facilities while meeting schedules and delivery dates which originate far from the plant floor is a constant headache. One plan which is an outgrowth of government (military) entry into the production and manufacturing business is *line of balance* [12].

Line of balance (LOB) is a means for measuring actual progress against a scheduled objective, employing the exception principle. There are four major phases to line of balance: the objective, the program, program progress, and comparison of program progress with the objective.

The initial step is the statement of the objective of the project in terms of number of units/time period, number of units to be delivered, scheduled completion date, or any other quantity/time combination. Naturally, there will be cases in which contract schedules are wholly unrealistic. In these cases contractors' estimates or other similar "objectives" may be used. These are plotted on an *objective chart.*

Phase two requires determination of the limiting and/or principal steps which must be accomplished en route to the objective. The graphical representation of this takes the form of a process flow chart with a horizontal time axis—a modified Gantt chart.

Evaluation of progress is performed in phase three. This is simply an inventory of the stock status for all limiting and/or principal steps identified in phase two. The status is depicted by a bar chart with a vertical axis of the same units as those on the objective chart. This chart is called a *progress chart.*

Striking the line of balance involves transferring points from the objective chart to the progress chart for the study date. A program that is exactly in phase results in a line of balance which would intersect every bar on the progress chart at (or near) its top.

Now consider an example of the application of the line-of-balance technique. A product that we are making has the production goals shown in Table 4.6. This information is plotted in Fig. 4.7 and represents the objective chart for the LOB study. Notice that this chart also records our performance against standard. The LOB study is to be performed at the beginning of month 6. Nine hundred units have been shipped during the first five months, while our objective was to ship 1400.

The next step in the LOB approach is to create a progress flow chart of

Table 4.6. PRODUCTION GOALS

Month	Planned	Cumulative
1	100	100
2	200	300
3	300	600
4	400	1000
5	400	1400
6	400	1800
7	400	2200
8	300	2500
9	100	2600
10	100	2700

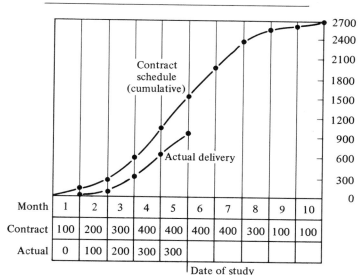

Month	1	2	3	4	5	6	7	8	9	10
Contract	100	200	300	400	400	400	400	300	100	100
Actual	0	100	200	300	300					

Date of study

Figure 4.7. LOB objective chart.

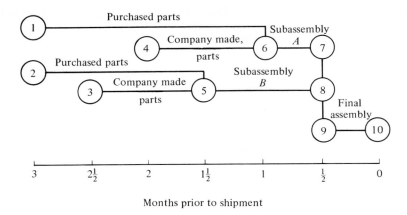

Figure 4.8. LOB progress flow chart.

the important steps involved in manufacturing our product. Figure 4.8 is an example of this type of chart. The chart tells us that we have two subassemblies feeding into final assembly. There are purchased parts and company-manufactured parts needed for each subassembly. This chart shows the expected time that it will take for a product in various stages of production to progress to product completion. For example, we expect that it will take three months from the time orders are placed for purchased parts until these parts find themselves in a finished product. The important control or information collection points are identified by numbers on the progress control chart. The number 1 corresponds to the placing of orders for purchased parts for subassembly *A*. There are ten identifiable points in our example.

Phase three identifies the status at each of the points identified on the progress flow chart for a particular point in time. For example, if our study is to be performed at the beginning of the sixth month of the production plan, we observe the cumulative number of units that have passed each point on the progress flow chart in the first five months. The information might be as shown in Table 4.7. We next plot the status on the progress chart as shown in Fig. 4.9. To evaluate the information just plotted, the line of balance is struck. The procedure for striking the line of balance is as follows:

1. Starting with the study date on the horizontal axis of the objective chart, mark off the lead time for each control point on the production plan.
2. Project up to the cumulative contract schedule curve from each point identified in step 1.

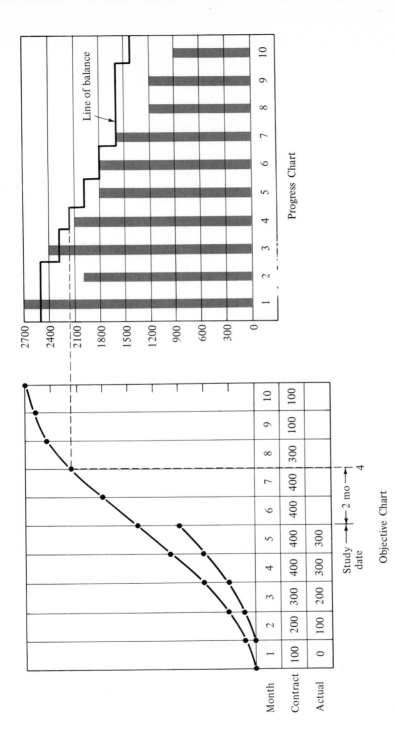

Figure 4.9. Striking the line of balance.

3. Project the points identified on the cumulative contract schedule horizontally to the progress chart.
4. Plot the line of balance.

Table 4.7. STATUS AT THE BEGINNING OF THE FIFTH MONTH FOR THE LOB PRODUCTION PLAN

Control Point	Description	Status
1	Order purchased parts for subassembly *A*	2700
2	Order purchased parts for subassembly *B*	2000
3	Begin manufacture of parts for subassembly *B*	2400
4	Begin manufacture of parts for subassembly *A*	2100
5	Begin subassembly *B*	1800
6	Begin subassembly *A*	1800
7	Finish subassembly *A*	1600
8	Finish subassembly *B*	1200
9	Begin final assembly	1200
10	Finish final assembly	900

Figure 4.9 demonstrates this procedure for control point 4. There is a two-month lead time associated with this point as shown in Fig. 4.8. Projecting, we find that the line of balance is struck at 2200 units for this point. Thus, if we expect to meet our production objective, 2200 units should have passed control point 4. Actually only 2100 have passed this point as shown on the progress chart. All bars which fall below the line of balance represent points which are behind the production plan; those above the line of balance indicate points which are ahead of schedule. From Fig. 4.9 we see that management should be made aware of lags at points 2, 4, 5, 8, 9, and 10. It appears that the purchasing department overlooked the placing of orders for the purchased parts for subassembly *B*. The final assembly is behind schedule, but it is dependent upon the completion of the two subassemblies. Subassembly *B* is behind schedule and is probably the source of the difficulties in this production plan. Iannone [6] presents a large-scale example of the use of LOB in the industrial environment.

4.4.2 PERT/cost. PERT/cost [7] has been proposed as an integrated management system designed to provide managers with the information they need to plan and control schedules and costs in projects. The system provides information in various levels of detail, thereby satisfying the needs of contractor management as well as the customer.

The PERT/cost system consists of a basic procedure for planning and control, and two closely related supplements set forth more advanced planning procedures. The basic PERT/cost procedure assists project managers by providing information in the varying levels of detail needed for planning schedules and costs, evaluating schedule and cost performance, and predicting and controlling time and cost overruns. The *time/cost option procedure*, a supplement, displays alternate time/cost/risk plans for accomplishing project objectives. The *resource allocation procedure*, also a supplement, determines the lowest cost allocation of resources among project tasks to meet a specified project duration.

The goal of the PERT/cost system is to set forth a procedure to identify critical schedule slippages and cost overruns in time for corrective action. The first step is to construct a standard PERT network consisting of the activities to be performed. After the network has been prepared and time estimates developed for the network activities, the manager establishes a schedule. This schedule is based on the critical path calculations, the directed dates, and the manager's judgment concerning the goals to be established for accomplishing the activities.

Once the schedule has been established, resource estimates to perform each cost-significant segment of the network (activity or group of activities) as scheduled are obtained. These estimates are then converted to total dollar estimates.

In the monitoring phase of PERT/cost, the actual costs and times are collected separately for each cost-significant segment of the network. These actual time and cost inputs are compared with the estimates to indicate the project status.

Standardized PERT/cost estimating forms are used to develop the type of data needed for preparing the time/cost network plan for a project. These forms contain information for estimating the time of the job as well as the manpower and material needed to perform the job. The breakdown of manpower into skill categories provides the data needed to indicate the manpower requirements of the project. The system converts man-hours and material to total dollars by applying the appropriate labor and overhead rates.

The PERT/cost system provides time, cost, and resource reports for various levels of management. The Time and Cost Status Report is a basic output of the PERT/cost system. It is designed to assist the project manager in evaluating overall time and cost progress and in pinpointing those activities which are causing schedule slippages or cost overruns, either actual or potential. The system permits this form to be printed in several degrees of data summation so that appropriate detail is presented to each level of management.

The "Manpower Requirements" report is a typical report that identifies the monthly manpower requirements to perform the project on schedule. The report is presented by total manpower and by individual manpower skills. An *activity slack* column provides the manager with the ranking of activities in their order of importance to completing the project on schedule.

The "Manpower Requirements" report is intended to point out those periods in the life of a project when manpower requirements for certain skill categories will exceed the availability or where substantial idle time appears in the plan. This report will assist the line manager in leveling out peaks and valleys in his manpower loading plan.

The "Rate of Expenditure" and "Cost of Work" reports are two summary reports, usually in graph form, which present the manager with the overall cost status of the project. The "Rate of Expenditure" report indicates the rate at which costs are budgeted and incurred over time. The "Cost of Work" report relates budgeted and actual costs to the amount of the work performed and indicates the estimated costs to complete the project. Together these reports establish periodic funding requirements and show the trend toward total cost overruns or underruns. Both reports are prepared as standard PERT/cost outputs.

Most proposals today stipulate that the contractor prepare only one time/cost plan to complete the proposed project by a directed completion date. Although this *directed-date* plan may be based primarily on a timed requirement, the factors of cost and technical risk are frequently of major importance in selecting a particular development plan. With only a single time/cost alternative to consider, neither the customer nor the contractor can determine that a directed-date plan is the *best* combination of time, cost, and technical risk for a particular project.

The time/cost option procedure calls for the preparation of three alternative time/cost plans for accomplishing the project. At least one of these plans will be prepared to meet the directed date. In addition to the directed-date plan, the procedure calls for a plan to accomplish the project in the shortest possible time and a plan for accomplishing the project in the time which will allow the contractor to achieve the project objectives in the most efficient manner.

In the development project, there frequently are various levels of resources that can be applied to each activity. It is important for the manager to recognize what effects the different applications will have on the total time and cost of a project, especially when a speed-up or stretch-out is being considered.

The resource allocation procedure identifies the specific allocation of resources for each activity that will yield the lowest total cost for one or more specified project durations. To do this, the procedure calls for alternate resource/time estimates for performing each activity in the project. The steps followed in this procedure are:

1. Construct network.
2. Obtain alternative time/cost estimates for each activity.
3. Select the lowest-cost alternative for each activity.

4. Calculate the critical path and compare to directed date.
5. If the critical path is too long, select higher-cost, shorter-time alternatives on critical path activities. These alternate points are picked where the ratio of increased cost to decreased time is least.
6. Repeat step 5 until length of critical path conforms to the directed date.

There are many forms of PERT/cost systems available. The government has created task forces to standardize the input to and output from these systems. References 1 and 4 give some examples of the types of reports expected from PERT/cost systems.

4.4.3 Resource allocation and multiproject scheduling. An early approach to a multiproject control method was offered by Mize [14]. He developed a noniterative heuristic model which schedules activities (jobs) for the several operating facilities of a multiproject organization when the objective is to minimize due-date slippage. The inputs to his scheduling model are the outputs from the individual project critical path analyses.

The method offered by Mize was one of the first heuristic approaches to be implemented by simulation. It also takes into account the dynamic relationships of activities to activities and project to project when schedule conflicts arise. The method generally is applicable to any program involving several projects competing for the same limited resources.

At approximately the same time that Mize was developing his model, the RAMPS (resource allocation and multiproject scheduling) algorithm was being completed [10]. RAMPS is another computerized model that handles several projects simultaneously. It schedules each activity so that project due dates are met, subject to stated resource constraints. Details of the RAMPS computational algorithm are not available, but it probably is a heuristic system based on juggling slack activities.

More recent research in this area has been performed by Fendley [5]. The development of his system was based on the concept of assigning due dates to incoming projects and then sequencing the activities of the projects toward meeting these due dates. He also used the heuristic approach, implemented by digital simulation, to solve the scheduling problem. Fendley concluded that giving priority to the activity with minimum slack-from-due-date (MSF rule) resulted in the best performance for his purposes. After determining that the MSF rule worked best, he used it to set realistic due dates by determining the amount of slippage that must occur to perform all projects with fixed resources.

Jordan and Mize [8] applied a simulation technique to the scheduling of multiengineering projects. They found that a rule based upon a combination of processing time and due date yielded the best results.

Wiest and Levy [17] give an interesting discussion of various approaches to the resource allocation problem.

EXERCISES

1. As a recent industrial engineering graduate you have decided, because of the unlimited opportunities, excellent pay, and fringe benefits (Pub Nights), to join the University of LeHigh Organization. More specifically, you have decided to work for the Buildings and Grounds Department because of their wide reputation as a progressive, dynamic organization.

Your first assignment is to develop quantitative methods of scheduling that will surplant existing methods of folklore and superstition. Together with Mr. Letters, the brilliant, lovable leader of Buildings and Grounds, you have developed the following list of activities that will lead to the completion of the Mert Library of Science and Engineering. Your job now is to develop an activity network for the project.

LIST OF ACTIVITIES

10—Contract award.
20—Site surveying and foundation excavation.
30—Site drainage.
40—Footings and foundation concrete.
50—Structural steel.
60—Foundation walls completed.
70—Finish concrete flooring and fire and waterproofing.
80—Rough plumbing.
90—Rough electrical.
100—Rough heating.
110—Masonry.
120—Roofing and flashing.
130—Set doors and window frames.
140—Excavation and pour sidewalks.
145—Lathing and brown coat of plaster.
150—Finish plaster
160—Glass and glazing.
170—Finish grade, topsoil, and seeding.
180—Tiles and marble.
190—Finish plumbing.
200—Finish heating.
210—Paint ceilings.
220—Floor coverings.
230—Paint walls.
240—Shelving and book stacks.
250—Window blinds.
260—Finish paint.
270—Completion of building.

2. *PERT Project:* Walker Portland Company Warehouse.
The Walker Portland Company is currently in the process of planning the

construction of a new warehouse facility to reduce the company's inventory storage problems. Most of the work will be performed by the company's own construction division, but certain portions of the project, such as electrical and plumbing work, will be subcontracted.

Assume you have been retained by the company as a consultant to aid in the planning and administrative control of the project. You have decided to use PERT (program evaluation and review technique) to perform this function. Before beginning on the PERT network, you called a meeting with the general foreman, estimator, and chief engineer, all from the company's construction division. Together, you have prepared a series of time estimates for the various separate portions of the project. These include a most optimistic, most pessimistic, and most likely time estimate. These separate jobs or activities and the three time estimates for each appear in Exhibit 4-1.

Exhibit 4-1

Activity Code	Activity	Optimistic Time Days*	Most Likely Time Days*	Pessimistic Time Days*
a	Obtain corporate management approval	2	5	8
b	Complete subcontractor negotiations	1	3	7
c	Grade building site and excavate for foundation	5	7	9
d	Procure structural steel for framework	10	15	20
e	Procure concrete for foundation	1	3	5
f	Procure exterior window and door frames	5	7	11
g	Procure supplies for exterior walls and roofing	1	3	5
h	Pour concrete for foundation	9	10	15
i	Pour building footings	4	5	10
j	Erect steel framework	8	11	14
k	Pour floor slab and lay concrete flooring	5	9	13
l	Erect exterior walls	21	24	27
m	Pour roof slab	9	12	15
n	Lay roofing	2	3	8
o	Electrical work—subcontracted	8	10	14
p	Plumbing—subcontracted	7	10	13
q	Install insulation and interior walls	10	15	20
r	Paint interior	3	5	9
s	Install fuel tank and heating system	8	10	16
t	Excavate and lay sewage drain	5	8	11
u	Driveway and parking lot—subcontracted	9	12	15
v	Backfill around building and grade	6	8	10
w	Clean up building and grounds	2	2	2
x	Obtain job acceptance	3	5	9

*Working days only, 5 days per week.

In addition to determining the list of activities, the estimator and foreman discussed in some detail how these activities should be sequenced since the list in Exhibit 4-1 does not necessarily indicate the order in which the work should be performed. During their discussion, you made the notes shown in Exhibit 4-2.

Exhibit 4-2

Corporate management approval must be obtained before any work or procurement of supplies can begin.

Procurement of materials and subcontracting negotiations can begin as soon as approval is obtained.

Grading can begin as soon as subcontracting negotiations are completed.

Foundation can be poured following excavation and after concrete arrives.

Building footings can be poured as soon as excavation work is completed.

Framework must be erected in conjunction with (simultaneously) pouring of building footings.

Floor slab can be poured at completion of foundation pouring. Extra time is allowed in estimates for safety reasons so that floor work does not interfere with framework construction.

Exterior walls can be put up following completion of steel framework and arrival of supplies.

Roof can be installed after steel framework is up. Roof slab pouring must precede roofing installation.

Electrical and plumbing work can commence as soon as steel framework is up.

Insulation and interior walls can be put up after electrical and plumbing work.

Interior painting must await installation of interior walls.

Work for sewage drain, driveway, and parking lot can begin as soon as building site is graded and excavated. Final backfilling and grading must await completion of these activities.

Installation of fuel oil tank and heating system must be started after the foundation has been poured. It must be completed before the interior walls are put in.

Clean up is the final work before job acceptance.

Company management has requested that your PERT analysis be included with the proposal which comes before them.

Prepare a PERT network representing the warehouse construction project. Identify each activity on the network. Also indicate the critical path showing your calculations which identify this path. What is the most likely time estimate for the project on the basis of current estimates, and what is the probability of completing it within this time period? It is anticipated that management will establish a scheduled completion date of three months, which is equal to sixty-five working days, after the time that the proposal is presented to them. What is the chance that the project will be completed within this scheduled time under current circumstances?

3. Petersen General Contractors is currently preparing a bid for a television station for the erection of a 225-foot television antenna tower and the construction of a building adjacent to the tower which will be used to house transmission and electrical equipment. Petersen is bidding only on the tower and its electrical

equipment, the building, the connecting cable between tower and building, and site preparation. Transmission equipment and other equipment to be housed in the building are not to be included in the bid and will be obtained separately by the television station. The site for the tower is at the top of a hill to minimize the required height of the tower, with the building to be constructed at a slightly lower elevation than the base of the tower and near a main road. Between the tower and building will be a crushed-gravel service road and an underground cable. Adjacent to the building, a fuel tank will be installed above ground on a concrete slab.

Prior to preparing the detailed cost estimates, Petersen's estimator met with the company's general foreman to go over the plans and blueprints for the job. In addition to preparing a cost estimate; the estimator was also preparing an estimate of the time it would take to complete the job. The television station management was very concerned about the time factor. It requested that bids be prepared on the basis of the most likely time for completing the job and also for the most optimistic and most pessimistic times for completing the job. During the conference between the estimator and general foreman, it was determined that the activities shown in Exhibit 4-3 would be necessary to complete the job. The estimator prepared time estimates for these activities as shown in Exhibit 4-3.

Exhibit 4-3. TELEVISION TOWER AND BUILDING CONSTRUCTION TIME AND COST ESTIMATE

Activity Code	Activity	Most Likely Time Days*	Optimistic Time Days*	Pessimistic Time Days*
a	Sign contract and complete subcontractor negotiations	5	5	5
b	Survey site	6	4	8
c	Grade building site and excavate for basement	8	6	10
d	Grade tower site	30	21	39
e	Procure structural steel and guys for tower	85	85	85
f	Procure electrical equipment for tower and connecting underground cable	120	120	120
g	Pour concrete for tower footings and anchors	42	25	59
h	Erect tower and install electrical equipment	38	25	51
i	Install connecting cable in tower site	8	4	12
j	Install drain tile and storm drain in tower site	35	18	52

Exhibit 4-3. (Cont.)

Activity Code	Activity	Most Likely Time Days	Optimistic Time Days	Pessimistic Time Days
k	Backfill and grade tower site	8	4	12
l	Pour building footings	29	21	37
m	Pour basement slab and fuel tank slab	14	11	17
n	Pour outside basement walls	34	30	38
o	Pour walls for basement rooms	9	7	11
p	Pour concrete floor beams	11	10	12
q	Pour main floor slab and lay concrete block walls	12	10	14
r	Pour roof slab	15	13	17
s	Complete interior framing and utilities	42	30	54
t	Lay roofing	3	2	4
u	Paint building interior, install fixtures, and clean up	19	13	25
v	Install main cable between tower site and building	35	25	45
w	Install fuel tank	3	2	4
x	Install building septic tank	12	8	16
y	Install drain tile and storm drain in building site	15	10	20
z	Backfill around building, grade, and surface with crushed rock	9	7	11
aa	Lay base for connecting road between tower and building	15	13	17
bb	Complete grading and surface connecting road	8	5	11
cc	Clean up tower site	5	3	7
dd	Clean up building site	3	2	4
ee	Obtain job acceptance	5	5	5

*Days shown are working days only.

In addition to determining the list of activities, the estimator and foreman discussed in some detail how these activities could be sequenced since the list of activities in Exhibit 4-3 did not necessarily indicate the order in which the work could be performed. In the course of the discussion, the estimator made the notes shown in Exhibit 4-4.

(a) Prepare a PERT network which portrays the plan for the construction project.
(b) Identify the critical path in the network. What is the most likely time estimate for the project?
(c) Under the most optimistic circumstances, would it be possible to complete

Exhibit 4-4

Survey work and procurement of the structural steel and electrical equipment for the tower can start as soon as contract is signed.

Grading of tower and building sites can begin when survey is completed.

After tower site is graded, footings and anchors can be poured.

After building site is graded and basement excavated, building footings can be poured.

Septic tank can be installed when grading and excavating of building site is done.

Construction of connecting road can start as soon as survey is completed.

Exterior and interior basement walls can be poured as soon as footings are in.

Basement floor and fuel tank slab should go in after basement walls.

Floor beams can go in after the basement walls and basement floor.

Main floor slab and concrete block walls go in after floor beams.

Roof slab can go on after block walls are up.

Interior can be completed as soon as roof slab is on.

Put in fuel tank any time after slab is in.

Drain tile and storm drain for building go in after septic tank.

As soon as tower footings and anchors are in and tower steel and equipment are available, tower can be erected.

Connecting cable in tower site, drain tile, and storm drain can be put in as soon as tower is up.

Main cable between building and tower goes in after connecting cable at tower site is in and basement walls are up.

Tower site can be backfilled and graded as soon as storm drain, connecting cable, and main cable are in.

Clean up tower site after backfilling and grading is done.

Backfill around building and grade after main cable is in and after storm drain is in.

Clean up building site after backfilling and grading is done.

the project within a 195-day period? If so, what portions of the job must be supervised most closely in order to achieve completion within this time?

4. Suppose you want to write a thesis. Construct an activity diagram of the various activities and events which would constitute such an endeavor from start to finish. Perform an analysis similar to that presented in this chapter for this situation.

5. Suppose that you wish to install a computer center in a company which previously had no computer. Construct an activity diagram which will describe all the activities and events that would constitute such an endeavor. Perform an analysis similar to that presented in this chapter for this situation.

6. Simulation has been proposed as a means of analyzing PERT networks. Write a computer program to perform a Monte Carlo simulation on a PERT network. Determine the probability that an activity will be on the critical path along with the distribution of the time to complete the project.

7. Ten plumbers and ten laborers are available for a sewerage project. The project is described as follows:

i	j	Plumbers	Laborers	Time
1	2	2	4	6
1	3	2	0	3
1	4	2	3	4
2	4	7	7	8
2	5	0	1	7
3	6	5	7	2
4	7	2	2	9
5	6	0	3	5
6	7	5	5	8

Plumbers and laborers must both work at the same time when assigned to the same activity. Prepare an activity schedule assuming that interruption of jobs is permitted if necessary.

REFERENCES

1. *AFSC PERT Cost System Cost Module.* Air Force Systems Command PERT Control Board, Headquarters AFSC, Washington, D. C., Vol. 1 (May, 1963).

2. Antill, J. M. and R. W. Woodhead, *Critical Path Methods in Construction Practice.* New York, John Wiley & Sons, Inc., 1966.

3. Battersby, A., *Network Analysis for Planning and Scheduling.* London, Macmillan and Co., Ltd., 1967.

4. *DOD and NASA Guide, PERT Cost Systems Design,* by the Office of the Secretary of Defense and the National Aeronautics and Space Administration. U. S. Government Printing Office, Washington, D. C. (June, 1962).

5. Fendley, L. G., "Toward the Development of a Complete Multi-Project Scheduling System." *Journal of Industrial Engineering*, Vol. 19, No. 10 (October, 1968), pp. 505–515.

6. Iannone, A. L., *Management Program Planning and Control with PERT, MOST, LOB.* Englewood Cliffs, N. J., Prentice Hall, Inc., 1967.

7. *An Introduction to the PERT/Cost System for Integrated Project Management.* Special Projects Office, Navy Department, 1961.

8. Jordan, L. F. and J. H. Mize, "Activity Assignment Procedures in an Engineering Organization." *Proceedings of the 21st AIIE Conference* (May, 1970).

9. Kelley, J. E., Jr., "Critical Path Planning and Scheduling: Mathematical Basis." *Operations Research*, Vol. 9, No. 3 (May–June, 1962).

10. Lambourn, S., "RAMPS—A New Tool in Planning and Control." *The Computer Journal* (January, 1963).

11. Levin, R. and C. Kirkpatrick, *Planning and Control with PERT/CPM.* New York, McGraw-Hill, Inc., 1966.

12. *Line of Balance*, Pamphlet NAVEX OSP1851, Office of Naval Material, 1958.

13. McMillan, C. and R. F. Gonzalez, *Systems Analysis, A Computer Approach to Decision Models.* Homewood, Ill., Richard D. Irwin, 1968.

14. Mize, Joe H., "A Heuristic Scheduling Model for Multi-Project Organizations." PhD. Dissertation, Purdue University, August, 1964.

15. Phillips, C. R., "Fifteen Key Features of Computer Programs for CPM and PERT," *Journal of Industrial Engineering* (January–February, 1964).

16. Ranck, N. A. and R. O. Farr, "Plannet Can Help PERT." *Aerospace Management* (February, 1972), pp. 18–21.

17. Wiest, J. and F. Levy, *Management Guide to PERT-CPM.* Englewood Cliffs, N. J., Prentice-Hall, Inc., 1969.

MAXIMAL FLOW AND SHORTEST PATH
ANALYSIS

In this chapter we will explore two related problems in network analysis. The first will be the determination of the shortest path through a network, and the second will interpret the network as a set of pipes with various flow capacities. The problem will be to maximize the flow in the network. This chapter will present algorithms to solve these problems along with some thoughts on the applications of the approaches. We will also take a detailed look at maximum flow analysis used to develop the optimum time/cost curve for the CPM model discussed in Chapter 3.

There are many good references to the material covered in this chapter, but the best overview in the author's opinion is the recent work of Elmaghraby [5, 6].

5

5.1 Shortest Path Analysis*

5.1.1 Introduction. We will consider solutions to the following three shortest-path problems for directed acyclic and cyclic networks:

1. Find the shortest path between the origin and the terminal point of a network.
2. Find the shortest paths from the origin to all the nodes in the network.
3. Find the shortest paths between all pairs of nodes.

We will also investigate the problem of finding the m shortest routes in a directed acyclic network, that is, we wish to rank the various paths through a network in ascending order according to their lengths. Finally, we will consider some applications of the shortest-path approach.

5.1.2 Acyclic directed networks. The PERT and CPM activity networks presented in Chapter 3 belong to a class of networks which are called *acyclic*

---

*This section is adapted from material appearing in Reference 6 and is used with the permission of TIMS and Dr. Elmaghraby.

directed networks. This means that all the paths are directed and contain no cycles or loops. Because of the special nature of these networks, we can always number the nodes in such a way that all activities lead from a smaller-numbered node to a node of larger number. Consider the network shown in Fig. 5.1(a). We can number the nodes as shown in Fig. 5.1(b) such that all

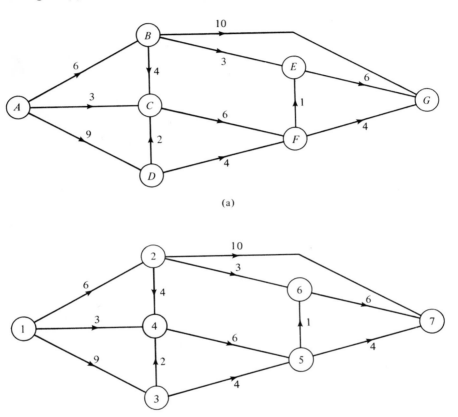

(a)

Figure 5.1. Node numbering for a directed acyclic network.

activities lead from smaller-numbered nodes to higher-numbered nodes. Because of this feature, the algorithms for finding the shortest paths in directed acyclic networks are quite simple.

Let us first consider the problem of finding the shortest path from node 1 to node 7 in Fig. 5.1(b). Thus we are seeking the shortest path from the source to the sink in this network. This problem can be solved by the following algorithm:

1. Label node 1 with the designation $m_1 = 0$.

2. Continue to label the remaining nodes (chosen in ascending order) according to the following formula:

$$m_j = \min_{i=1,2,\ldots,j-1} (m_i + d_{ij})$$

where d_{ij} is the distance between node i and node j. d_{ij} can be assumed to approach infinity for those combinations of i and j which do not have paths in the network of interest.

3. When the last node, n, has been labeled, m_n is the shortest path through the network. The path is determined by tracing backwards from node n to all nodes such that $m_i + d_{ij} = m_j$ for $j = n,\ n - 1,\ n - 2, \ldots, 1$. That is, any path satisfying the equality $m_i + d_{ij} = m_i$ is a candidate for being on the shortest path from the origin to the terminal of the network.

Now let us apply this algorithm to the network shown in Fig. 5.1(b).

Step 1:

$$m_1 = 0$$

Step 2:

$$m_2 = \min (m_1 + d_{12}) = \min (0 + 6) = 6$$
$$m_3 = \min (m_1 + d_{13}) = \min (0 + 9) = 9$$
$$m_4 = \min (m_1 + d_{14}, m_2 + d_{24}, m_3 + d_{34})$$
$$= \min (0 + 3, 6 + 4, 9 + 2) = 3$$
$$m_5 = \min (m_3 + d_{35}, m_4 + d_{45}) = \min (9 + 4, 3 + 6) = 9$$
$$m_6 = \min (m_2 + d_{26}, m_5 + d_{56}) = \min (6 + 3, 9 + 1) = 9$$
$$m_7 = \min (m_2 + d_{27}, m_5 + d_{57}, m_6 + d_{67})$$
$$= \min (6 + 10, 9 + 4, 9 + 6) = 13$$

Step 3: Since node 7 has been labeled, we conclude that the shortest path through the network is of length 13. Checking for the path, we find that activities (5, 7), (2, 6), (4, 5), (1, 4), (1, 3), and (1, 2) satisfy the equality $m_i + d_{ij} = m_j$. Thus the shortest path from 1 to 7 will be (1, 4), (4, 5), and (5, 7).

It is interesting to note that the algorithm just presented also solves the problem of finding the shortest path from the origin to any node in the network. The m labels give the lengths of such paths. The actual paths are determined from the set of activities satisfying the equality $m_i + d_{ij} = m_j$. For the example shown in Fig. 5.1(b), $m_6 = 9$; therefore, the shortest path from node 1 to node 6 is 9. The path includes activities (1, 2) and (2, 6). Similarly, $m_3 = 9$, so the shortest path from node 1 to node 3 is of length 9 and includes only activity (1, 3).

A simple modification to the algorithm will allow us to determine the shortest path between all pairs of nodes. Due to the special nature of directed acyclic networks we observe that a node can only have a path to a node of higher number. Therefore, it is possible for us to have a path from node 2 to node 5, but not from node 5 to node 2. To find the shortest path from nodes k to h, $h > k$, consider the portion of network including all nodes of higher number than k and apply the same procedure outlined above with $m_k = 0$.

For example, if we were interested in the shortest paths from node 3 to nodes 4, 5, 6, and 7, we would first modify the network as shown in Fig. 5.2 and then apply our algorithm:

Step 1:

$$m_3 = 0$$

Step 2:

$$m_4 = \min (m_3 + d_{34}) = \min (0 + 2) = 2$$
$$m_5 = \min (m_3 + d_{35}, m_4 + d_{45}) = \min (0 + 4, 2 + 6) = 4$$
$$m_6 = \min (m_5 + d_{56}) = \min (4 + 1) = 5$$
$$m_7 = \min (m_5 + d_{57}, m_6 + d_{67}) = \min (4 + 4, 5 + 6) = 8$$

Step 3:

Activities (3, 4), (3, 5), (5, 6), and (5, 7) satisfy the equality $m_i + d_{ij} = m_j$; so the results of our calculations are summarized below.

Shortest Path from Node 3 to	Length	Includes Activities
node 4	2	(3, 4)
node 5	4	(3, 5)
node 6	5	(3, 5)(5, 6)
node 7	8	(3, 5)(5, 7)

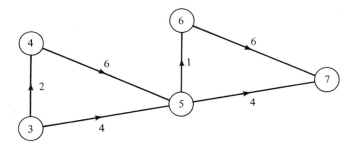

Figure 5.2. Modified acyclic network.

The shortest path between all pairs of nodes can be determined in a similar way.

5.1.3 Cyclic directed networks.

In this section we will consider the more general directed network which includes cycles. There are numerous algorithms [2, 7, 11] available for finding the shortest path from the origin to the terminal of the network. We will concentrate on one of these algorithms credited to Dijkstra [3]. This is probably the most efficient and flexible of the algorithms developed to solve this problem. The algorithm takes advantage of the fact that, if j is a node on the minimal path from the origin to the terminal of a network, then we automatically know the shortest path from the origin to j. As we see, the algorithm constructs minimal paths from the origin to other nodes in order of increasing minimum path length until the terminal node is reached.

Dijkstra subdivided the nodes into three sets:

A. Nodes for which the minimum path from the origin is known; nodes are added to this set in order of increasing minimum path length from the origin.

B. Nodes from which the next node to be added to set A will be selected; the set includes all nodes that are connected to at least one node of set A but have not yet qualified for A.

C. The remaining nodes.

The arcs are also subdivided into three sets:

I. Arcs which are part of minimal paths from the origin to the nodes in set A.

II. Arcs from which the next arc to be placed in set I will be chosen; one and only one arc from this set will lead to each node in set B.

III. The remaining arcs (rejected or not yet considered).

Initially, all nodes are considered to be in set C and all arcs are in set III. We then transfer the origin node to set A and perform the following steps.

Step 1: Consider all arcs connecting the node just transferred to set A with nodes j in sets B or C. If node j is in set B, test whether the use of the arc yields a shorter path from the origin to j than the known path that uses the arc now in set II. If this is not the case, the arc is rejected; however, if the arc yields a shorter path from the origin to j, it replaces the corresponding arc in set II, and the replaced arc is rejected. If the node j belongs to set C, it is added to set B and the arc is added to set II.

Step 2: Every node in set B can be connected to the origin in only one way if we consider the arcs from set I and one from set II. In this sense, each node in set B has a distance from the origin. The node with the minimum distance is transferred from set *B* to set *A*, and the corresponding arc is moved from set II to set I. Next we return to step 1 and repeat the process until the terminal node is transferred to set *A*.

As an example of the procedure, consider the network shown in Fig. 5.3. The procedure can be applied as follows. At the start, the sets will

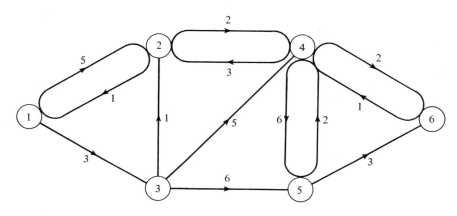

Figure 5.3. Cyclic directed network.

appear as:

Set A	B	C	I	II	III
1		2, 3, 4, 5, 6			(1, 2)(2, 1)(1, 3)(3, 2)
					(2, 4)(4, 2)(3, 4)(3, 5)
					(4, 5)(5, 4)(5, 6)(4, 6)
					(6, 4)

When we apply step 1 of the procedure, nodes 2 and 3 move to set *B*, and arcs (1, 2) and (1, 3) move to set II, yielding:

Set A	B	C	I	II	III
1	2, 3	4, 5, 6		(1, 2)(1, 3)	(2, 1)(3, 2)(2, 4)(4, 2)
					(3, 4)(3, 5)(4, 5)(5, 4)
					(5, 6)(4, 6)(6, 4)

In step 2 we survey the distance from the origin of those nodes in set *B* and find that node 2 has a distance of 5 while node 3 has a distance of 3. Thus node 3 is added to set *A* and (1, 3) is moved to set I:

Set A	B	C	I	II	III
1, 3	2	4, 5, 6	(1, 3)	(1, 2)	(2, 1)(3, 2)(2, 4)(4, 2) (3, 4)(3, 5)(4, 5)(5, 4) (5, 6)(4, 6)(6, 4)

Nodes 2, 4, and 5 are connected to node 3. Node 2 is in set *B*, so a comparison must be made to determine if the new route to node 2 is better than the existing path. The path (1, 2) is 5 units long, while the path including (3, 2) would be 4 units in length. Thus (3, 2) replaces (1, 2) in set II, and path (1, 2) is rejected (rejected paths are shown in set III with a slash through them). Nodes 4 and 5 are not in set *B*, so they are added; paths (3, 4) and (3, 5) are added to set II.

Set A	B	C	I	II	III
1, 3	2, 4, 5	6	(1, 3)	(3, 2)(3, 4)(3, 5)	(2, 1)(2, 4)(4, 2)(1, 2) (4, 5)(5, 4)(5, 6)(4, 6) (6, 4)

In the next step we find that node 2 has the shortest route to the origin ($1 \rightarrow 2 = 4$, $1 \rightarrow 4 = 8$, $1 \rightarrow 5 = 9$).

Set A	B	C	I	II	III
1, 3, 2	4, 5	6	(1, 3)(3, 2)	(3, 4)(3, 5)	(2, 1)(2, 4)(1, 2) (4, 2)(4, 5)(5, 4) (5, 6)(4, 6)(6, 4)

Since node 4 is in set *B*, the paths including (3, 4) and (2, 4) must be compared to determine the best route to node 4. Activity (3, 4) is rejected.

Set A	B	C	I	II	III
1, 3, 2	4, 5	6	(1, 3)(3, 2)	(2, 4)(3, 5)	(2, 1)(4, 2)(4, 5)(5, 4)(1, 2) (5, 6)(4, 6)(6, 4)(3, 4)

The distances from the origin to nodes 4 and 5 are compared, and it is found that node 4 is closer to the origin.

Set A	B	C	I	II	III
1, 3, 2, 4	5	6	(1, 3)(3, 2)(2, 4)	(3, 5)	(2, 1)(4, 2)(4, 5)(5, 4)(3, 4) (5, 6)(4, 6)(6, 4)(1, 2)

After the next step of our procedure we have:

Set A	B	C	I	II	III
1, 3, 2, 4	5, 6		(1, 3)(3, 2)(2, 4)	(3, 5)(4, 6)	(2, 1)(4, 2)(5, 4)~~(1, 2)~~ (5, 6)(6, 4)~~(3, 4)(4, 5)~~

Finally, the distances of nodes 5 and 6 are compared, and it is found that node 6 is closer to the origin; so the node is added to set *A*. Since node 6 is the terminal node, the algorithm is terminated.

Set A	B	C	I	II	III
1, 3, 2, 4, 6	5		(1, 3)(3, 2)(2, 4)(4, 6)	(3, 5)	(2, 1)(4, 2)(5, 4)~~(1, 2)~~ (5, 6)(6, 4)~~(3, 4)(4, 5)~~

The algorithm yielded a distance of 8 units from nodes 1 to 6 along the path (1, 3)(3, 2)(2, 4), and (4, 6).

Now consider the problem of determining the shortest path from the origin to each of the nodes in the network, and we see that Dijkstra's algorithm gives the desired result. All elements in set *A* have their shortest paths calculated. In the example we just investigated, set *A* included all the nodes except node 5. By continuing the application of Dijkstra's algorithm, we can force all the nodes in the network into set *A* and thus solve the problem of interest.

For our example the algorithm will yield:

Set A	B	C	I	II	III
1, 3, 2, 4, 6, 5			(1, 3)(3, 2)(2, 4) (4, 6)(3, 5)		(2, 1)(4, 2)(5, 4)~~(1, 2)~~ (5, 6)(6, 4)~~(3, 4)(4, 5)~~

Thus we determine the following results:

Origin to Node	Distance	Path
2	4	(1, 3)(3, 2)
3	3	(1, 3)
4	6	(1, 3)(3, 2)(2, 4)
5	9	(1, 3)(3, 5)
6	8	(1, 3)(3, 2)(2, 4)(4, 6)

The last unanswered question for the directed cyclic network is finding the shortest path between all pairs of nodes. One approach to this problem would be the successive applications of one of the shortest-path algorithms such as Dijkstra's method, with each node in the system assumed to be the origin of the network. This is an inefficient approach to this problem, so an alternate will be proposed. This method is called the *revised cascade method* [7] and could have been applied to the case of the directed acyclic network discussed in Section 5-1.2. It hinges on the operation:

$$d_{ik} = \min (d_{ik}; d_{ij} + d_{jk})$$

in which the distance d_{ik} between nodes i and k is compared to the sum $d_{ij} + d_{jk}$ for some intermediate node j. The minimum of these two quantities replaces d_{ik} in the matrix of distances D.

The operation is performed as follows. Let j be fixed at $j = 1, 2, \ldots, n$. For each value of j, say $j = j_0$, do the operation for each entry d_{ik} with $k \neq j_0 \neq k$ and $k \neq j_0 \neq i$. The matrix thus generated is the shortest-distance matrix. To determine the shortest routes, a route matrix is generated while the shortest-distance matrix is being constructed.

$$r_{ik} = \begin{cases} r_{ik} & \text{if } d_{ik} \leq d_{ij} + d_{jk} \\ r_{ij} & \text{otherwise} \end{cases}$$

The matrix $R = [r_{ik}]$ is the route matrix. The initial R matrix has entries of $r_{ik} = k$ for each nonzero d_{ik} entry.

Consider the network shown in Fig. 5.4. The D and R matrices for this network are:

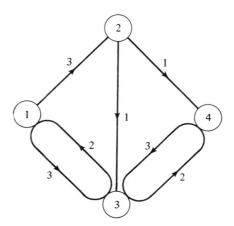

Figure 5.4. Cyclic directed network.

$$D = \begin{array}{c} \\ 1 \\ 2 \\ 3 \\ 4 \end{array} \begin{array}{cccc} 1 & 2 & 3 & 4 \\ \begin{bmatrix} 0 & 3 & 3 & \infty \\ \infty & 0 & 1 & 1 \\ 2 & \infty & 0 & 2 \\ \infty & \infty & 3 & 0 \end{bmatrix} \end{array}, \qquad R = \begin{array}{c} \\ 1 \\ 2 \\ 3 \\ 4 \end{array} \begin{array}{cccc} 1 & 2 & 3 & 4 \\ \begin{bmatrix} - & 2 & 3 & - \\ - & - & 3 & 4 \\ 1 & - & - & 4 \\ - & - & 3 & - \end{bmatrix} \end{array}$$

where the ∞ indicates paths that do not exist. Now let $j_0 = 1$ in the equation:

$$d_{ik} = \min (d_{ik}; d_{ij_0} + d_{j_0 k})$$

and consider each entry d_{ik} with $i \neq 1 \neq k$ and $k \neq 1 \neq i$.

$$d_{23} = \min (d_{23}; d_{21} + d_{13}) = \min (1; \infty + 3) = 1$$
$$r_{23} = 3$$
$$d_{24} = \min (d_{24}; d_{21} + d_{14}) = \min (1; \infty + \infty) = 1$$
$$r_{24} = 4$$
$$d_{32} = \min (d_{32}; d_{31} + d_{12}) = \min (\infty; 2 + 3) = 5$$
$$r_{32} = r_{31} = 1$$
$$d_{34} = \min (d_{34}; d_{31} + d_{14}) = \min (2; 2 + \infty) = 2$$
$$r_{34} = 4$$
$$d_{42} = \min (d_{42}; d_{41} + d_{12}) = \min (\infty; \infty + 3) = \infty$$
$$r_{42} = -$$
$$d_{43} = \min (d_{43}; d_{41} + d_{13}) = \min (3; \infty + 3) = 3$$
$$r_{43} = 3$$

Thus:

$$D = \begin{bmatrix} 0 & 3 & 3 & \infty \\ \infty & 0 & 1 & 1 \\ 2 & 5 & 0 & 2 \\ \infty & \infty & 3 & 0 \end{bmatrix}, \qquad R = \begin{bmatrix} - & 2 & 3 & - \\ - & - & 3 & 4 \\ 1 & 1 & - & 4 \\ - & - & 3 & - \end{bmatrix}$$

It should be noted that whenever d_{ij_0} or $d_{j_0 k} = \infty$, d_{ik} cannot change. Therefore, only d_{32} was a candidate for change. We wasted effort by evaluating d_{23}, d_{24}, d_{34}, d_{42}, and d_{43}.

Now let $j_0 = 2$. Since d_{21} and $d_{42} = \infty$, then d_{31}, d_{41}, and d_{43} cannot change. So we will only evaluate d_{13}, d_{14}, and d_{34}.

$$d_{13} = \min (d_{13}; d_{12} + d_{23}) = (3; 3 + 1) = 3$$
$$r_{13} = 3$$

$$d_{14} = \min (d_{14}; d_{12} + d_{24}) = (\infty; 3 + 1) = 4$$
$$r_{14} = r_{12} = 2$$
$$d_{34} = \min (d_{34}; d_{32} + d_{24}) = (2; 5 + 1) = 2$$
$$r_{34} = 4$$

Thus:

$$D = \begin{bmatrix} 0 & 3 & 3 & 4 \\ \infty & 0 & 1 & 1 \\ 2 & 5 & 0 & 2 \\ \infty & \infty & 3 & 0 \end{bmatrix}, \qquad R = \begin{bmatrix} - & 2 & 3 & 2 \\ - & - & 3 & 4 \\ 1 & 1 & - & 4 \\ - & - & 3 & - \end{bmatrix}$$

When we let $j_0 = 3$:

$$d_{12} = \min (d_{12}; d_{13} + d_{32}) = \min (3, 3 + 5) = 3$$
$$r_{12} = 2$$
$$d_{14} = \min (d_{14}; d_{13} + d_{34}) = \min (4; 3 + 2) = 4$$
$$r_{14} = 2$$
$$d_{21} = \min (d_{21}; d_{23} + d_{31}) = \min (\infty; 1 + 2) = 3$$
$$r_{21} = r_{23} = 3$$
$$d_{24} = \min (d_{24}; d_{23} + d_{34}) = \min (1; 1 + 2) = 1$$
$$r_{24} = 4$$
$$d_{41} = \min (d_{41}; d_{43} + d_{31}) = \min (\infty; 3 + 2) = 5$$
$$r_{41} = r_{43} = 3$$
$$d_{42} = \min (d_{42}; d_{43} + d_{32}) = \min (\infty; 3 + 5) = 8$$
$$r_{42} = r_{43} = 3$$

Thus:

$$D = \begin{bmatrix} 0 & 3 & 3 & 4 \\ 3 & 0 & 1 & 1 \\ 2 & 5 & 0 & 2 \\ 5 & 8 & 3 & 0 \end{bmatrix}, \qquad R = \begin{bmatrix} - & 2 & 3 & 2 \\ 3 & - & 3 & 4 \\ 1 & 1 & - & 4 \\ 3 & 3 & 3 & - \end{bmatrix}$$

Finally, let $j_0 = 4$:

$$d_{12} = \min (d_{12}; d_{14} + d_{42}) = \min (3; 4 + 8) = 3$$
$$r_{12} = 2$$
$$d_{13} = \min (d_{13}; d_{14} + d_{43}) = \min (3; 4 + 3) = 3$$
$$r_{13} = 3$$

$$d_{21} = \min(d_{21}; d_{24} + d_{41}) = \min(3; 1 + 5) = 3$$
$$r_{21} = 3$$
$$d_{23} = \min(d_{23}; d_{24} + d_{43}) = \min(1; 1 + 3) = 1$$
$$r_{23} = 3$$
$$d_{31} = \min(d_{31}; d_{34} + d_{41}) = \min(2; 2 + 5) = 2$$
$$r_{31} = 1$$
$$d_{32} = \min(d_{32}; d_{34} + d_{42}) = \min(5; 2 + 8) = 5$$
$$r_{32} = 1$$

Therefore, the final D and R matrices are:

$$D = \begin{bmatrix} 0 & 3 & 3 & 4 \\ 3 & 0 & 1 & 1 \\ 2 & 5 & 0 & 2 \\ 5 & 8 & 3 & 0 \end{bmatrix}, \quad R = \begin{bmatrix} - & 2 & 3 & 2 \\ 3 & - & 3 & 4 \\ 1 & 1 & - & 4 \\ 3 & 3 & 3 & - \end{bmatrix}$$

To interpret these matrices, enter matrix D to determine the minimum distance between two nodes; then matrix R is used to trace the path. For example, let us find the shortest distance from node 4 to node 2. Entering matrix D we find $d_{42} = 8$; then the shortest distance is 8. We find path by entering matrix R. $r_{42} = 3$, $r_{32} = 1$, and $r_{12} = 2$; so the route from node 4 to node 2 is 4–3–1–2.

Thus the revised cascade method gives us a generalized method for investigating the shortest paths between any two nodes in a network. This approach can be used for analyzing cyclic and acyclic networks. It also can be used to find the shortest path from the origin to terminal or from the origin to any other node in the network. It is a generalized algorithm that can be used to answer all the questions posed in Sections 5.1.2 and 5.1.3. But the specialized algorithms presented in Sections 5.1.2 and 5.1.3 are more efficient than the revised cascade method for the restricted conditions for which the specialized algorithms are applicable.

5.1.4 The k shortest paths from the origin to the terminal.

The problem of finding the k shortest paths between nodes is an interesting and important concept. For example, it might not be feasible to use the shortest route. Thus some path such as the second or third shortest might be adopted. Sensitivity analysis is another important use for this approach. For instance, we might be interested in what it will cost us to deviate from the optimum shortest-path solution.

There have been a number of attempts [1, 13, 14] at a generalized algorithm to solve this problem for the cyclic directed graph. Elmaghraby [6] discusses the use of one of these in his recent paper. We will not attempt to solve the general problem in this book, but we will concentrate on the much simpler problem of finding the k shortest paths in a directed acyclic network.

This problem can be handled by a slight modification of the step used in finding the shortest path between two nodes. Elmaghraby [6] suggests the notation $\min_1$ to denote the minimum, $\min_2$ to denote the second minimum, and so on. The general step of the procedure is as follows. For each node j, consider the set of nodes i which are connected to j. The kth shortest distance is found using the following equation:

$$m_j^{(k)} = \min_k_{\substack{\text{all } i \\ \text{connected} \\ \text{to } j}} (m_i^{(r)} + d_{ij}) \quad \text{for} \quad 1 < r < k, \quad k = 1, 2, \ldots, n$$

where $m_j^{(k)}$ is the kth shortest path to node j.

To illustrate, let us determine the three shortest paths for the network illustrated in Fig. 5.1(b). For node 1, $m_1^{(1)} = 0$ since we assume node 1 occurs at time 0. $m_1^{(2)}$ and $m_1^{(3)}$ are undefined since there is only one possible time.

We are now ready to apply our procedure:

node 2:	$m_2^{(1)} = 6$;	$m_2^{(2)}$ and $m_2^{(3)}$ are undefined	
node 3:	$m_3^{(1)} = 9$;	$m_3^{(2)}$ and $m_3^{(3)}$ are undefined	
node 4:	$m_4^{(1)} = 3$;	$m_4^{(2)} = 10$;	$m_4^{(3)} = 11$
node 5:	$m_5^{(1)} = 9$;	$m_5^{(2)} = 13$;	$m_5^{(3)} = 16$
node 6:	$m_6^{(6)} = 9$;	$m_6^{(2)} = 10$;	$m_6^{(3)} = 14$
node 7:	$m_7^{(1)} = 13$;	$m_7^{(2)} = 15$;	$m_7^{(3)} = 16$

As an example of the calculations involved, consider node 7 which is connected to nodes 2, 5, and 6.

$$m_7^{(1)} = \min_1 (m_2^{(1)} + d_{27}; m_5^{(1)} + d_{57}; m_6^{(1)} + d_{67})$$
$$= \min_1 (16, 13, 15) = 13$$
$$m_7^{(2)} = \min_2 (m_2^{(1)} + d_{27}; m_2^{(2)} + d_{27}; m_5^{(1)} + d_{57}; m_5^{(2)} + d_{57}; m_6^{(1)}$$
$$+ d_{67}; m_6^{(2)} + d_{67}) = \min_2 (16, \text{undefined}, 13, 17, 15, 16) = 15$$
$$m_7^{(3)} = \min_3 (m_2^{(1)} + d_{27}; m_2^{(2)} + d_{27}; m_2^{(3)} + d_{27}; \ldots ; m_6^{(3)} + d_{67})$$
$$= 16$$

5.1.5 Some applications of the shortest-path algorithms. In this section we will consider some applications for the techniques just presented.

AN EQUIPMENT REPLACEMENT MODEL

Consider a company which is preparing its equipment replacement policy for the next n years. The company can buy a piece of equipment now (year 1) and keep it until year j (where $j \leq n$). In year j another piece of equipment is bought which is kept until year k. The process continues until year n is reached. We could calculate a cost c_{ij} associated with buying equipment in year i and selling it in year j. The cost figure would include the purchase price, the salvage value, and the running and maintenance costs. If $n = 6$ and equipment must be kept at least two years (except equipment purchased in year 5 and which is phased out in year 6), Fig. 5.5 shows a net-

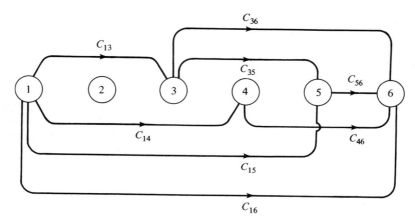

Figure 5.5. Network model of an equipment replacement model.

work representation of the equipment replacement situation. This is a directed acyclic network. The shortest path from node 1 to node 6 represents the minimum cost replacement policy. If the shortest path included paths (1, 4) and (4, 6), this would imply that equipment should be purchased in year 1, kept for 4 years, and then sold. More equipment would then be purchased in year 4 and sold at the end of the planning horizon in the 6th year.

A DISTRIBUTION MODEL

A familiar distribution model is the so-called *transportation model*. In this model there are a number of sources of supply and a number of destinations that must be supplied. The sources of supply are often considered to be warehouses, and the destinations are considered to be customers. Associated with each pair of sources and destinations is a per-unit distribution cost, c_{ij}. The model finds the optimal number of units to ship between each source i and each destination j, x_{ij}, which will minimize the distribution costs $\sum_i \sum_j c_{ij} x_{ij}$.

A modification of this model is known as the *transshipment problem*. In this model it is assumed that it is possible to ship products between the sources and destinations by way of intermediate points. For example, if one wanted to ship a product from Trenton, New Jersey, to Flagstaff,

Arizona, it might be cheaper to truck the product from Trenton to New York, then fly it from New York to Phoenix, and finally truck it from Phoenix to Flagstaff, than it would be to truck it directly from Trenton to Flagstaff. Actually there are a number of possible ways of getting products between Trenton and Flagstaff, including such things as shipping the product to the west coast by boat and then trucking the product back to Flagstaff. Shortest-path analysis can be used to find the shortest or cheapest path between any two distribution points. If there is only one source and one destination, the transshipment problem is a direct application of the shortest-path analysis discussed in this section. If there are a number of sources and destinations, shortest-path analysis could be applied to each pair of sources and destinations. After the most economical routing is found between each pair of points, this information can be fed into the transportation model discussed earlier. The transportation model would then be used to optimize the entire system.

Shortest-path analysis is a useful tool for any organization for which distribution represents a significant portion of their costs. It is useful for the distribution media themselves. Railroads, trucking firms, and airlines can make use of these techniques for their scheduling purposes.

A Reliability Model

Elmaghraby [5] suggests the following analogy between the so-called *most-reliable-route* problem and the shortest-path problem. In the most-reliable-route problem, the arcs of the network are used to represent the components of a system, and the nodes represent the junction points among the components. The input must follow one and only one path from the source to the sink of the network. The problem is to find the most reliable path, where we define reliability as the probability of nonfailure. The reliability of a path through the network is defined as the product of the individual reliabilities of the arcs on the path.

This problem can be reformulated as a shortest-path problem if we take the log of each of the reliabilities in the system. Since the reliability will be maximized when its log is maximized, we can modify the network as discussed in the next paragraph dealing with the relationship between longest- and shortest-path problems.

The Relationship Between Longest- and Shortest-Path Problems

There are two approaches to the longest-path problems. First, we could modify the algorithms by replacing all the minimum operators with maximum operators. The other alternative is to take the negative of all the arc transmittances and to apply the shortest-path algorithms described in this section. These algorithms will find the shortest (most negative) path of the modified network, but this turns out to be the longest path in the original network. These approaches should prove interesting for analyzing the activity networks discussed in Chapters 3 and 4. In the activity network, we are interested in the longest path in a directed acyclic network. Thus the algorithms in this chapter give us the opportunity to analyze activity networks more fully.

5.2 Maximal Flow Methods

5.2.1 Introduction. An important question that can be asked about networks is: "What is the maximum possible flow between two points in a network?" There are a number of physical interpretations of this problem, e.g., telephone messages in the Bell System's network, fuel oil being shipped by pipe from the Gulf coast to other areas of the country, and traffic flowing on a highway network between two or more cities. Recently, network flow techniques have emerged as a powerful tool for analyzing problems formerly analyzed by other research techniques. Many problems, particularly those previously solved by linear programming techniques, have been formulated as network flow models. Usually there is less computational time involved in solving the network flow model than is necessary using other operations research techniques. However, as we will see from our discussion of the maximum flow model of the project cost model introduced in Chapter 3, the conversion to a network flow model can be delicate. There has been great activity in the literature dealing with network flow methods. Ford and Fulkerson [8] wrote the first book on the subject. Hu [10] in his recent text has attempted to report the new developments in the area since Ford's and Fulkerson's text. Elmaghraby [5, 6] also presents an interesting summary of recent work in the area. Hadley [9] and Moder and Phillips [12] present interesting discussions of the elementary concepts of network flow analysis.

5.2.2 Maximal flows; a simple algorithm. In this section we will consider an intuitive approach for finding the maximum flow in a network with a single source and a single sink. We will also assume that there is conservation of flow at each node of the network; that is, the flow into each node must equal the flow out of each node. Obviously, the conservation-of-flow restriction does not hold for the source, where commodities only flow out, and the sink, where commodities only flow in. It is interesting to note, however, that the amount flowing out of the source will equal the amount flowing into the sink because of the conservation of flow at all other nodes. The flow problem also has flow constraints implied for each arc of the network. The flow capacities can, however, be assumed to be infinite for certain of the arcs in the network.

The flow network can be formulated as a linear programming model. Assume that there are n nodes in a network, and node 1 is the source and node n is the sink. In our network we will let x_{ij} denote the flow along the branch connecting nodes i and j, and c_{ij} will equal the capacity in that branch. Thus we have flow capacity constraints for each branch of the form:

$$0 \leq x_{ij} \leq c_{ij} \quad \text{for all } ij$$

We also have constraints that guarantee the conservation of flow at each node:

$$\sum_i x_{ik} = \sum_j x_{kj}, \qquad k = 2, 3, \ldots, n - 1$$

or

$$\sum_i x_{ik} - \sum_j x_{kj} = 0, \qquad k = 2, 3, \ldots, n - 1$$

Our goal is to maximize the total flow in the network. The objective is thus to maximize the flow out of the source $\sum_j x_{1j}$ or into the sink $\sum_i x_{in}$. The linear programming model for this system can be written as

$$\text{Maximize } Z = \sum_j x_{ij}$$

subject to:

$$\sum_i x_{ik} - \sum_j x_{kj} = 0, \qquad k = 2, 3, \ldots, n - 1$$
$$0 \leq x_{ij} \leq c_{ij} \quad \text{for all } ij$$

In the network shown in Fig. 5.6(a), the numbers on the branches correspond to the capacities of branches, i.e., c_{ij}. The linear programming model for this network is:

$$\text{Maximize } Z = x_{12} + x_{13}$$

subject to:

$$x_{12} - x_{23} - x_{24} = 0$$
$$x_{13} + x_{23} - x_{34} - x_{35} - x_{36} = 0$$
$$x_{24} + x_{34} - x_{46} = 0$$
$$x_{35} - x_{56} = 0$$
$$0 \leq x_{12} \leq 8$$
$$0 \leq x_{13} \leq 4$$
$$0 \leq x_{23} \leq 3$$
$$0 \leq x_{24} \leq 3$$
$$0 \leq x_{34} \leq 6$$

$$0 \le x_{35} \le 2$$
$$0 \le x_{36} \le 2$$
$$0 \le x_{46} \le 6$$
$$0 \le x_{56} \le 3$$

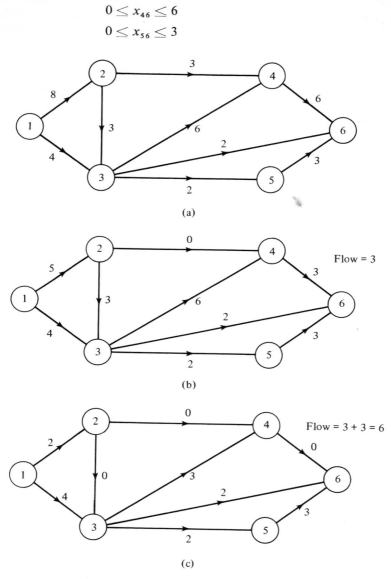

(a)

(b)

(c)

Figure 5.6. An example of the intuitive approach to network flow analysis.

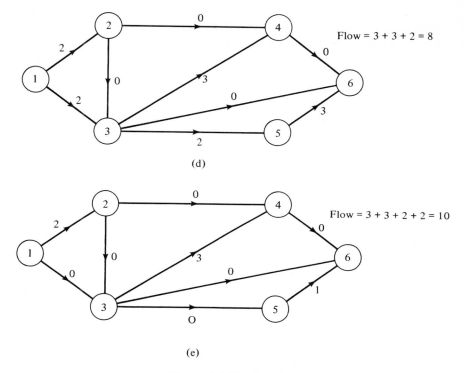

(d)

Flow = 3 + 3 + 2 = 8

(e)

Flow = 3 + 3 + 2 + 2 = 10

Figure 5.6 Continued.

Thus a standard linear programming approach can be used to solve this maximum flow problem. It is, however, inefficient to use linear programming algorithms for networks involving a large number of nodes and arcs.

We will now consider an alternate and simpler method for maximizing the flow through a network. Consider a network such as that shown in Fig. 5.6(a). An intuitive approach to finding the maximum flow is the following:

Step 1: Start at the source and move along branches of positive capacity until we reach the sink. Consider the capacities of the branches on the path from the source to the sink. Let M denote the minimum of these capacities. Next deduct M from each of the capacities for the branches on the path. The remainder on each branch can be viewed as the excess branch capacity.

Step 2: Repeat the above processes on the new network capacities until there is no longer a path from the source to the sink which can carry flow. The maximal flow is the sum of all of the flows found at each step.

We will apply our procedure to the network shown in Fig. 5.6(a). First consider the path 1–2–4–6. The minimum capacity along the path is 3, which occurs on the branch from node 2 to node 4. Thus 3 units of flow can be sent along path 1–2–4–6, and the residual capacities can be calculated by reducing every unit along this path by 3. The revised network is shown in Fig. 5.6(b). Next the path 1–2–3–4–6 is considered from Fig. 5.6(b). The maximum quantity that can flow along this path is 3 units, as restricted by the branch from node 2 to node 3. After the capacities are modified the network will appear as shown in Fig. 5.6(c), and 6 units have flowed through the network. Next it is observed that 2 units can flow along the path 1–3–6, which yields Fig. 5.6(d). Finally, 2 units can flow through the path 1–3–5–6, which yields the network shown in Fig. 5.6(e). Examining Fig. 5.6(e) we can find no other path that can accommodate a positive flow; so our procedure terminates with a total flow of 10 units.

The obvious question to ask is whether the maximal flow for this network is 10? The answer is yes, but we must consider ourselves lucky because, had we chosen our flow paths in a slightly different order, we might not have achieved the flow of 10. Figure 5.7 illustrates one such sequence. The flows were chosen as follows:

Figure	Path	Flow	Total Flow
5.7(a)	1–2–3–6	2	2
5.7(b)	1–3–4–6	4	6
5.7(c)	1–2–4–6	2	8
5.7(d)	1–2–3–5–6	1	9

Examining Fig. 5.7(e), we see no path that will accommodate a positive flow. Therefore, we would assume the maximum flow in the network to be 9 units. The algorithm has failed us. It is not tolerable to have a method that yields different results as a function of the arbitrary ordering of the paths selected. The algorithm can be modified to allow fictional flows in the wrong direction. That is, step 2 of our approach can be modified to allow flow in the wrong direction as long as the net flow remains nonnegative. If x'_{ij} and x'_{ji} denote the simultaneous flows in the same branch of a network, but in opposite directions, then the net flow x_{ij} satisfies the following equations:

$$c_{ij} \geq x_{ij} = x'_{ij} - x'_{ji} \geq 0$$
$$x'_{ji} \leq x'_{ij}$$

Viewing Fig. 5.7(e), we see that with the concept of negative flow, 1 unit can flow along the path 1–2–4–3–5–6 which will reduce the flow in arc

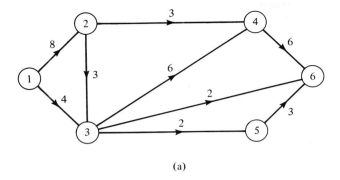

(a)

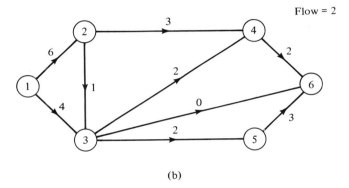

Flow = 2

(b)

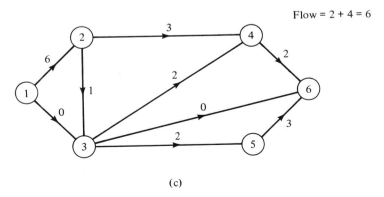

Flow = 2 + 4 = 6

(c)

Figure 5.7. An example of where the intuitive network flow approach fails.

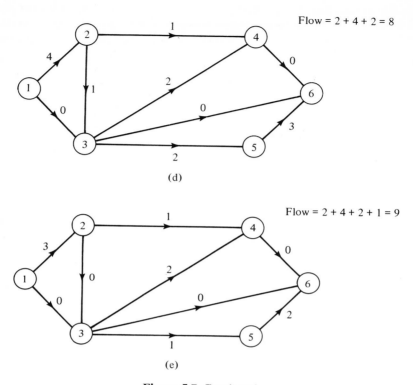

Figure 5.7 Continued.

(3, 4) to 1, which is still feasible. It is therefore possible to obtain the optimal solution to the network flow problem using this intuitive approach. But this technique is inconvenient for complicated networks, so in the next section we will consider a systematic labeling procedure which will do away with some of the objections to this method. In a complicated network it is difficult to keep track of all the paths from the source to the sink. The consideration of backward flows also complicates the situation because it introduces many new paths for consideration.

5.2.3 Maximal flow—a labeling procedure. Ford and Fulkerson's [8] well-known maximum-flow labeling procedure will be discussed in this section. The labeling procedure is different from the intuitive procedure in that it *fans out* from the source to the sink, rather than searching particular paths from the source to the sink. We will first define some nomenclature. Let c_{ij} and c_{ji} represent the capacities in the path from node i to node j. Note that the algorithm does not preclude the possibility of flow in both directions along a path. If we have only one directional flow between two nodes, i and j, then $c_{ij} > 0$ and $c_{ji} = 0$. Next we will define the excess capacities, d_{ij} and d_{ji}, for each branch:

$$d_{ij} = c_{ij} - x_{ij} + x_{ji}$$

for all ij

$$d_{ji} = c_{ji} + x_{ij} - x_{ji}$$

x_{ij} and x_{ji} are the flows in each direction along the branch (i, j). Initially, all flows are assumed to be zero. We then label each branch with its excess capacity. The goal of the procedure is to increase flow in the network. We begin at the source and consider all nodes connected to the source by branches which have a positive excess capacity. Next we will assume as before that the source is node 1 and the sink is node n. Let the index s designate a node that is connected to the source, and the entire set will be designated by S. We then label each node $s \in S$ on the network with two pieces of information (A_s, B_s), where:

A_s = excess capacity from the source to node s

B_s = the node from which the flow came

Thus:

$$(A_s, B_s) = (d_{1s}, 1) \quad \text{for} \quad s \in S$$

If node n was labeled during this first labeling, we move to the last step of the algorithm which is a method for increasing the flow in the network. Usually the sink is not reached in this step, so we proceed as follows. From the set of nodes $s \in S$ we choose a node. Then we look for nodes not yet labeled which are adjacent to this node and have a positive excess capacity. If there are none, we choose another node from S and repeat the process. If unlabeled nodes can be reached, we will designate them by the index t and call the set T. We now are in a position to label the node $t \in T$ as follows:

$$A_t = \min (d_{st}, A_s)$$

$$B_t = s$$

The label A_t is thus equal to the minimum excess capacity of the two branches from the source to node s and node s to node t. The label B_t indicates the node from which we moved to label node t. This labeling process is carried out for all the nodes $s \in S$. If the sink is labeled, we proceed to the last part of the algorithm.

The algorithm continues in the same manner as the previous step. If we have a set of labeled nodes $u \in U$, we search for nodes which are not labeled and are connected to nodes in U with paths that have positive excess capacity. The process is repeated for each $u \in U$. If we find a set of unlabeled nodes $v \in V$, they are labeled as follows:

$$A_v = \min\,(d_{uv}, A_u)$$
$$B_v = u$$

The algorithm continues, and in a finite number of steps, one of the following conditions occurs:

1. The sink is labeled.
2. The sink has not been labeled and no other node can be labeled.

If condition 1 is reached, we can increase the flow in the system. If condition 2 is reached, the existing flow is the maximum flow, and the algorithm terminates.

We will now consider how to increase the flow if condition 1 is reached. The label, A_n, indicates the positive excess capacity that exists from the source to the sink for the path indicated by this labeling step; thus it indicates how much the flow can be increased. It is simple to find the path of flow because the second label òn the node, B, indicates the preceding node leading to a node. Hence, we trace the path backwards from the sink.

Let d_{kl} represent the excess capacities of the branches in the path that allowed us to increase the flow by A_n. The new excess capacities become:

$$d'_{kl} = d_{kl} - A_n$$
$$d'_{lk} = d_{lk} + A_n$$

All the capacities not on the path connecting the source and sink remain unchanged. The next step is to repeat the entire labeling process with the new excess capacities. In a finite number of steps, we will reach condition 2 if the flow in the system is not infinite. The flow in each branch is:

$$x_{ij} = c_{ij} - d_{ij} \quad \text{if} \quad c_{ji} = 0$$

If both c_{ij} and $c_{ji} > 0$, then either:

$$x_{ij} = c_{ij} - d_{ij} \quad \text{and} \quad x_{ji} = 0$$

or

$$x_{ji} = c_{ji} - d_{ji} \quad \text{and} \quad x_{ij} = 0$$

depending upon which of the $c_{ij} - d_{ij}$ and $c_{ji} - d_{ji}$ is positive.

To demonstrate the algorithm, consider the network discussed in the previous section and shown in Fig. 5.6(a). Initially we assume that all flows are zero; so the network will have the excess capacities as shown in Fig. 5.8(a). We start at node 1 and find all the nodes connected to 1 with positive excess capacity. Nodes 2 and 3 meet these conditions. Thus, $A_2 = d_{12} = 8$ and $B_2 = 1$. Likewise, $A_3 = d_{13} = 4$ and $B_3 = 1$. These labels are shown on Fig. 5.8(a). Next node 2 is chosen; all nodes that have not been labeled and are connected to node 2 are considered. Node 4 meets these conditions. The labels for node 4 are calculated as follows:

$$A_4 = \min(d_{24}, A_2) = \min(3, 8) = 3$$
$$B_4 = 2$$

Similarly, the labels for the unlabeled nodes that can be reached from node 3 are developed. Nodes 5 and 6 are labeled (2, 3) and (2, 3). Node 6 has been labeled so the sink has been reached. Therefore, the flow can be increased by $A_6 = 2$ units. Next we locate the path along which these 2 units flowed. $B_6 = 3$, so we look at B_3 which equals 1. Therefore, the path of interest is 1–3–6. The new excess capacities for this path can now be calculated.

$$d'_{13} = 4 - 2 = 2 \qquad d'_{31} = 0 + 2 = 2$$
$$d'_{36} = 2 - 2 = 0 \qquad d'_{63} = 0 + 2 = 2$$

The modified network is now shown in Fig. 5.8(b). The process is repeated and the labels shown in Fig. 5.8(b) are developed. The sink once again has been labeled, and the algorithm proceeds as shown in Fig. 5.8(a) through 5.8(e). In Fig. 5.8(e) it is observed that the sink cannot be labeled. Therefore the maximum flow has been found and is equal to the sum of A_n realized at each step of our procedure, and:

$$\text{Maximal flow} = 2 + 3 + 2 + 3 = 10$$

We have not yet considered why this algorithm works. First we will have to define a *cut* of a network as a set of directed branches of the network such that every path of directed branches from the source to the sink contains at least one branch of the set. There are many cuts for a network, but the number is always finite. The cut concept is important because the maximal flow cannot be greater than the branch capacities of any cut. Now if we examine the last network created by the application of this algorithm, we observe that we have a disjoint set of nodes with respect to nodes that can and cannot be labeled.

Now consider the set of branches connecting those nodes which can be labeled and those which cannot be labeled. The set of branches is a cut of the network. We will now show that the sum of the capacities of this cut is

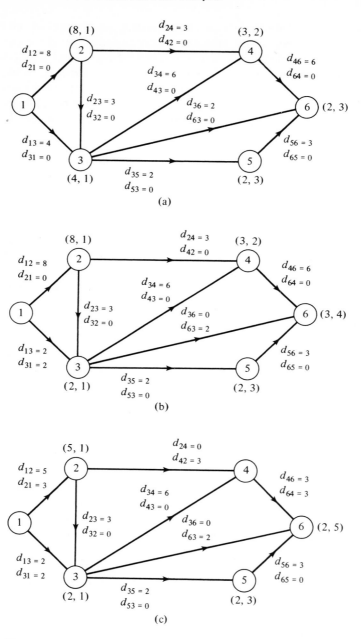

Figure 5.8. An example of the application of the Ford–Fulkerson algorithm.

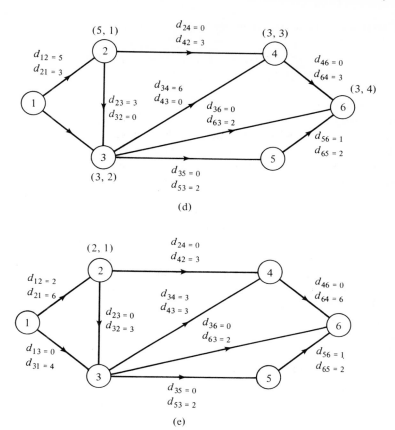

Figure 5.8. Continued.

equal to the maximal flow in the network. Substituting the excess capacity equations into the conservation-of-flow equations at each node, we get:

$$\sum_{\substack{j=1 \\ j \neq i}}^{n} (c_{ij} - d_{ij}) = 0 \quad \text{for} \quad i = 2, \ldots, n-1$$

and for the source we get:

$$\sum_{j=2}^{n} (c_{1j} - d_{1j}) = \sum_{j=2}^{n} x_{ij} = \text{total flow}$$

If we sum these equations for the labeled nodes, remembering that the source is always a labeled node, we have:

$$\sum_{\substack{i \text{ for labeled} \\ \text{nodes}}} \sum_{\substack{j=1 \\ j \neq i}}^{n} (c_{ij} - d_{ij}) = \sum_{j=2}^{n} x_{ij} = \text{total flow}$$

If both i and j are elements of the labeled set, then the preceding equation will contain the terms $c_{ij} - d_{ij}$ and $c_{ji} - d_{ji}$ which will cancel each other. Therefore, all terms where j is a node in the set labeled "nodes" will cancel, and only terms for the j's which are in the unlabeled set will be present. But for the case where i is in the labeled set and j is in the unlabeled set, $d_{ij} = 0$ because the two sets are disjoint. Therefore, we arrive at the equation:

$$\sum_{\substack{i \text{ for} \\ \text{labeled} \\ \text{nodes}}} \sum_{\substack{j \text{ for} \\ \text{unlabeled} \\ \text{nodes}}} c_{ij} = \text{total flow}$$

Thus, the sum of the capacities on the cut separating the labeled from the unlabeled nodes equals the total flow in the system. We know that the flow cannot exceed the sum of the capacities in a cut, and we have found a cut in which the total flow equals the capacities of the branches in a cut. The sum of the capacities for this cut must equal the maximum flow in the network. Therefore, we can be confident that Ford and Fulkerson's labeling technique yields the maximum flow through the network.

What has been proved here is that the max-flow–min-cut theorem exists:

Max-Flow–Min-Cut Theorem: If the sums of the capacities of the branches of every cut of a network are calculated, then the minimum of these sums is equal to the maximal possible flow in the network.

The Ford and Fulkerson method is just a systematic procedure for finding the minimum cut. The theorem can be illustrated by considering the network originally introduced in Fig. 5.6(a). The cuts of this network are as follows:

Cut	Sum of Capacities
(1, 2)(1, 3)	12
(1, 3)(2, 3)(2, 4)	10
(1, 2)(2, 3)(3, 6)(3, 5)	21
(2, 4)(3, 4)(3, 6)(3, 5)	13
(4, 6)(3, 6)(3, 5)	10
(4, 6)(3, 6)(5, 6)	11

The minimum cut of this network is 10, which equals the maximum flow calculated previously. The minimum actually occurs for two cuts (1, 3) (2, 3)

(2, 4) and (4, 6) (3, 6) (3, 5). The labeling procedure was blocked by the first of these.

The labeling algorithm is an efficient approach for finding the maximum flow through a network. It is computationally more efficient than a linear programming approach to this problem and much more systematic and dependable than the intuitive approach. Its main disadvantage is that it is cumbersome to modify the flows on the network diagram. Therefore, the approach is not very appropriate for computer solution. A matrix approach is available for solving the system.

5.2.4 Maximal flow—a matrix approach.* This section shows how a labeling technique such as the one just presented can be converted into a matrix or tableau type of calculation. The importance of this section is not limited to the algorithm presented; the approach of converting labeling processes into matrix approaches will be useful for other network flow methods.

A network of n nodes can be represented by an nth-order matrix of capacities $||c_{ij}||$. The diagram could be completely avoided, but it could, however, be drawn from the matrix because the c_{ij} indicate the nodes joined by branches and their orientations. When $c_{ij} = 0$, there is no flow allowed from i to j. When $c_{ij} > 0$, there is a path from node i to node j, and it has a capacity of c_{ij}. Using a tableau approach, Hadley [9] shows that you can find the maximal flow in a network without ever using a diagram. The solution procedure is to construct a series of excess capacity matrices $||d_{ij}||$. The labeling procedure is easily performed if we use a tableau such as in Table 5.1.

Table 5.1. MATRIX FOR MAXIMAL FLOW ANALYSIS.

Node	1	2	j	N	Labels	Stage
1	—	d_{12}	d_{1j}	d_{1n}	A_1 B_1	
2	d_{21}	—	d_{2j}	d_{2n}	A_2 B_2	
i	d_{j1}	d_{j2}	d_{ij}	d_{in}	A_i B_i	
Total flow	d_{n1}	d_{n2}	d_{nj}	d_{nn}	A_n B_n	

Given the excess capacities, the labels A_j and B_j are easily calculated from the tableau. We first consider row 1 of the tableau and locate columns having $d_{1j} > 0$. For such columns we set $A_j = d_{1j}$ and $B_j = 1$. Assuming that the sink has not been labeled, we move on to the row of the lowest index,

---

*This section is adapted from Reference 9 and is used with the permission of Addison-Wesley Publishing Co., Inc.

say i, that was labeled at the first step. We are searching for d_{ij} which is greater than zero and for which j has not been labeled. When we find one we set $A_j = \min(d_{ij}, A_i)$ and $B_j = i$. If no value is assigned to A_n, we consider other rows of the tableau that were labeled in the first stage of labeling and repeat the procedure. We continue until the sink is labeled or the first stage has been exhausted. We then move to the rows labeled during the second stage of labeling and we repeat the process. In a finite number of steps we either assign a positive value to A_n or we find it impossible to label the sink. When A_n is assigned a value, the flow can be increased. If the sink is not labeled, the flow is maximum.

When the sink is labeled we must construct a new tableau. As before we trace the path of flow using the B_j labels. Assume $B_n = t$. This means that the last row was reached from row t; thus the excess capacity is represented by $d'_{tn} = d_{tn} - A_n$. Now suppose $B_t = s$, which means that row t was reached from row s. Therefore, d_{st} must be modified and will be $d'_{st} = d_{st} - A_n$. Eventually row 1 will be reached, which will complete the set of elements that must be reduced by a value A_n. For each element d_{kl} that is decreased by A_n, the element d_{lk} is increased by A_n. As we remember from the labeling procedure, an increase in flow in one direction increases the excess capacity in the opposite direction. All other entries remain unchanged as they did on the diagram. At this point we are ready to begin our process anew. The entries in our new tableau have a one-to-one correspondence, and the excess capacities shown in the figures are used in the labeling procedure just discussed. The procedure continues until the sink cannot be labeled.

We will now use the tableau approach to find the maximal flow for the network discussed throughout this chapter and represented in Fig. 5.6(a). The initial tableau is given in Table 5.2. In this tableau $d_{ij} = c_{ij}$ because there is no flow in the system. At stage 1 rows 2 and 3 can be labeled. Moving to row 2 we find that row 4 can be labeled. Next using row 3, rows 5 and 6 can be labeled. We observe that:

$$A_6 = \min(d_{36}, A_3) = \min(2, 4) = 2$$

Table 5.2. MATRIX FLOW ANALYSIS, STEP 1.

Nodes	1	2	3	4	5	6	Labels		Stage
1	—	8	④	0	0	0	A	B	
2	0	—	3	3	0	0	8	1	1
3	[0]	0	—	6	2	②	4	1	1
4	0	0	0	—	0	6	3	2	2
5	0	0	0	0	—	3	2	3	2
0	0	0	[0]	0	0	—	2	3	2

Thus the flow can be increased by 2 along the traced path. We find that row 6 was reached from row 3, so we circle d_{36} to indicate that its capacity must be reduced. B_3 indicates that row 3 was reached from row 1; so d_{13} is also circled. This completes the path; but since d_{36} and d_{13} will be reduced by 2, d_{63} and d_{31} will be increased by 2.

The elements to be increased are marked by a square. The new tableau is next constructed as shown in Table 5.3. Notice that all circled elements

Table 5.3. MATRIX FLOW ANALYSIS, STEP 2.

Nodes	1	2	3	4	5	6	Labels		Stage
1	—	(8)	2	0	0	0	A	B	
2	[0]	—	3	(3)	0	0	8	1	1
3	2	0	—	6	2	0	2	1	1
4	0	[0]	0	—	0	(6)	3	2	2
5	0	0	0	0	—	3	2	3	2
2	0	0	2	[0]	0	—	3	4	3

have been decreased by $A_6 = 2$ and all elements marked with a square are increased by $A_6 = 2$. All other elements remain unchanged. The procedure continues as shown in Tables 5.3 through 5.6. It is interesting to note that

Table 5.4. MATRIX FLOW ANALYSIS, STEP 3.

Nodes	1	2	3	4	5	6	Labels		Stage
1	—	5	(2)	0	0	0	A	B	
2	3	—	3	0	0	0	5	1	1
3	[2]	0	—	6	(2)	0	2	1	1
4	0	3	0	—	0	3			
5	0	0	[0]	0	—	(3)	2	3	2
5	0	0	2	3	[0]	—	2	5	3

Table 5.5. MATRIX FLOW ANALYSIS, STEP 4.

Nodes	1	2	3	4	5	6	Labels		Stage
1	—	(5)	0	0	0	0	A	B	
2	[3]	—	(3)	0	0	0	5	1	1
3	4	[0]	—	(6)	0	0	3	2	2
4	0	3	[0]	—	0	(3)	3	3	3
5	0	0	2	0	—	1			
7	0	0	2	[3]	2	—	3	4	4

Table 5.6. MATRIX FLOW ANALYSIS, STEP 5.

Nodes	1	2	3	4	5	6	Labels		Stage
1	—	2	0	0	0	0	A	B	
2	6	—	0	0	0	0	2	1	1
3	4	3	—	3	0	0			
4	0	3	3	—	0	0			
5	0	0	2	0	—	1			
10	0	0	2	6	2	—			

these tables correspond to networks shown in Fig. 5.8. Finally, the flow in each arc is found in a manner similar to that discussed in Section 5.2.3.

In this section we have shown that it is possible to convert a network flow labeling process into a tableau form. The tableau approach is far superior for solving flow problems using the computer.

5.2.5 Some examples of the use of network flow models.

Most of the important applications of network flow models involve the conversion of linear programming models, and this will be discussed in the next section. Elmaghraby [4], however, has suggested a number of interesting situations in which the algorithm as presented in this chapter is directly applicable.

For example, suppose we wish to transport goods from a number of sources, $S_1, S_2, \ldots, S_m$, to a number of destinations, $D_1, D_2, \ldots, D_n$. There are a number of units a_i available at each source and a number of units d_j needed at each destination. Now distributions can take place only over certain routes, and these routes have capacity c_{ij}. The problem is to determine a feasible distribution pattern if one exists. Note that we are not trying to minimize cost as is often the case in problems of this sort.

Figure 5.9 represents a network approach for modeling this system. The sources are represented by nodes S_1 through S_m and the destinations by nodes D_1 through D_n. For each permissible route (i, j), an arc with capacity c_{ij} is shown joining nodes S_i and D_j. Next a master source, S, and a master terminal, T, are added to the flow network. The paths (S, S_i) which have capacity a_i limit the amount of product sent from S_i. The needs of each destination, D_j, are limited in a similar manner by the arcs (D_j, D) whose capacities are d_j. We next apply network flow methods to the model. If the flow from D equals the sum of the d_2, then the solution satisfies the conditions of the problem. If not, it is impossible to ship the desired commodities from sources S_i to the destinations D_j using the permissible routes.

Elmaghraby [4] suggests a number of machine assignment problems which can be solved by these methods. For example:

1. There are m machines in a job shop and n jobs to perform. Each job can be handled only by certain machines, and each machine can be assigned only one job. What is the maximum number of jobs that can be assigned?

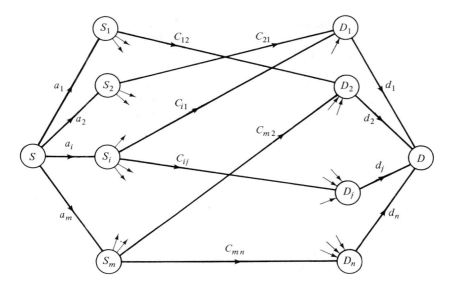

Figure 5.9. Network model of a transportation system.

2. There are m machines in a job shop and n jobs to perform. There is a measure of efficiency assigned to each job on each machine. Our goal is to assign the jobs in such a way that the lowest efficiency realized is as large as possible. To accomplish this, an arbitrary assignment is made. The lowest efficiency is observed and all assignments with that efficiency or lower are considered to be illegal assignments. A flow network of the legal assignments is made and another feasible solution is found. Once again, the lowest efficiency is observed and assignments with that efficiency or lower are eliminated from consideration. The process continues until the flow method can no longer find a feasible assignment. The lowest efficiency in the last feasible assignment is then the answer to the problem.

Ford and Fulkerson [8] also present a number of interesting flow examples which the interested reader is encouraged to survey.

5.3 The Relationship Between Linear Programming and Maximal Flow

5.3.1 Introduction. We have briefly discussed the possibility of converting a linear programming model into a maximal flow formulation. The solution of the maximal flow model is often a more efficient algorithm from a computational standpoint. In this section we will consider in detail the maximal flow model for the CPM project cost model discussed in Chapter 3.

5.3.2 Linear programming concepts. To illustrate the linear programming concepts that we will need in this section, let's consider a linear programming model for finding the critical path in an activity network.

The model is especially interesting because a flow concept is used to develop the linear programming model. Consider the model shown in Fig. 5.10. Now the activity diagram is viewed as a flow network in which a hypothetical unit of flow leaves the source nodes and enters node 5 which is the

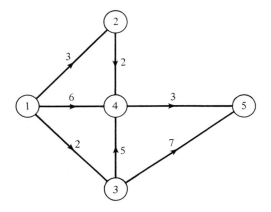

Figure 5.10. Sample activity network.

sink. Nodes 2, 3, and 4 can be viewed as transshipment nodes, and conservation of flow must be observed at each of these nodes. The time shown on the activities is then interpreted as the time for sending a unit of flow along the activity. When we consider a project network in this manner, the problem of finding the critical path is equivalent to finding the network path which requires the maximum time to move from the source to the sink. If we apply the argument just discussed, the linear programming model for the network shown in Fig. 5.10 becomes:

Maximize $f(x) = 3x_{12} + 2x_{13} + 6x_{14} + 2x_{24} + 5x_{34} + 7x_{35} + 3x_{45}$
subject to:

$$x_{12} + x_{13} + x_{14} = 1$$
$$-x_{12} + x_{24} = 0$$
$$-x_{13} + x_{34} + x_{35} = 0$$
$$-x_{14} - x_{24} - x_{34} + x_{45} = 0$$
$$-x_{35} - x_{45} = -1$$
$$x_{12}, x_{13}, x_{14}, x_{24}, x_{34}, x_{35}, x_{45} \geq 0$$

In this model, $x_{ij} = 1$ will denote the presence of flow in activity i–j and $x_{ij} = 0$ will indicate that there is no flow. The first and last constraints indicate that a unit of flow leaves node 1 and arrives at node 5. The remaining constraints show that there is conservation of flow at nodes 2, 3, and 4. Thus any set of x_{ij} which satisfies these constraints indicates a path from the source to the sink of our network. The path will be indicated by those x_{ij} which are equal to 1. The path 1–4–5 would be indicated by x_{14} and x_{45} being equal to 1 while all other x_{ij}'s would be 0. Now since the x_{ij}'s for those units on the path equal 1, the objective function gives the sum of the activity times for the chosen path. When we maximize the objective function we will find the critical path. This is obviously an inefficient approach for finding the critical path, but it will prove to be an interesting application of the duality theory of linear programming.

From the well-known duality theorem of linear programming it can be shown that for every linear programming model there is a corresponding linear programming model with some very interesting properties. These models are referred to as the *primal* and *dual* models. The relationship between these two models is given by the following theorem.

Duality Theorem: Given a linear programming model, call it the *primal*, there is always a related linear programming problem, called the *dual*, which is defined as follows:

Primal	*Dual*
Maximize $f(x) = C^T x$	Minimize $g(y) = B^T y$
Subject to:	Subject to:
$Ax \le B$	$A^T y \ge C$
$x \ge 0$	$y \ge 0$

where C and x are $n \times 1$ vectors, y and B are $m \times 1$ vectors, and A is an $m \times n$ matrix.

If there exists a solution, x^*, which gives a finite maximum value to $f(x)$, there is a related solution, y^*, for which $g(y)$ equals $f(x)$. The solution to the primal gives the solution to the dual, and vice versa. If any of the primal variables are unrestricted in sign, then the corresponding dual constraints are equalities; if the primal constraints are equalities, then the corresponding dual variables are unrestricted in sign.

The relationship between the primal and dual is further clarified by another famous theorem of duality theory known as the *complementary slackness theorem*.

Complementary Slackness Theorem: At optimal, if a dual variable y_j is greater than zero, then it must be true that the jth inequality in the primal is satisfied as an equality. Also, if some x_i is greater than zero, then the corresponding ith dual constraint must be satisfied as an equality.

We will now apply duality theory to the example we have been considering. We should note that since all the primal constraints are equalities, the dual variables will be unrestricted in sign. The dual is as follows:

$$\text{Minimize } g(y) = y_1 - y_5$$

subject to:

$$y_1 - y_2 \geq 3$$
$$y_1 - y_3 \geq 2$$
$$y_1 - y_4 \geq 6$$
$$y_2 - y_4 > 2$$
$$y_3 - y_4 > 5$$
$$y_3 - y_5 > 7$$
$$y_4 - y_5 > 3$$
$$-\infty < y_i < \infty \quad \text{for} \quad i = 1, 2, 3, 4, 5$$

Now since all the dual constraints involve only two variables, they can be solved by inspection in terms of y_1. To see this, consider the following formulation of the constraints:

$$y_2 < y_1 - 3 \qquad\qquad y_2 = y_1 - 3$$
$$y_3 < y_1 - 2 \qquad\qquad y_3 = y_1 - 2$$

$$\left.\begin{array}{l} y_4 < y_1 - 6 \\ y_4 < y_2 - 2 = y_1 - 5 \\ y_4 < y_3 - 3 = y_1 - 7 \end{array}\right\} \qquad y_4 = y_1 - 7$$

$$\left.\begin{array}{l} y_5 < y_3 - 7 = y_1 - 9 \\ y_5 < y_4 - 3 = y_1 - 10 \end{array}\right\} \qquad y_5 = y_1 - 10$$

Since y_5 varies directly with y_1 and $g(y)$ is the difference between y_1 and y_5, y_1 can be assigned an arbitrary value without affecting $g(y)$. Thus we can set $y_1 = 0$; then to minimize $g(y) = y_1 - y_4 = -y_4$, we must minimize $-y_4$ and at the same time satisfy the above constraints. We can see by inspection that this is achieved by the solution indicated above, i.e., $y_1^* = 0$, $y_2^* = -3$, $y_3^* = -2$, $y_4^* = -7$, and $y_5^* = -10$, and the optimal value of the dual objective function is $g(y^*) = y_1^* - y_5^* = 0 - (-10) = 10$. The solution of these inequalities is very similar to the solution procedure using the forward-pass rules discussed in Chapter 3. If this solution is substituted into our dual formulation, we find that the third, fourth, and sixth equations are satisfied as inequalities. The complementary slackness theorem tells us that the corresponding primal variables x_{14}, x_{24}, x_{35} are zero while the others must be nonnegative. Substituting these results into the primal constraints, we get:

$$x_{12} + x_{13} = 1$$
$$-x_{12} = 0$$
$$-x_{13} + x_{34} = 0$$
$$-x_{34} + x_{45} = 0$$
$$-x_{45} = -1$$

This suggests an optimal solution to the primal of $(x^*) = (x_{12}^*, x_{13}^*, x_{14}^*, x_{24}^*, x_{34}^*, x_{35}^*, x_{45}^*) = (0, 1, 0, 0, 1, 0, 1)$ and $f(x^*) = 3 \times 0 + 2 \times 1 + 6 \times 0 + 2 \times 0 + 5 \times 1 + 7 \times 0 + 3 \times 1 = 10$. Therefore, it can be concluded that the critical path is composed of activities (1, 3), (3, 4), and (4, 5) and is of length 10.

5.3.3 CPM and the network flows. In Section 3.3.3 we introduced the idea of using linear programming and network flow methods to solve the time/cost tradeoff problem introduced in the CPM model. In this section we will develop these concepts further and develop a network flow model to solve the CPM system.

In Section 3.3.3 a linear programming formulation of the CPM problem was given. We will be using a slightly different formulation in this section. First, we will consider the objective function. The cost of each activity will be described as follows:

$$\text{Activity } (i, j) \text{ direct cost} = K_{ij} - C_{ij}x_{ij}$$

where: K_{ij} is the y intercept of the cost curve for activity (i, j),
       C_{ij} is the *absolute value* of the slope of the cost curve for activity
         (i, j), and
       x_{ij} is the duration of activity (i, j).

Our objective is to minimize the sum of all the activity costs which is expressed as follows:

$$\text{Total direct project costs} = \sum_i \sum_j (K_{ij} - C_{ij}x_{ij})$$

Since the sum of the K_{ij} terms is a constant, the minimization of the total direct project costs can be accomplished by if we maximize the following expression:

$$\sum_i \sum_j C_{ij}x_{ij}$$

The linear programming formulation becomes:

$$\text{Maximize } f(x) = \sum_i \sum_j C_{ij}x_{ij} \tag{5-1a}$$

subject to:

$$T_i + x_{ij} - T_j \leq 0 \qquad \text{for all } ij \tag{5-1b}$$
$$x_{ij} \leq D_{ij} \qquad \text{for all } ij \tag{5-1c}$$
$$-x_{ij} \leq -d_{ij} \qquad \text{for all } ij \tag{5-1d}$$
$$T_n - T_1 \leq \tau \tag{5-1e}$$

where: T_k denotes the earliest expected time for node k to be realized,
D_{ij} is the normal duration for activity (i, j),
d_{ij} is the crash duration for activity (i, j), and
τ is a parametric variable which is used to restrict the project duration time.

The first set of constraints (5-1b) apply to each activity (i, j) and state that the difference between the earliest node times, T_i and T_j, must be at least x_{ij}, which is the scheduled duration of activity (i, j). Equations (5-1c) and (5-1d) imply that x_{ij} is constrained to lay between its normal and crash times. Finally, Eq. (5-1e) guarantees that the time between the realization of the start node, T_1, and the realization of the sink node, T_n, is less than or equal to the project duration τ.

The formulation for the network shown in Fig. 5.10 becomes:

$$\text{Maximize } f(T, x) = C_{12}x_{12} + C_{13}x_{13} + C_{14}x_{14} + C_{24}x_{24}$$
$$+ C_{34}x_{34} + C_{35}x_{35} + C_{45}x_{45}$$

subject to:

$$T_1 + x_{12} - T_2 \leq 0$$
$$T_1 + x_{13} - T_3 \leq 0$$
$$T_1 + x_{14} - T_4 \leq 0$$
$$T_2 + x_{24} - T_4 \leq 0$$
$$T_3 + x_{34} - T_4 \leq 0$$
$$T_3 + x_{35} - T_5 \leq 0$$
$$T_4 + x_{45} - T_5 \leq 0$$
$$x_{12} \leq D_{12}$$
$$x_{13} \leq D_{13}$$
$$x_{14} \leq D_{14}$$
$$x_{24} \leq D_{24}$$
$$x_{34} \leq D_{34}$$

$$x_{35} \le D_{35}$$
$$x_{45} \le D_{45}$$
$$-x_{12} \le -d_{12}$$
$$-x_{13} \le -d_{13}$$
$$-x_{14} \le -d_{14}$$
$$-x_{24} \le -d_{24}$$
$$-x_{34} \le -d_{34}$$
$$-x_{35} \le -d_{35}$$
$$-x_{45} \le -d_{45}$$
$$T_5 - T_1 \le \tau$$

As discussed in Chapter 3, this problem could be solved by the simplex method, which finds the optimum schedule for different values of τ. The simplex procedure will prove less efficient than the flow algorithm that will now be developed.

The first step in the development of the flow algorithm to solve this system is to find the dual of the problem just presented. We will use different dual variables for each set of equations in the primal problem. f_{ij}, v_{ij}, and w_{ij} correspond to Eqs. (5-1b), (5-1c), and (5-1d), respectively, while the dual variable for Eq. (5-1e) is y. We note that all the dual constraints are equalities since all the primal variables are unrestricted in sign, and the dual variables are all sign-constrained since the primal constraints are all inequalities. The dual formulation becomes:

$$\text{Minimize } g(f, v, w, y) = \tau y + \sum_i \sum_j D_{ij} v_{ij} - \sum_i \sum_j d_{ij} w_{ij} \qquad (5\text{-}2a)$$

subject to:

$$\sum_j (f_{ij} - f_{ji}) = \begin{cases} y; & i = 1 \\ 0; & i \ne 1, n \\ -y; & i = n \end{cases} \qquad (5\text{-}2b)$$

$$f_{ij} + v_{ij} - w_{ij} = C_{ij} \qquad \text{for all } ij \qquad (5\text{-}2c)$$

$$f_{ij}, v_{ij}, w_{ij}, y \ge 0 \qquad \text{for all } i, j$$

Applying this formulation to the model shown in Fig. 5.10, we get:

$$\begin{aligned} \text{Minimize } g(f, v, w, y) = \; & \tau y + D_{12} v_{12} + D_{13} v_{13} + D_{14} v_{14} \\ & + D_{24} v_{24} + D_{34} v_{34} + D_{35} v_{35} \\ & + D_{45} v_{45} - d_{12} w_{12} - d_{13} w_{13} \\ & - d_{14} w_{14} - d_{24} w_{24} - d_{34} w_{34} \\ & - d_{35} w_{35} - d_{45} w_{45} \end{aligned}$$

subject to:

$$f_{12} + f_{13} + f_{14} = y$$
$$-f_{12} + f_{24} = 0$$
$$-f_{13} + f_{34} + f_{35} = 0$$
$$-f_{14} - f_{24} - f_{34} + f_{45} = 0$$
$$-f_{35} - f_{45} = -y$$
$$f_{12} + v_{12} - w_{12} = C_{12}$$
$$f_{13} + v_{13} - w_{13} = C_{13}$$
$$f_{14} + v_{14} - w_{14} = C_{14}$$
$$f_{24} + v_{24} - w_{24} = C_{24}$$
$$f_{34} + v_{34} - w_{34} = C_{34}$$
$$f_{35} + v_{35} - w_{35} = C_{35}$$
$$f_{45} + v_{45} - w_{45} = C_{45}$$
$$\text{all } f_{ij}, v_{ij}, w_{ij}, y \geq 0$$

The first set of constraint Eq. (5-2b) has a flow interpretation. The equations imply that there is a flow of y out of the source node and an equal flow into the sink. All intermediate nodes have conservation of flow. It will now be shown that the variables v_{ij} and w_{ij} can be expressed as a function of the flow variables, f_{ij}, and the activity cost slopes, C_{ij}. Thus, constraint Eq. (5-2c) will be expressed as constraints on the flow in the network activities.

Since $D_{ij} \geq d_{ij}$, our dual formulation tells us that v_{ij} and/or w_{ij} must be zero in an optimal solution. For example, suppose that $C_{ij} - f_{ij}$ is greater than zero. Then $v_{ij} - w_{ij} > 0$ since equation set (5-2c) states that

$$f_{ij} + v_{ij} - w_{ij} = C_{ij}$$

This implies that $v_{ij} < w_{ij}$. Now, since $D_{ij} > d_{ij}$, minimizing $(D_{ij}v_{ij} - d_{ij}w_{ij})$ in our object function implies that w_{ij} must equal zero. Using similar arguments, we can verify the following statements:

$$C_{ij} - f_{ij} > 0 \longrightarrow w_{ij} = 0$$
$$C_{ij} - f_{ij} = 0 \longrightarrow v_{ij} = w_{ij} = 0$$
$$C_{ij} - f_{ij} < 0 \longrightarrow v_{ij} = 0$$

Therefore, we can eliminate the v_{ij} and w_{ij} variables by expressing them in terms of C_{ij} and f_{ij}. We observe that:

$$v_{ij} = \text{maximum } (0, C_{ij} - f_{ij}), \quad \text{and}$$
$$w_{ij} = \text{maximum } (0, f_{ij} - C_{ij})$$

Substituting this result into Eq. (5-2a), we obtain the objective function:

$$\text{Minimize } G(y, f_{ij}) = \tau y + \sum_i \sum_j [D_{ij} \max (0, C_{ij} - f_{ij})$$
$$- d_{ij} \max (0, f_{ij} - C_{ij})]$$

Although we have successfully eliminated v_{ij} and w_{ij} from our formulation, unfortunately the resulting problem is now a nonlinear programming problem because the objective function is nonlinear. But the function

$$H_{ij} = D_{ij} \max (0, C_{ij} - f_{ij}) - d_{ij} \max (0, f_{ij} - C_{ij})$$

is piecewise linear because, when

and when
$$f_{ij} < C_{ij}, \quad \text{then} \quad H_{ij} = D_{ij}(C_{ij} - f_{ij})$$

$$f_{ij} > C_{ij}, \quad \text{then} \quad H_{ij} = d_{ij}(C_{ij} - f_{ij})$$

Since $D_{ij} \geq d_{ij}$, we can also conclude that H_{ij} is convex. Thus, even though the new objective function is nonlinear, it can be handled by linear programming methods if each of the variables f_{ij} is broken into two pieces corresponding to the two pieces of its cost curve. Therefore, f_{ij} can be replaced by two variables as follows:

where
$$f_{ij} = f_{ij1} + f_{ij2}$$

Thus,
$$0 \leq f_{ij1} \leq C_{ij} \quad \text{and} \quad 0 \leq f_{ij2} \leq \infty$$

$$H_{ij} = -D_{ij} f_{ij1} - d_{ij} f_{ij2}.$$

Since the coefficient of f_{ij2} is algebraically greater than the coefficient of f_{ij2}, it follows that, in minimizing the dual objective function, if $f_{ij2} > 0$, then $f_{ij1} = C_{ij}$. Also, if $f_{ij1} < C_{ij}$, then $f_{ij2} = 0$. Now using the preceding arguments, we can reformulate the dual linear programming using the following definitions:

$$C_{ij1} = C_{ij} \quad \text{and} \quad C_{ij2} = \infty$$
$$d_{ij1} = D_{ij} \quad \text{and} \quad d_{ij2} = d_{ij}$$
$$\text{Minimize } G(d_{ijk}, y) = \tau y - \sum_i \sum_j \sum_k f_{ijk} d_{ijk}$$

subject to:

$$\sum_j \sum_k (f_{ijk} - f_{jik}) = \begin{cases} y; & i = 1 \\ 0; & i \neq 1, n \\ -y; & i = n \end{cases}$$

$$0 \leq f_{ijk} \leq C_{ijk}; \quad \text{all } ijk$$
$$y \geq 0$$

This formulation has the following network flow interpretation. First, we enlarge the network by paralleling each original activity by a second activity labeled $ij1$ and $ij2$, respectively. Each flow, f_{ijk}, has a maximum capacity of C_{ijk} and an objective function coefficient of d_{ijk}. For example, the network shown in Fig. 5.10 would be modified as shown in Fig. 5.11. The problem now is to find a total flow of value y from nodes 1 to n that will minimize G, the objective function. In the next section we will consider a labeling procedure that has been suggested to solve this problem.

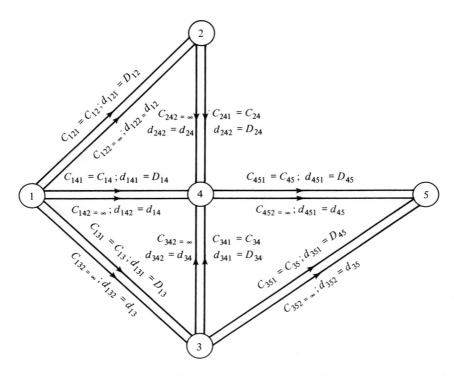

Figure 5.11. Network flow interpretation of the CPM system.

5.3.4 A network flow algorithm to solve the CPM model.

The algorithm discussed in this section is similar to the labeling process discussed in Section 5.2. The method was developed by Ford and Fulkerson [8]. The algorithm is a systematic search for a path from the source to the sink of the network. An event is considered to be in one of three states: (1) unlabeled, (2) labeled and scanned, and (3) labeled and unscanned. An interesting feature of this algorithm is that once a node has been labeled and scanned, it can be ignored for the rest of the labeling cycle. This feature makes the algorithm very efficient. As was the case in Section 5.2, the labeling of a node i corresponds to locating a path from the source to event i that can be part of the flow path we are seeking. Enough information is carried along in the labels so that, if the terminal event is labeled, the resulting flow change along the path can be made.

If we do not obtain a breakthrough, the flow through the network is maximal, and the minimal cut-set determines what event times must be reduced and by how much they may be reduced before the next labeling cycle is begun. The event time reductions determine the total project duration reduction as well as the reduction in the individual activities. Thus, the increase in the project direct cost can be calculated. In this algorithm, we assume that the source is numbered 1, the terminal event is numbered n, and all the intermediate nodes are labeled so that for every activity (i, j), $i < j$. Now we define:

$$\bar{d}_{ijk} = T_i + d_{ijk} - T_j$$

Activities for which $\bar{d}_{ijk} = 0$ are called *admissible*. Observe that $\bar{d}_{ij1}$ is the negative of the total slack for activity (i, j) when this activity is scheduled at its normal time ($d_{ij1} = D_{ij}$), and $\bar{d}_{ij2}$ is interpreted in a similar manner when the activity is scheduled at its crash time.

To start the algorithm, we:

1. Let τ = project duration when all activities are scheduled at their normal duration, D_{ij}.
2. Let all $f_{ijk} = 0$.

We are now ready to begin the first of two labeling steps. The label is of the form $[i, k^{+ \text{ or } -}, e(j)]$. The first label, i, denotes the node from which the labeling is done. The second label can take any of four values, 1^+, 1^-, 2^+, or 2^-. It describes the path and direction of the labeling. The numerical value denotes the path and the plus sign shows that the flow is from i to j; the minus sign implies flow from j to i. $e(j)$ equals the largest permissible flow change along the path from the source to j.

FIRST LABELING

The source node is always labeled $[-, -, e(j) = \infty]$. Event j can be labeled from node i if activity $(i, j, 2)$ is admissible. The event would be labeled $[i, 2^+, e(j) = \infty]$. If a breakthrough occurs during this labeling, terminate the procedure; otherwise go to the second labeling.

SECOND LABELING

The labels from the first stage of this process are retained and all events revert to the unscanned state. When we scan a labeled event i, the labeling rules are that we can label a node if either of the following conditions are met:

1. Activity (i, j, k) is admissible and $f_{ijk} < C_{ijk}$. The event j is labeled $[i, k^+, e(j)]$ where $e(j) = \min [e(i), C_{ijk} - f_{ijk}]$.
2. Activity (j, i, k) is admissible and $f_{jik} > 0$. The event j is labeled $[i, k^-, e(j)]$ where $e(j) = \min [e(i), f_{jik}]$. In this case $j < i$.

If a breakthrough occurs at the completion of this labeling, change the flow by adding or subtracting $e(n)$ along the flow path as indicated by the labels.

If a nonbreakthrough occurs, identify the following subsets of activities which comprise the minimal cut-set of the network, consistent with the sign of the superscript or k in the event label.

$$S_1 = [(ijk) \,|\, i \text{ labeled}, j \text{ unlabeled}, \bar{d}_{ijk} < 0] \quad \text{and let}$$
$$\Delta_1 = \min_{S_1} [-\bar{d}_{ijk}]$$
$$S_2 = [(ijk) \,|\, i \text{ unlabeled}, j \text{ labeled}, \bar{d}_{ijk} > 0] \quad \text{and let}$$
$$\Delta_2 = \min_{S_2} [\bar{d}_{ijk}]$$

Then define $\Delta = \min (\Delta_1, \Delta_2)$. Next change the event times, T_i, by substracting Δ from all T_i corresponding to unlabeled i. Discard all labels and return to labeling step 1.

Each new set of event times, T_i, yields a new point on the project time/cost curve. Let $T(\tau)$ equal the total project direct costs for a schedule having a duration of τ. Then:

$$T(\tau) = \sum_i \sum_j k_{ij} - \sum_i \sum_j C_{ij} x_{ij}$$

where:

$$x_{ij} = \min [D_{ij}, (T_j - T_i)]$$

We will now apply this algorithm to the network shown in Fig. 5.11. The values of C_{ij}, d_{ij}, and D_{ij} are assumed to be as follows:

Activity	C_{ij}	d_{ij}	D_{ij}
1, 2	2	1	2
1, 3	5	2	3
1, 4	6	4	6
2, 4	∞	2	2
3, 4	4	3	5
3, 5	3	5	7
4, 5	7	2	4

The labeling procedure is shown in Fig. 5.12(a) through 5.12(j). The nodes on these diagrams contain two pieces of information. The left-hand number is the earliest scheduled time, T_i, for the node, and the right-hand number is the node number. Each flow activity is labeled with four pieces of information. The upper left-hand number is f_{ij1}, and the lower left-hand number is f_{ij2}. $\bar{d}_{ij1}$ is the upper right-hand number, and $\bar{d}_{ij2}$ is the lower right-hand number.

We should also note that for simplicity of drawing, each flow activity is shown as one path rather than the two paths that actually exist. Figure 5.12(a) shows the initial conditions for the flow labeling procedure. All flows are equated to 0 and the T_i are calculated assuming that all activities are scheduled at their normal times. Next, the $\bar{d}_{ijk}$ values are calculated using the

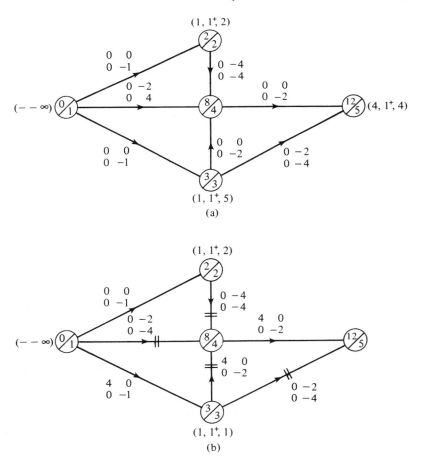

Figure 5.12. CPM network flow labeling procedure.

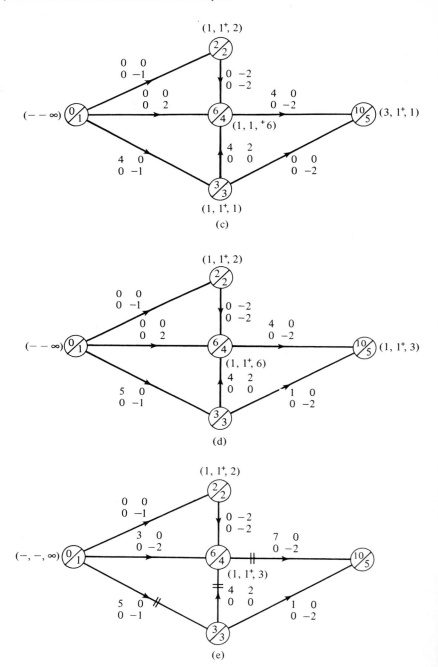

Figure 5.12. Continued.

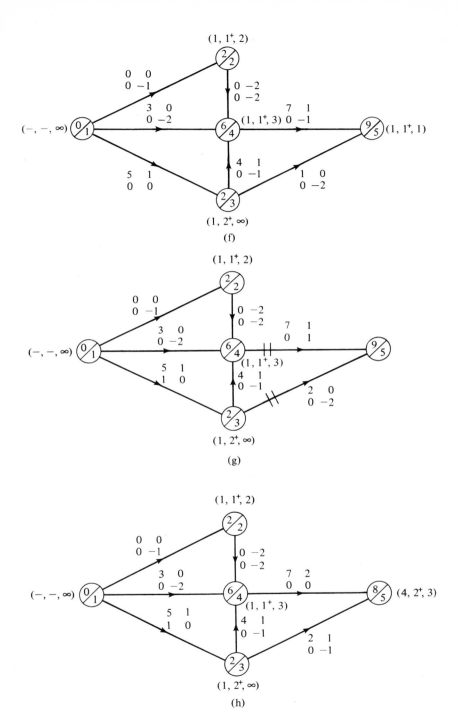

Figure 5.12. Continued.

153

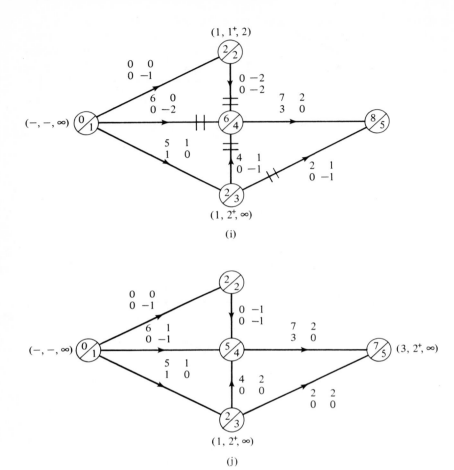

Figure 5.12. Continued.

formula:

$$\bar{d}_{ijk} = T_i + d_{ijk} - T_j$$

For example, $\bar{d}_{142}$ equals -4 because T_1 and T_4 equal 0 and 8, respectively; the crash time for this activity, d_{142}, is 4. We can now begin our labeling procedure.

The first labeling allows us to label only node 1, since there is no path from the source which has a $\bar{d}_{ij2}$ value that equals zero. Moving to the second labeling, we find that nodes 2 and 3 can be labeled from node 1 along path 1 because $\bar{d}_{121}$ and $\bar{d}_{131}$ equal zero, i.e., they are admissible. Next, node 2 is scanned but no node can be labeled from this node. When we scan node 3 we find that node 4 can be labeled. Note that $e(4) = \min [e(3), C_{241} - f_{241}]$ $= \min (5, 4 - 0) = 4$. We get a breakthrough when we label node 5 from node 4. The next step is to augment the flow in the path 1–3–4–5. We see that $e(5) = 4$ has been added to f_{131}, f_{341}, and f_{451} in Fig. 5.12(b). Figure 5.12(b) can now be labeled. We find that only nodes 1, 2, and 3 can be labeled; so a nonbreakthrough condition exists. The minimal cut-set for this figure includes activities (1, 4), (2, 4), (3, 4), and (3, 5). All these activities belong to S_1; so:

$$\Delta_1 = \min (-\bar{d}_{141}, -\bar{d}_{142}, -\bar{d}_{241}, -\bar{d}_{242}, -\bar{d}_{341}, -\bar{d}_{342}, -\bar{d}_{351}, -\bar{d}_{352}) = 2$$

Since S_2 is a null set, Δ_2 will be undefined, and

$$\Delta = \min (\Delta_1, \Delta_2) = 2$$

All the earliest times for unlabeled nodes are reduced by Δ. Thus, T_4 and T_5 become 6 and 10, respectively, as in Fig. 5.12(c). These reductions cause $\bar{d}_{141}$, $\bar{d}_{142}, \bar{d}_{241}, \bar{d}_{242}, \bar{d}_{341}, \bar{d}_{342}, \bar{d}_{351}$, and $\bar{d}_{352}$ to be increased by Δ as shown in Fig. 5.12(c). We have now reduced the duration of the project to ten days by reducing activity (3, 4) to its crash time. This would cost $8. The labeling procedure continues with breakthroughs shown in Figs. 5.12(c) and 5.12(d). The next nonbreakthrough occurs in Fig. 5.12(e). In analyzing this case, we find that the project duration can be reduced to 9 units if activities (1, 3) and (4, 5) are decreased while activity (3, 6) is increased. The cost is $5 + 7 - 4 = \$8$. In Fig. 5.12(e), there is a tendency to try to label node 3 from node 4; but this is illegal because, although $\bar{d}_{342}$ indicates that the second flow arc from node 3 to node 4 is admissible, there is no flow in the path, $f_{342} = 0$. The first labeling procedure is able to label node 3 in Fig. 5.12(f) because $\bar{d}_{132} = 0$. The analysis continues as shown in Fig. 5.12(f) through 5.12(j). In Fig. 5.12(j), the first labeling process was successful in labeling node 5; so the process ends. We can assume that it is impossible to reduce the project beyond seven days. The results of the algorithm are summarized in Table 5.7.

Table 5.7. CPM TIME/COST SCHEDULE.

	Time of Activity							Incremental	Total
Duration	(1, 2)	(1, 3)	(1, 4)	(2, 3)	(3, 4)	(3, 5)	(4, 5)	Cost	Cost
12	2	3	6	2	5	7	4	0	0
10	2	3	6	2	3	7	4	8	8
9	2	2	6	2	4	7	3	8	16
8	2	2	6	2	4	6	2	10	26
7	2	2	5	2	3	5	2	13	39

Thus, we can see that flow methods solve project time/cost problems with far fewer calculations than would be necessary using linear programming methods. This problem is typical of those handled by flow methods in that the mathematical argument involved in converting to a flow method is tedious. But once the conversion is made, the calculations are much easier to perform than those using conventional solution procedures.

5.3.5 Other network models. A number of network flow algorithms have been developed in recent years. The most general of these is the solution to the general minimal cost circulation problem. The solution procedure leads to the so-called *out-of-kilter* algorithm. This algorithm was also developed by Ford and Fulkerson [8] and discussed recently by Elmaghraby [6]. Other applications of this method have been to the personnel assignment problem [9], the transportation problem [8], production-inventory problems [6], and shortest paths in undirected networks [6]. Most of these models are just special cases of the general minimal cost circulation problem.

5.4 Extensions and New Areas of Study

Hu [10] and Elmaghraby [5, 6] present many new ideas in the area of network flow and shortest-path analysis which are impossible to cover within the confines of this chapter. Hu discusses the multiterminal maximal flows, multiterminal shortest chains, multicommodity flow, and flows in continua. Elmaghraby [5, 6] summarizes many of Hu's ideas in a concise and readable manner. He also considers the problem of nonlinear cost curves for the general minimal cost algorithm. Ford and Fulkerson [8] still represent the prime source of background in this area.

EXERCISES

For each of the following acyclic networks find:

  (a) Shortest path between the origin and terminal nodes.
  (b) Shortest paths from the origin to all nodes in the network.
  (c) Shortest paths between all parts of nodes.
  (d) Three shortest paths from the source to the sink.

1. *Arc*   *Duration*
  1–2   7
  1–3   6
  1–4   10
  2–3   2

2–4	4
3–4	2
3–5	7
4–5	3

2.

Arc	Duration
1–2	3
1–3	1
2–3	2
2–4	1
3–4	8
3–6	12
3–5	9
4–5	2
4–6	10
5–6	6

3.

Arc	Duration	Arc	Duration
1–2	6	5–7	3
1–3	9	6–7	6
2–3	6	6–8	4
2–4	2	7–8	3
2–5	5	7–9	8
3–4	8	8–9	4
3–6	3		
4–5	1		
4–7	7		

For each of the following cyclic networks find:

(a) Shortest path between the origin and terminal nodes.
(b) Shortest paths from the origin to all nodes in the network.
(c) Shortest paths between all pairs of nodes.

4.

Arc	Duration	Arc	Duration
1–2	4	3–5	8
1–4	2	4–2	1
2–4	3	4–3	5
2–3	6	4–5	7
3–4	2		

5.

Arc	Duration
1–2	10
1–3	2
1–4	11
2–3	4
3–2	3
2–4	3
3–4	10

6. Develop a computer program for finding the shortest path between all pairs of nodes in a directed acyclic network. Test your routine on Exercises 1 to 3.

7. Develop a computer program to apply Dijkstra's method for finding the shortest path in a directed cyclic network.

8. Develop a computer program to apply the revised cascade method for finding the shortest path between all pairs of nodes in a directed cyclic network.

9. Develop a procedure for finding the m longest paths in an activity network. Apply your method to Exercise 5, Chapter 3.

10. Given the following reliability network, find the most reliable route through the network:

Arc	Reliability
1–2	0.80
1–3	0.80
2–3	0.70
2–4	0.90
3–2	0.90
3–4	0.90
3–5	0.75
4–3	0.95
4–5	0.80

11. Consider the following equipment replacement model. A company is planning its equipment policy for the next six years. C_{ij} is the cost of buying the equipment in year i and selling it in year j. No equipment can be kept more than three years. Find the best replacement policy if:

$$C_{12} = 800, C_{13} = 1400, C_{14} = 1700, C_{23} = 900, C_{24} = 1300,$$
$$C_{25} = 1700, C_{34} = 1000, C_{35} = 1200, C_{36} = 2000, C_{45} = 800,$$
$$C_{46} = 1200, \text{ and } C_{56} = 700$$

How much better is this policy than the second and third best methods?

Use the intuitive, labeling, and matrix methods for analyzing each of the following flow networks.

12. | Arc | C_{ij} |
|-----|----------|
| 1–2 | 6 |
| 1–3 | 4 |
| 2–3 | 3 |
| 2–4 | 9 |
| 3–4 | 1 |
| 3–5 | 4 |
| 4–5 | 6 |

13. | Arc | C_{ij} |
|-----|----------|
| 1–2 | 10 |
| 1–3 | 11 |
| 2–3 | 6 |
| 2–4 | 7 |
| 3–4 | 4 |

3–5	5
3–6	3
4–5	4
4–6	5
5–6	5

14. Develop a computer program using the matrix method to find the maximum flow in a network.

15. For the distribution problem discussed in Section 5.3.5:

$$a_1 = 10 \qquad d_1 = 10 \qquad c_{12} = 5$$
$$a_2 = 10 \qquad d_2 = 10 \qquad c_{13} = 5$$
$$a_3 = 20 \qquad d_3 = 5 \qquad c_{14} = 2$$
$$d_4 = 10 \qquad c_{21} = 10$$
$$c_{22} = 5$$
$$c_{24} = 6$$
$$c_{31} = 8$$
$$c_{32} = 8$$
$$c_{33} = 8$$
$$c_{34} = 8$$

Is there a feasible distribution pattern available?

16. The matrix below shows the jobs that can be handled by machines in a job shop. If each machine can handle only two jobs, what is the maximum number of jobs that can be processed?

Job/Machine	*A*	*B*	*C*	*D*
a	✓		✓	
b	✓	✓		✓
c	✓			✓
d	✓			✓
e	✓	✓		
f	✓		✓	
g	✓			
h	✓	✓	✓	
i	✓		✓	
j	✓		✓	
k	✓	✓		
l	✓			

17. In Section 5.2.5, a method for maximizing the minimum level of efficiency in an assignment problem was discussed. Apply this approach to the following assignment matrix.

Job/Person	1	2	3	4	5
1	10	9	6	5	1
2	2	3	8	4	5
3	7	6	2	1	5
4	4	8	8	4	3
5	4	5	2	4	2

18. For the dual formulation of the CPM model, verify that if $C_{ij} - f_{ij} = 0$, then v_{ij} and $w_{ij} = 0$; and verify that if $C_{ij} - f_{ij} < 0$, then $v_{ij} = 0$.

19. Use the Ford–Fulkerson CPM algorithm to verify that the CPM example discussed in Section 3.3.1 was analyzed correctly.

20. Use the Ford–Fulkerson CPM algorithm to analyze Exercise 9 in Chapter 3.

21. A telephone network is an example of a multiterminal network. Such networks have their source and sink selected from a given set of nodes. In the telephone network a call may originate at any node (city) or terminate at any node (city). Maximum flow analysis can be used between any two nodes. Show that one must solve $\binom{n}{2}$ maximum flow problems in the case of the telephone network.

22. A city must assign n families to n available low-income houses. Certain assignments are impossible because the houses are too small for the family. Suggest a model that will aid the city.

23. A company wishes to develop an optimum replacement policy for some pollution equipment it needs for the next seven years. Assume p_t will be the purchase price at the beginning of year t, and $P_1 = 200$, $P_2 = 210$, $P_3 = 225$, $P_4 = 250$, $P_5 = 270$, $P_6 = 290$, $P_7 = 300$.

   The salvage value of a 1-year-old piece of equipment is 100, and it is 50 for a 2-year-old piece of equipment. Older equipment is worthless. The operating costs for equipment in use for K consecutive periods is $Y_1 = 50$, $Y_2 = 75$, $Y_3 = 100$, $Y_4 = 150$, $Y_5 = 200$, $Y_6 = 300$, $Y_7 = 450$. Can you help the company solve this problem?

24. One of the classical problems in mathematical programming is the *caterer problem*. A catering firm requires R_j fresh napkins at the start of day j, where $j = 1, \ldots, T$. Normal laundering takes one full day at B cents a napkin; rapid laundering takes overnight and costs C cents a napkin. The caterer wants a plan for purchasing and laundering napkins to minimize costs and to meet all fresh napkin requirements. Will network modeling help in the solution of this problem?

REFERENCES

1. Clarke, S., A. Kirkonian, and J. Rausen, "Computing the n Best Loopless Paths in a Network." *Journal of SIAM*, Vol. 2, No. 4 (December, 1963), 1096–1102.

2. Dantzig, G. B., "On the Shortest Route Through a Network." *Management Science*, Vol. 6, No. 2 (January, 1960), 187–190.

3. Dijkstra, E. W., "A Note on Two Problems in Connection with Graphs." *Numerische Mathematik*, Vol. 1 (1959), 269–271.

4. Elmaghraby, S. E., *The Design of Production Systems*. New York, Van Nostrand-Reinhold Publishing Corp., 1966.

5. Elmaghraby, S. E., *Network Models in Management Science*. Springer-Verlag Lecture Series on Operations Research, 1970.

6. Elmaghraby, S. E., "Theory of Networks and Management Science I and II." *Management Science*, Vol. 17, Nos. 1 and 2 (September and October, 1970).

7. Floyd, R. W., "Algorithm 97, Shortest Path." *Communications of A.C.M.*, Vol. 5 (1962), 345.

8. Ford, L. R. and D. R. Fulkerson, *Flows in Networks*. Princeton, N.J., Princeton University Press, 1962.

9. Hadley, G., *Linear Programming*. Reading, Mass., Addison-Wesley Publishing Co., Inc., 1962.

10. Hu, T. C., *Integer Programming and Network Flows*. Reading, Mass., Addison-Wesley Publishing Co., Inc., 1969.

11. Minty, G. J., "A Comment on the Shortest-Route Problem." *Operations Research*, Vol. 5, No. 5 (October, 1957).

12. Moder, J. J. and C. R. Phillips, *Project Management with CPM and PERT*. New York, Van Nostrand-Reinhold Publishing Corp., 1964.

13. Pollack, M., "Solutions of the *K*th Best Route Through a Network—A Review." *Journal of Mathematical Analysis and Applications*, Vol. 3, No. 3 (December, 1961).

14. Pollack, M., "The *K*th Best Route Through a Network." *Operations Research*, Vol. 9, No. 4 (July–August, 1967).

FLOWGRAPHS

Elmaghraby [6] stated: "The distinguishing aspect of a system—as opposed to an individual component—is the interaction among its various components. The mode of this interaction, its genre and direction determine the structure of the system. The engineer in his analysis or synthesis of systems is ultimately concerned with such structure. Insight into the cause-and-effect relationships and into the performance of the system as a whole can be gained only through such knowledge. In fact, the determination of the structure of the system is a prelude to the framing of the problem for quantitative analysis."

In this chapter, we will explore flowgraph analysis as a tool of systems analysis, keeping Elmaghraby's charge in mind. The original concept of flowgraphs was worked out by Shannon [23] in a report dealing with analog computers. However, the primary credit for the development and refinement of the theory is given to Mason [19]. Lorens [17] presents the most comprehensive treatment of flowgraphs in his paperback volume.

6

6.1 Definition of Flowgraph Analysis

A *flowgraph* is a graphical representation of the relationships among variables. It simultaneously displays all the relationships (equations) among the variables of a given system. Basically, a flowgraph is a way of writing linear algebraic equations. The flowgraph may be solved for the variables in the system (represented by nodes) in terms of the relationships that exist between the variables (represented by branches and called transmittances). A flowgraph then consists of a set of nodes and branches, where the nodes represent variables and the branches indicate that a relationship exists between the nodes which the branches connect.

Now let us consider the simple flowgraph in Fig. 6.1 which expresses the relationship $B = RA$. In this flowgraph A and B are *nodes; R* is a *transmittance.* The arrow's direction is important because it shows that A is an *independent variable* while node B is a *dependent variable.* The transmittance R expresses the relationship between A and B. If we reverse the direction of the arrow, the flowgraph will appear as in Fig. 6.2. This yields the equation $A = (1/R)B$, which, of course, is consistent with the equation $B = RA$. However, the relationship between the dependent and the independent variables has been changed, and the operation is known as *path inversion.*

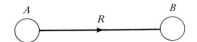

Figure 6.1. Simple flowgraph element.

Figure 6.2. Path inversion of a simple flowgraph element.

The purpose of a flowgraph is to depict a system graphically and to study the system by analyzing the flowgraph. To accomplish the latter, methods have been developed that permit direct manipulations to be performed on flowgraphs. These manipulations change the graph but do not alter the relationships depicted by the graph. The fundamental property associated with a flowgraph is: The value associated with a node is equal to the sum of the transmittances of all incoming branches times the value associated with the node from which the branch originated. Examples of this property are illustrated in Fig. 6.3. Figure 6.3(a) can be stated mathematically as $y = ex + fz$ and represents addition. Figure 6.3(b) also demonstrates addition and can be stated as $y = x(g + h)$. $y = fx - ex$ is the equation represented in Fig. 6.3(c) and thus demonstrates subtraction in flowgraph format. Multiplication is shown in Fig. 6.3(d) and is stated mathematically as $y = ex$. Figure 6.3(e) can be stated mathematically as $y = x/e$ and thus demonstrates division. Another common relationship in flowgraph representation is shown in Fig. 6.3(f). Whenever a transmittance is not shown, we assume that a unit transmittance exists, and this expresses the relationship $y = x$.

In writing the equation for a node, we assume that the node in question is the dependent variable. However, the same node may also be an independent variable if it has branches emanating from it. Consider the flowgraph shown in Fig. 6.4. Node x can be considered as both a dependent and an independent variable as shown by the following equations:

$$x = es + ft + gu$$

and

$$y = hx = hes + hft + hgu$$

We often find it desirable, especially in the detailed examination of a flowgraph system, to introduce a new node separated by a unit transmittance. This is illustrated by Fig. 6.5. The value of x is unchanged by the addition

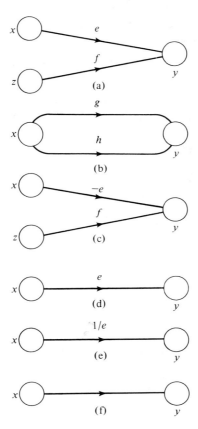

Figure 6.3. Basic mathematical manipulations represented in flowgraphs.

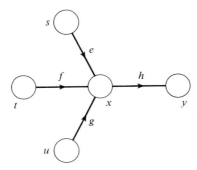

Figure 6.4. Flowgraph demonstrating an element which is both an independent and a dependent variable.

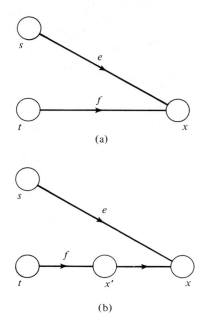

Figure 6.5. Introduction of a new node into flowgraph network.

of node x' in Fig. 6.5(b), but the transmittance e makes a contribution to x and not to x'.

$$x = x' + es = ft + es$$

and

$$x' = ft$$

In this manner, we can split a complicated node into a series of simpler ones by the use of unit transmittances. This operation is demonstrated in Fig. 6.6. Node y is split, but Fig. 6.6(a) and Fig. 6.6(b) are equivalent. The importance of the operation is that any complex node (a node which has multiple inputs and outputs) can be converted into the following two basic types of nodes:

1. *Distributive node.* A distributive node is one at which at most one transmittance terminates and at least two originate, as shown in Fig. 6.7(a).
2. *Contributive node.* A contributive node is one at which at least two transmittances terminate and at most one originates, as shown in Fig. 6.7(b).

We will use these concepts later when we describe the operation of path inversion.

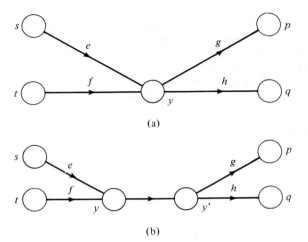

(a)

(b)

Figure 6.6. Conversion of a complex flowgraph node to basic types of nodes.

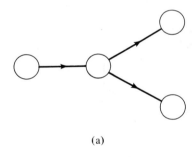

(a)

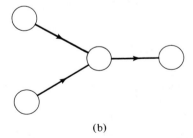

(b)

Figure 6.7. Distributive and contributive nodes.

Figure 6.8 illustrates two additional important components of a flowgraph:

1. A *path* is a series of branches which join two nodes and do not pass through any node more than once. There are three paths from node 1 to node 5, in Fig. 6.8: the paths are *a* to *g*, *a* to *c* to *h*, and *b* to *h*. We note that *a* to *d* to *e* to *g* is not a path because the route passes through node 2 twice. The value of a path is the product of the transmittances along the path.
2. A *loop* is a series of branches which lead from a node and eventually return to the node without passing through any other node more than once. In Fig. 6.8, *e* to *d* and *f* are the loops. The value of a loop is equal to the product of the transmittances around the loop.

We will now discuss methods for solving flowgraph models.

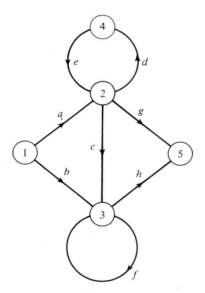

Figure 6.8. Sample flowgraph.

6.2 Methods of Solution of Flowgraphs

6.2.1 Topological equations. A topological equation of flowgraph theory will now be discussed. Though the fundamental meaning of the topological equation is simple, its implications are profound. It provides the basis for solving complex flowgraphs. In Section 6.1, our reference to a loop was really to *first-order loops*, which are a consecutive path of arrows leading from a node and returning to the same node. There are n*th-order loops* also, which can be described as n nontouching first-order loops. The *value of an nth-order loop* is equal to the product of the transmittances for the n nontouching loops. The loops in Fig. 6.8 combine to make a second-order loop, since they have no common nodes. A topological equation, H, for closed flowgraphs can now be defined. A *closed flowgraph* is one composed entirely of loops. The topological equation takes the following form: the sum of the loops formed in the following manner will always equal zero in a closed flowgraph.

$$H = 1 - L_1 + L_2 - L_3 + \cdots + (-1)^i L_i + \cdots = 0$$

where L_i is the sum of the ith-order loops. We will derive a form of this equation in the next section. The topological relationship is demonstrated for the flowgraph shown in Fig. 6.9 as follows:

$$\text{Loop I } (xyzw) = 1 \times 5 \times \tfrac{1}{5} \times \tfrac{1}{4} = \tfrac{1}{4}$$
$$\text{Loop II } (xyzuvw) = 1 \times 5 \times 3 \times 3 \times \tfrac{1}{15} \times \tfrac{1}{4} = \tfrac{3}{4}$$
$$L_1 = \tfrac{1}{4} + \tfrac{3}{4} = 1$$

and

$$H = 1 - L_1 = 1 - 1 = 0$$

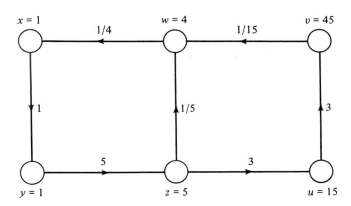

Figure 6.9. Demonstration of the topological relationship in flowgraphs.

As another example, the topological equation states that the variables in the flowgraph shown in Fig. 6.10 must satisfy the following equation:

$$H = 1 - (af + bg + ch + di + ej) + (afch + afdi + afej + bgdi$$
$$+ bgej + chej) - (afchej) = 0$$

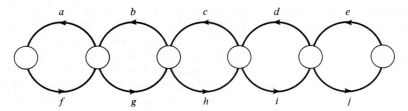

Figure 6.10. Calculation of the sum of the loops in a closed flow-graph.

The topological equation can be used for open flowgraphs if we convert them to closed graphs with a fictitious $1/T$ transmittance, where T is the equivalent transmittance of the open graph. Let us now consider the solution of the open flowgraph shown in Fig. 6.11(a). The solution of the flowgraph is defined as the equivalent transmittance, T, from node x to node w, as shown in Fig. 6.11(b). If the flowgraph in Fig. 6.11(b) is closed with a transmittance from node w to node x, the value of this transmittance from the topological equation $T' = 1/T$, as shown in Fig. 6.11(c), is $H = 1 - T(T') = 0$. Thus, we can solve the graph shown in Fig. 6.11(a) for the transmittance between nodes x and w by closing the graph as shown in Fig. 6.11(d) and solving for T.

$$\text{(Topological equation):} \quad H = 0$$

$$\text{First-order loops:} \quad a, b, \frac{c}{T},$$

$$\text{Second-order loops:} \quad \frac{ac}{T},$$

and

$$1 - a - b - \frac{c}{T} + \frac{ac}{T} = 0.$$

Thus:

$$T = \frac{[c(1-a)]}{(1-a-b)}$$

This is a useful technique. When we close the flowgraph with a variable transmittance $1/T$, the system is constrained. However, open flowgraphs need not be closed because the topological equation can be put in the following form so as to handle open flowgraphs:

$$T = \frac{[\sum (\text{path} \times \sum \text{nontouching loops})]}{\sum \text{loops}}$$

This equation, known as *Mason's rule*, is stated as follows: Write down the product of transmittances along each path from the independent to the depen-

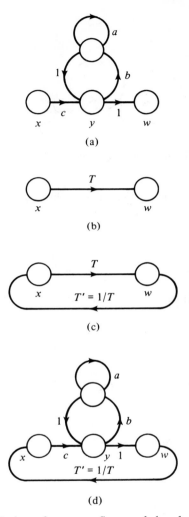

(a)

(b)

(c)

(d)

Figure 6.11. Solution of an open flowgraph by closing the graph.

dent variable. Multiply its transmittance by the sum of the nontouching loops to that path. Sum these modified path transmittances and divide by the sum of all the loops in the open flowgraph.

Consider the flowgraph shown in Fig. 6.12:

Path from x to $w = c$

$$\sum \text{nontouching loops} = 1 - (\sum \text{first-order nontouching loops})$$
$$+ (\sum \text{second-order nontouching loops})$$
$$- \cdots = 1 - a$$

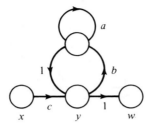

Figure 6.12. Solution of an open flowgraph by means of the topological equation.

$$\sum \text{loops in system} = 1 - (\sum \text{first-order loops})$$
$$+ (\sum \text{second-order loops}) - \cdots = 1 - (a + b)$$

and

$$T = \frac{[c(1 - a)]}{(1 - a - b)}$$

This is consistent with the result we obtained by closing the flowgraph.

EXAMPLE:

Consider the following riddle which was suggested by Happ [10]:

A baby is $\frac{1}{4}$ her father's age less 3 years, while her brother is 3 times her age minus $\frac{1}{2}$ his own. Dad is twice Mom's age minus twice Brother's age, while Mom is Dad's age minus twice the baby's age.

The flowgraph for this system is shown in Fig. 6.13. It can be seen that this graph represents the following set of linear equations:

$$\text{Baby} = \tfrac{1}{4} \text{Dad} - 3$$
$$\text{Bro.} = -\tfrac{1}{2} \text{Bro.} + 3 \text{Baby}$$
$$\text{Mom} = \text{Dad} - 2 \text{Baby}$$
$$\text{Dad} = 2 \text{Mom} - 2 \text{Bro.}$$

If we desire to find Dad's age from this flowgraph, we could apply Mason's rule to find the equivalent transmittance between nodes, 1 year, and Dad. This transmittance, T, will represent the coefficient in the following equation:

$$\text{Dad} = T \cdot (1 \text{ year})$$

Thus, T will be Dad's age.

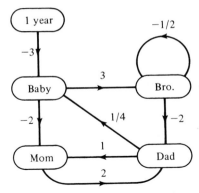

Figure 6.13. Flowgraph representation of an age puzzle.

To apply Mason's rule we must locate all the loops and paths in the system:

LOOPS IN THE TOTAL SYSTEM

First-order loops:

No.	Path	Value
1—Baby to Mom to Dad to Baby		$(-2)(2)(\frac{1}{4}) = -1$
2—Mom to Dad to Mom		$(2)(1) = 2$
3—Bro. to Bro.		$(-\frac{1}{2}) = -\frac{1}{2}$
4—Baby to Bro. to Dad to Baby		$(3)(-2)(\frac{1}{4}) = -\frac{3}{2}$

Second-order loops:

Nontouching first-order loops	Value
No. 1 and No. 3	$(-2)(2)(\frac{1}{4})(-\frac{1}{2}) = \frac{1}{2}$
No. 2 and No. 3	$(2)(1)(-\frac{1}{2}) = -1$

Third-order loops:

None

$$\sum \text{loops in the system} = 1 - (-1 + 2 - \tfrac{1}{2} - \tfrac{3}{2}) + (\tfrac{1}{2} - 1) = \tfrac{3}{2}$$

PATHS FROM 1-YEAR TO DAD

	Path	Value
Path 1	1-year to Baby to Mom to Dad	$(-3)(-2)(2) = 12$

First-order nontouching loops to path 1:

Bro. to Bro.

$$\sum \text{loops not touching path } 1 = 1 - (-\tfrac{1}{2}) = \tfrac{3}{2}$$

Path 2	1-year to Baby to Bro. to Dad	$(-3)(3)(-2) = 18$

First-order nontouching loops to path 2:

None

$$\sum \text{loops not touching path } 2 = 1$$

APPLYING MASON'S RULE

$$\text{Dad's age} = \frac{(12)(\tfrac{3}{2}) + (18)(1)}{(\tfrac{3}{2})} = 24 \text{ years}$$

By similar analysis:

$$\text{Mom's age} = \frac{(6)(\tfrac{3}{2}) + (18)(1)}{(\tfrac{3}{2})} = 18 \text{ years}$$

$$\text{Baby's age} = \frac{(-3)(-\tfrac{3}{2})}{(\tfrac{3}{2})} = 3 \text{ years}$$

$$\text{Bro.'s age} = \frac{(-9)(-1)}{(\tfrac{3}{2})} = 6 \text{ years}$$

The author does not wish to comment on the social implications of this problem.

6.2.2 Why does Mason's rule work? Mason's rule has been proved by a number of authors, e.g., Mason [19], Lorens [17], and Robichaud [22]. The following discussion follows closely that of Robichaud.

Each variable of a flowgraph can be defined by a linear equation of the form:

$$x_k = f(x_1, x_2, \ldots, x_n)$$

or else it is an independent variable. A matrix T of branch transmittances can be written as a square matrix of order n, where n is the number of nodes in the flowgraph:

$$T = \begin{bmatrix} t_{11} & t_{12} & t_{1n} \\ t_{21} & t_{22} & t_{2n} \\ & & \\ t_{n1} & t_{n2} & t_{nn} \end{bmatrix} \tag{6-1}$$

If we write X for the column matrix of the variables $x_1, x_2, \ldots, x_n$ and Y for the column matrix of dependent variables ($y_K = x_K - f_K$), we have the system of equations represented in the flowgraph:

$$AX = Y \tag{6-2}$$

Where A is defined by:

$$A = I - T = \begin{bmatrix} 1 - t_{11} & -t_{12} & -t_{1n} \\ -t_{21} & 1 - t_{22} & -t_{2n} \\ -t_{31} & -t_{32} & -t_{3n} \\ & & \\ -t_{n1} & -t_{n2} & 1 - t_{nn} \end{bmatrix} \tag{6-3}$$

and I is an identity matrix of order n.

In terms of the elements of the matrix A, the determinant of the system of equations will be written:

$$D = \sum (-1)^{K+K'} (a_{aa'} \times a_{bb'} \times a_{cc'} \times \ldots \times a_{nn'}) \tag{6-4}$$

where each term contains n elements, one and only one taken from each row and each column, and where K and K' are the number of inversions of the numbers $1, 2, \ldots, n$ in the sequences of indices $a, b, c, \ldots, n$ and $a'\, b'\, c'\, \ldots n'$. We must take the sum for all possible choices of the elements of the matrix.

We will now show that the terms in this sum consist of all the possible products of transmittances of nonintersecting loops. Consider any term in Eq. (6-4). This term contains the element a_{ij}. It also necessarily contains another element of row j, and this element can only be a_{ji} or a_{jk}. If it is a_{ji}, we can find next an element a_{kl}, since one must choose an element in rows and columns other than i and j. If it is an element a_{jk}, we can find an element from row k which has to be a_{ki} or a_{kl}. Proceeding in this manner for each term in Eq. (6-4) and changing the order of the factors, we can write for the indices one or several closed sequences:

$$ij, jk, kl, lx, \ldots, yp, pi \tag{6-5}$$

$$ij, ji, \ldots, rr, \ldots, nm, mp, pn, \ldots, qu, uq \tag{6-6}$$

The number of inversions K' then becomes $K' = (K + n + M)$, n being the number of factors in each term and M the number of sequences.

The matrix A corresponding to the flowgraph in Eq. (6-3) will now be considered. Considering the sequences in Eqs. (6-5) and (6-6), and by decomposing into several terms containing elements of the first diagonal of the matrix, we obtain the following expressions:

$$t_{ij}t_{jk}t_{kl}t_{lx} \cdots t_{yp}t_{pi} \tag{6-7}$$

$$t_{ij}t_{ji} \cdots (1) \cdots t_{nm}t_{mp}t_{on} \cdots t_{qu}t_{uq} \tag{6-8}$$

$$t_{ij}t_{ji} \cdots t_{rr} \cdots t_{nm}t_{mp}t_{pn} \cdots t_{qu}t_{up} \tag{6-9}$$

Equations (6-8) and (6-9) are derived from the sequence suggested in Eq. (6-6) because a_{rr} is on the first diagonal and equals $(1 - t_{rr})$. We determine the sign in front of each product by taking into account the minus signs of the transmittances; hence:

$$(-1)^{K + K' + n} \tag{6-10}$$

which we rewrite by substituting the value found previously for K':

$$(-1)^{K + K + n + M + n} = (-1)^{2(K + n) + M} = (-1)^M \tag{6-11}$$

Each of the Eqs. (6-7), (6-8), and (6-9) can be recognized as a loop transmittance or a product of loop transmittances, since the sequences in the indices start at a node and end at the same node. If one line contains several loops, the nonintersection is assured by the initial restriction on the choice of the elements of each term. We observe that the sign in front of each term depends uniquely on the number of loops as was shown in Eq. (6-11). If we add the unity term obtained from the product of the elements on the principal diagonal, we can, by regrouping the terms, rewrite the determinant:

$$D = 1 - L_1 + L_2 - L_3 + \cdots \tag{6-12}$$

The numerator of the expression for a variable such as x_2, for example, if there exists a nonzero element in the Y vector in the first equation only, is given by Cramer's rule as the determinant:

$$\begin{bmatrix} a_{11} & x_0 & a_{13} & \cdots & a_{1n} \\ a_{21} & 0 & a_{23} & \cdots & a_{2n} \\ \cdots & & & & \\ \cdots & & & & \\ a_{n1} & 0 & a_{n3} & \cdots & a_{nn} \end{bmatrix} \tag{6-13}$$

By writing the terms of the determinant in the same manner as before, we obtain closed sequences for the indices and one open sequence, starting with 2 and ending with 1, which will be associated with the element x_0:

$$x_0 a_{1k} a_{kj} a_{j2} \cdots a_{rr} \cdots a_{ps} a_{sq} a_{qp} \qquad (6\text{-}14)$$

Using the matrix of the flowgraph in Eq. (6-3) and writing the factors corresponding to the preceding indices, once again realizing that the a_{rr} term equals $(1 - t_{rr})$, we get:

$$x_0 t_{1k} t_{kj} t_{j2} \cdots t_{rr} \cdots t_{ps} t_{sq} t_{qp} \qquad (6\text{-}15)$$

$$x_0 t_{1k} t_{kj} t_{j2} \cdots (1) \cdots t_{ps} t_{sq} t_{qp} \qquad (6\text{-}16)$$

In all terms obtained, there will be a path transmittance for a path from node 1 to node 2, and each of these transmittances will be multiplied by a product of loop transmittances. We note that these loops cannot intersect the path considered because of the initial restriction on the choice of the elements in the Eq. (6-4) of the determinant. Combining all these results, we then can finally write Mason's rule realizing that the value of the determinant is equal to the sum of loops in the system.

6.2.3 Other methods of solution.

The topological approach to the solution of flowgraphs is by no means the only approach, but it is the principal approach that will be used in this text. Four other techniques will now be briefly discussed.

TOPOLOGICAL EQUIVALENCE

Stepwise reduction of flowgraphs is often referred to as *topological equivalence*. For example, Fig. 6.14 shows the rules pertinent to loop reduction. Figure 6.14(a) is equivalent to Fig. 6.14(b), and Fig. 6.14(c) is equivalent to Fig. 6.14(d). These relationships are quite easy to show mathematically. For example, the equivalence of Fig. 6.14(c) and Fig. 6.14(d) is developed as follows:

$$y = x + Ly$$

$$z = y$$

$$x = y(1 - L)$$

and

$$y = x/(1 - L)$$

hence:

$$z = x/(1 - L)$$

Examples of the use of topological equivalence can be found in References 3, 17, and 22.

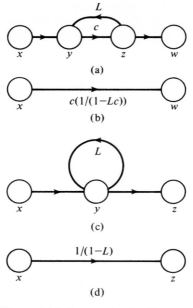

Figure 6.14. Loop rules for flowgraphs.

NODE-REDUCTION METHODS

By applying procedures similar to Gauss–Jordan reduction procedures, researchers have developed procedures for reducing flowgraphs by successively removing nodes. One such procedure is based upon the following formula:

$$t'_{ij} = t_{ij} + \frac{t_{ik}t_{kj}}{1 - t_{kk}}$$

where the indices i and j refer to residual nodes, and k refers to the node which will be removed in a particular step.

We can demonstrate the procedure by solving the problem shown in Fig. 6.15(a) for the transmittance between nodes 1 and 4. As a first step, we will remove node 3. The resulting flowgraph is shown in Fig. 6.15(b). We calculate the new transmittance for the path from node 2 to node 4 as follows:

$$i = 2, \quad j = 4, \quad \text{and} \quad k = 3$$

$$t_{24} = 0, \quad t_{23} = c, \quad t_{34} = f, \quad \text{and } t_{33} = e$$

Therefore:

$$t'_{24} = 0 + \frac{cf}{1 - e}$$

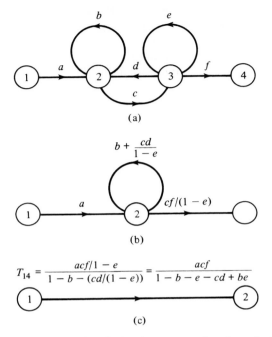

Figure 6.15. Flowgraph solution by means of node reduction procedures.

The final step of the reduction is to remove node 2 as shown in Fig. 6.15(c). This is, of course, the same result that you would expect from applying Mason's rule.

This method is more cumbersome to use in hand solutions of flowgraphs, but it is much easier to program for computer solution. It is the procedure used in the computer program presented in Appendix 1.

APPROXIMATION TECHNIQUES.

Approximation techniques for flowgraphs have been developed [5, 10] which reduce the graph by eliminating the insignificant paths and loops to provide a more concise model of a system, consistent with a preassigned level of accuracy. In theory, we significantly reduce the difficulty in recognizing the principal interactions of variables and solving the flowgraph by using the approximate flowgraph.

COMPUTER TECHNIQUES

Computer programs have been developed for most computers to solve flowgraphs. These programs use matrix, topological equivalence, and topological methods for solving the flowgraphs. Robichaud [22] presents a chapter describing the development of a simple program using topological equivalence methods.

In Appendix I we present a computer program that can be used to analyze open flowgraphs. The input format for this program is as follows:

Card 1   (Identification card)
    Col. 1–80   Title of the program
Card 2   (Source nodes)
    Col. 1–2   Node no. of first source node
    Col. 3–4   Node no. of second source node

         .   .
    Col. 79–80   Node no. of fortieth source node
Card 3   (Transmittance Cards)
    Col. 1–2   Node no. of node preceding the transmittance
    Col. 3–4   Node no. of node terminating the transmittance
    Col. 5–14   Transmittance

A blank card marks the end of the problem, and problems may be stacked. The node numbers are limited to the range 1–99. The output of the program is the equivalent transmittance between each source node and all other nodes in the system.

Let's consider the use of this program for the example shown in Fig. 6.13. The input to the program is as follows:

Fig. 6.13, 1 = 1 year, 2 = Baby, 3 = Brother, 4 = Mom, 5 = Dad
1
1 2   −3.0
2 3   3.0
2 4   −2.0
3 3   −0.5
3 5   −2.0
4 5   2.0
5 4   1.0
5 2   0.25

The output of the program is shown in Table 6.1. We see that the results derived in Section 6.2.1 are verified.

Table 6.1. COMPUTER OUTPUT FOR A SAMPLE FLOWGRAPH.

SOLUTION OF FLOWGRAPH
FIG. 6.13, 1=1 YEAR, 2=BABY, 3=BROTHER, 4=MOM, 5=DAD
INPUT DATA

FROM	TO	TRANSMITTANCE
1	2	−3.0000
2	3	3.0000
2	4	−2.0000
3	3	−0.5000
3	5	−2.0000
4	5	2.0000
5	4	1.0000
5	2	0.2500

1 SPECIFIED AS A SOURCE
RESULTING EFFECTIVE TRANSMITTANCES

FROM	TO	TRANSMITTANCE
1	2	0.30000000E 01
1	3	0.60000000E 01
1	4	0.18000000E 02
1	5	0.24000000E 02

END OF PROBLEM
END OF DATA—END OF PROGRAM

6.3 Path Inversion

Path inversion is equivalent to inverting the relative independent-dependent relationship between the two variables. This is often an important tool in the solving of flowgraphs. For a flowgraph to be solvable, every node with a known value must be an independent variable. When we set up the original graph, path inversion is often used to meet this restriction. Inversion has meaning for only two types of paths: a path which goes from a strictly independent node to a strictly dependent node, and a path which forms a loop. We define a strictly independent node as one that has no transmittances flowing into it. The strictly dependent node is one that has no transmittances flowing out of it. The rules for inverting paths through contributive and distributive nodes are as follows.

PATH INVERSION AT A CONTRIBUTIVE NODE

Consider the flowgraph shown in Fig. 6.16(a). If we wish to invert the path from node w to node y, we reverse the direction of the transmittances $y - z$ and $z - w$. As shown in Fig. 6.16(a):

$$w = c(ax + by)$$

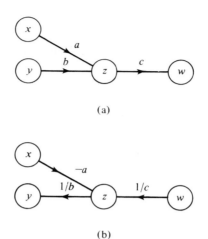

(a)

(b)

Figure 6.16. Path inversion at a contributive node.

Solving for y, we obtain:

$$y = \frac{w - cax}{cb}$$

In general, the rule for path inversion at a contributive node is: Reverse the direction of the arrows in the through path; replace each transmittance in the through path with its reciprocal; change the sign of the unchanged transmittances.

PATH INVERSION AT A DISTRIBUTIVE NODE

The rule for path inversion at a distributive node is: Reverse the direction of the arrows in the through path; replace each transmittance in the through path by its reciprocal; leave the other transmittance unchanged. The graph in Fig. 6.17(a) is equivalent to the one in Fig. 6.17(b).

We note that path inversion will *never* change a contributive node to a distributive node, or vice versa. It should also be pointed out that, in practice, the nodes are not really split, but only handled as if they were split. (Refer to Fig. 6.6.) This point is illustrated in Fig. 6.18, where Fig. 6.18(a) is equivalent to Fig. 6.18(b).

EXAMPLE:

Consider the following example from the field of engineering economy:

A person deposited twice as much in a bank this year as he did last year. The bank pays 4% per year. In three years the value in the bank will be $1000. How much did he deposit this year?

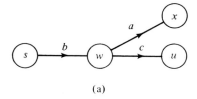

(a)

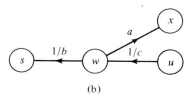

(b)

Figure 6.17. Path inversion at a distributive node.

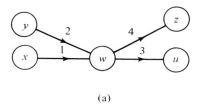

(a)

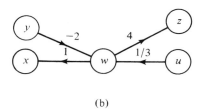

(b)

Figure 6.18. Path inversion at a complex node.

The flowgraph is shown in Fig. 6.19; the node numbers represent the year number, and the 1.04 transmittance between nodes represents the 4% interest rate paid by the bank. Notice that the unknown X is an independent variable in Fig. 6.19. Path inversion must be performed in order to convert node X to a dependent node. Figure 6.20 shows the graph after the path from $1000 to X has been inverted. The solution to the graph in Fig. 6.20 is:

$$X = \$\,1000\left(\frac{(1/1.04)^3}{(1 - (\tfrac{1}{2})(-1.04))}\right)$$

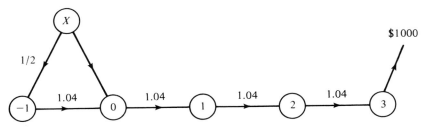

Figure 6.19. Flowgraph of an engineering economy problem.

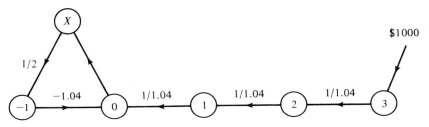

Figure 6.20. Figure 6.19 after path inversion.

6.4 Additional Topics to be Considered in Flowgraph Analysis

A topic which has been studied to some extent is that of sensitivity analysis. Usually three types of sensitivity are considered. If we consider the sensitivity of some dependent variable Q to another variable P, the types of sensitivities are:

1. Ratio sensitivity—which is the ratio of the two variables, or Q/P.
2. Rate sensitivity—which is the rate at which a variable will change with respect to a change in another variable, or dQ/dP.
3. Fractional sensitivity—which is the fractional change of the dependent variable with respect to a fractional change in another variable, or $(dQ/Q)/(dP/P)$.

Happ [10], Lynch [18], and Lorens [17] give brief descriptions of the development and use of sensitivity analysis related to flowgraph analysis.

We have thus far discussed flowgraph analysis with respect to the analysis of linear systems of variables. Lorens [17] discusses the use of flowgraphs for the analysis of matrix systems and nonlinear systems. The nonlinear system is first linearized and then solved by the procedures that we have been discussing. For example, a nonlinear function $x = f(x_1, x_2, x_3)$ can be linearized as follows:

$$\partial x = \frac{\partial f}{\partial y_1}\, \partial y_1 + \frac{\partial f}{\partial y_2}\, \partial y_2 + \frac{\partial f}{\partial y_3}\, \partial y_3$$

6.5 Applications of Flowgraphs for System Modeling*

Possible uses of flowgraph analysis in the area of system modeling will now be explored [25].

6.5.1 Economic systems. Flowgraphs of economic systems to be modeled are based on linear or approximately linear relations among the variables of the system. The flowgraph models of economic systems have several useful properties:

1. The functional relationships can be evaluated or approximated by elementary rules.
2. The pattern of functional interdependence discloses the relations among variables at a glance.
3. The sensitivity, stability, and other figures of merit of the system can be evaluated by well-known techniques.

We will use two models of economic systems to demonstrate the effectiveness of this approach.

THE NATIONAL ECONOMY

The dynamics of economic systems are similar to those of feedback circuits; hence, the techniques for systems analysis apply. An example of a linear system is the model for the national economy [18] shown in Fig. 6.21. The interrelation of the variables as well as the forces that control the economy are postulated. The stabilizing effects that government spending can have are included in this model of the national economy. Simplified models or more complex models of the national economy can be obtained by modification of the flowgraph in Fig. 6.21. The variables in the system are:

X_c—gross national product
X_p—private investment
X_r—excess government spending
X_d—dollars for production

*Portions of this section have been adapted from Reference 25 and are used with the permission of AIIE.

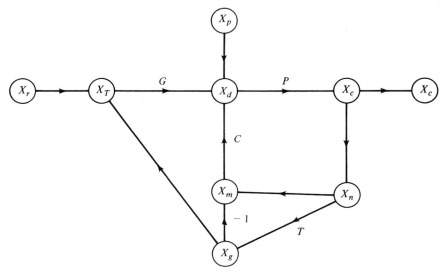

Figure 6.21. A simplified flowgraph model of national economy.

X_g—government income
X_n—gross income to the consumer
X_m—net income to the consumer
X_t—total government dollars invested

The parameters of the system are:

G—government spending rate
P—business production rate
C—consumer spending rate
T—tax rate

The system equation may be written directly from the flowgraph as:

$$X_c = \frac{X_r GP + X_p P}{1 - ((PC(1 - T)) + TGP)}$$

By using the path inversion and sensitivity techniques discussed in this chapter, we can determine the influence of changes in the parameters of the system on the sensitivity and stability of the system for various levels of government spending, for example.

CORPORATE ECONOMY

The economy of a corporation is of interest. A simplified flowgraph model is given in Fig. 6.22. The nodes are:

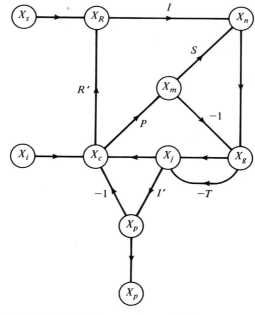

Figure 6.22. A simplified flowgraph model of corporate economy.

X_s—customer-funded research and grants
X_i—new capital investment
X_p—return to the investors
X_c—total capital to invest
X_r—total research funds
X_m—cost of goods sold
X_n—gross income
X_g—gross profit
X_j—net profit after taxes

Then the parameters of the system are:

R'—fraction of capital used in research
I—income from research
I'—fraction of net profit returned to investors
S—return on manufactured products
P—fraction of capital in products and inventories, etc.
T—tax rate

We can find the interactions among various corporate activities from the flowgraph model. Similarly, sensitivity and stability of the economy of the corporation may be evaluated. For example, the mathematical relationship for the investor income may be written directly from the flowgraph as:

$$X_p = \frac{X_s(I(1-T)I') + X_iI'(1-T)(PS - P + R'I)}{I - (1-T)(1-I')(R'I - P + PS)}$$

6.5.2 Sequential event systems. Present and projected factors influencing the financial status of a corporation depend on past performance. Many of these factors depend on combinations of exponential growth and decay, or the evaluation of exponential and linear factors. Flowgraphs of sequential events yield models that are at once mathematically rigorous and visually enlightening.

EXPONENTIAL GROWTH AND DECAY: COMPOUND INTEREST

A model for amortization of a loan is given in Fig. 6.23. Interest charges are compounded on the unpaid balance for a specific payment period. For a certain loan, the principal is X_0, while X_1 is the amount remaining after the

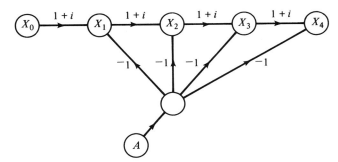

Figure 6.23. A model for amortization of a loan.

first payment, A. Thus:

$$X_1 = (1 + i)X_0 - A$$

Similarly, the outstanding debt after n equal payment periods is denoted by:

$$X_n = X_0(1 + i)^n - A((1 + i) + (1 + i)^2 + (1 + i)^3 + \cdots + (1 + i)^{n-1})$$

The periodic payment A to repay the debt in n equal payments is desired. Then $X_n = 0$; therefore A must be expressed as a dependent node. We accomplish this directly on the flowgraph by inverting a path from X_n to A. The resulting expression, after setting $X_n = 0$, is:

$$A = \frac{X_0(1 + i)^n}{(1 + i) + (1 + i)^2 + (1 + i)^3 + \cdots + (1 + i)^{n-1}}$$

THE HOSKOLD FORMULA

Another problem that we can use flowgraphs to solve is the economic justification for applied research. This is a comparison of recovery with exponential growth compared to linear return, and it is solved with the Hoskold formula [11]:

$$P = \frac{D}{R + R'/(1 + R')^n}$$

where: P—present worth of the income the project will yield if successful,

D—average annual incremental income yielded if the project is successful,

R'—average net return on capital invested in the enterprise,

R—current rate of interest for low-risk investment, and

n—number of years within which the research costs must be recovered.

Figure 6.24 shows the flowgraph representation for this formula. Notice the comparison between the two types of investment, as demonstrated by the loop through P.

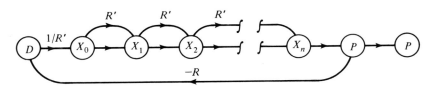

Figure 6.24. A flowgraph model of the Hoskold formula.

6.5.3 Probabilistic models. Since flowgraphs represent a multiplicative system, the technique is effective for investigating systems involving independent probabilistic events.

RELIABILITY ENGINEERING

Reliability engineering and probability analysis are very adaptable to flowgraph analysis. Much has been written in this area. For an excellent collection of examples, the reader is referred to the work by Happ [10].

Happ considers the following example of a recoverable missile casing.

Define:

Events

i = the first flight
f = any flight
g = a successful flight
b = an unsuccessful flight
d = nonrecovery after g
e = nonrecovery after b

Probabilities

S = a successful flight
R = a recovery after g
T = a recovery after b

Applying the topology equation to the system as shown in Fig. 6.25 yields the following probabilities:

$$g = (S)/(1 - (1 - S)T - SR)$$
$$b = (1 - S)/(1 - (1 - S)T - SR)$$
$$f = 1/(1 - (1 - S)T - SR)$$
$$d = ((1 - R)S)/(1 - (1 - S)T - SR)$$
$$e = ((1 - T)(1 - S))/(1 - (1 - S)T - SR)$$

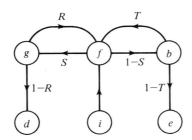

Figure 6.25. A flowgraph model of recoverable rocket casing.

Burroughs [1] has performed extensive analysis on this problem. He has performed extensive sensitivity analysis and determined various conditional probabilities. For example, the probability of a second successful flight, given that you have already had one successful flight, can be found by modifying the flowgraph as shown in Fig. 6.26. The graph is then solved from g' to g'' to yield:

$$P_{g'g''} = \frac{RS}{1 - (1 - S)(T)}$$

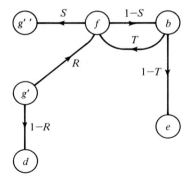

Figure 6.26. Modified flowgraph of the recoverable rocket casing.

6.5.4 Other applications. Three approaches to the solution of waiting-line problems have been considered by Gue [8], Pritsker [21], and Huggins [15]. Howard considered applications to inventory control [13]. Probability models have received much attention in recent literature [15, 16]. Howard [14], Sittler [24], and Huggins [15] have suggested the applicability of flow-graph theory to Markov systems. Carroll [4] applied the technique to information systems analysis.

EXERCISES

1. Evaluate the equivalent transmittance between x and y for each flowgraph in Fig. 6.27.

2. Consider the following engineering problem discussed by Flagle [7]. Define:

$$x_1 = \text{the weight of the rocket}$$
$$x_2 = \text{the propulsion thrust}$$
$$x_3 = \text{the payload}$$

Also, we assume that the state of the art is such that with all other factors held constant:

    (1) One unit increase in total weight requires one unit increase in propulsion thrust.

    (2) One unit increase in thrust requires $\frac{2}{3}$ unit increase in total weight.

    (3) One unit of total weight costs 4 units of money, whereas one unit of thrust costs 10 units of money.

Determine:

(a) The cost of increasing the payload one unit.

(b) The cost of increasing the net thrust one unit.

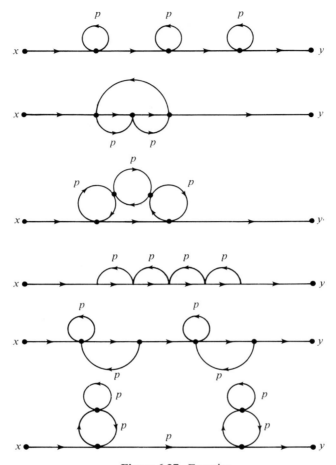

Figure 6.27. Exercise.

(c) The decrease in payload that must be accepted if the net thrust is to be increased by one unit without increasing the cost.

3. For the rocket casing system modeled in Figure 6.25, find:
   (a) Probability of recurrence of node b; of node f.
   (b) Sensitivity of the probability of reaching node d with respect to S; to R; to T.

4. Show that the flowgraph in Fig. 6.14(a) is equivalent to the one in Fig. 6.14(b).

5. Three honest guys and a monkey have a bunch of coconuts, say N. They go to bed and decide to divide these up in the morning. During the night, A gets up, takes one-third of the coconuts and leaves two-thirds to the others. But there is an extra coconut which he gives to the monkey. Next B gets up and

does exactly the same; next *C*. Next morning, the remaining coconuts are split and *A*, *B*, and *C* get exactly *M* coconuts. Draw the flowgraph, and relate *N* with *M* by flowgraphs.

6. Review one of your favorite courses taken in the past. Give three specific examples of how flowgraphs could have been used to your advantage in that course.

7. If first-order loops are denoted by L_1, L_2, L_3, etc., define $M_1 = 1 - L_1$, $M_2 = 1 - L_2$, etc. Express the gain formula in terms of M_i. Explain when *M* formulation is preferable to the *L* formulation.

8. Write the equations associated with the flowgraphs shown in Fig. 6.28.

9. For the system modeled in Fig. 6.21, solve for the excess government income (x_g) necessary for a stable GNP. (*Hint:* Solve for x_g in terms of x_p and GNP.)

10. Expand the corporate model shown in Fig. 6.22 to make it more realistic.

11. The total cost of *X* units can be simply expressed as $TC_x = FC + (VC)(X)$, where:

$$TC_x = \text{total cost for } X \text{ units}$$
$$FC = \text{fixed cost}$$
$$VC = \text{variable cost per unit of } X$$

Consider two possible methods of making the units with different fixed and variable costs.

(a) Show that the break-even point (the number of units of production where it would be profitable to choose the alternative approach to the problem) is equal to $(FC_2 - FC_1)/(VC_1 - VC_2)$. (*Hint:* The break-even point is the value of *X* at which the two methods give the same total cost.)

(b) Given:
$$VC_1 = \$10$$
$$VC_2 = \$5$$
$$FC_1 = \$100$$
$$FC_2 = \$200$$

Show that the break-even point is 20 units and interpret the result.

12. (a) If you deposit \$100 per year in the bank for four years, in a bank paying 4% per year, how much would your money be worth at the end of the four years? (Assume the money is deposited at the end of the year.)

(b) If the bank pays 2% each half year, would you expect more money in the bank than under the first condition?

(c) In part (a), if you left your money in the bank for ten more years without adding to it or withdrawing from it, how much would it be worth?

13. (a) If you borrow \$1000 at 1% per month and agree to repay at the rate of \$50 per month (end-of-month payments), how much do you still owe of the original \$1000 after 4 months?

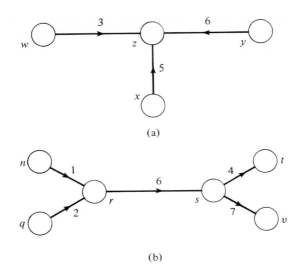

(a)

(b)

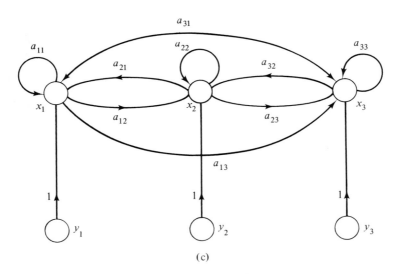

(c)

Figure 6.28. Exercise.

(b) If $1000 is deposited in a three-year annuity paying 10% per year, how much would you expect the three equal annual payments to be?

14. Write the specifications of a Fortran computer program to solve flowgraphs by topological methods.

15. A genetic model: There are two types of genes: G and B. Everyone is born with two genes, one from each parent. The various combinations are GG, GB,

and *BB*. A person has green eyes if and only if the combination *GG* is present. Otherwise the eyes will be brown. Genes are selected at random from the parents and have no effect on the selection of mates. Assume that for a gene selected at random the probability that it will be *G* is *p*.

(a) If your paternal grandfather has a gene pair *BB*, what is the probability that your eyes will be green?

(b) If you observe twice as many brown-eyed people as green-eyed people, what is your estimate of *p*?

16. With emphasis on pollution control, the city has decreed that if a particular pollution control device is not working at eight in the morning, the plant must send all its employees home and close for the day. When the device is turned on there is a probability of 0.5 that the equipment will not work. If the machine survives the first hour of operation there is a 0.8 probability that it will survive each successive hour. The repair crew finds the fault in the first hour 70% of the time, and it takes two hours 30% of the time. Should we try to turn on the equipment at 5 a.m., 6 a.m., 7 a.m., or 8 a.m. if we want to maximize the probability of it being in operation at 8 a.m?

17. A doctor's office has three chairs, one for conference and two for waiting. The doctor divides his time into 15-minute intervals. At the end of each time interval there is a 0.80 probability that the conference will end. During an interval, 50% of the time no patients arrive and 50% of the time one patient arrives. Describe this system using a flowgraph. What is the probability that a patient will be turned away because there are insufficient chairs?

REFERENCES

1. Burroughs, J. L. and W. W. Happ, "Flowgraph Techniques for Reliability Engineering," in Goldberg, M. F. and J. Voccaro, *Physics of Failure in Electronics.* Washington, U. S. Department of Commerce, Office of Technical Services AD434/329 (1964), 375–423.

2. Busaker, R. G. and T. L. Saaty, *Finite Graphs and Networks—An Introduction with Applications.* New York, McGraw-Hill Book Company, 1965.

3. Chan, S. P., *Introductory Topological Analysis of Electrical Networks.* New York, Holt, Rinehart and Winston, Inc., 1969.

4. Carroll, J. M., "A Methodology for Information Systems Analysis," *The Journal of Industrial Engineering.* Vol. 18, No. 11 (November, 1967), 650–657.

5. Compton, J. B. and W. W. Happ, "Flowgraph Models of Integrated Circuits and Film-Type System," *Proceedings of the Aerospace Electro-Technology Conference.* Phoenix, Arizona, 1964.

6. Elmaghraby, S. E., *The Design of Production Systems.* New York, Van Nostrand Reinhold Publishing Corp., 1966.

7. Flagle, G. D., W. H. Huggins, and R. H. Roy, "Operations Research and Systems Engineering." Baltimore, Maryland, John Hopkins University Press, 1960.

8. Gue, R.,"Signal Flowgraphs and Analog Computation in the Analysis of Finite Queues," *Operations Research*, Vol. 14, No. 2 (March–April, 1966).

9. Hall, Arthur D., *A Methodology for Systems Engineering*. Princeton, N.J., Nostrand Company, Inc., 1962.

10. Happ, W. W., *Application of Flowgraph Techniques to Solutions of Reliability Engineering Problems*. Technical Report, Hughes Aircraft Company, Semiconductor Division, Report No. 9403–31/219.

11. Heyel, Carl, *Handbook of Industrial Research Management*. New York, Van Nostrand-Reinhold Publishing Corp., 1963.

12. Holt, C. C., F. Modigliani, J. F. Muth, and H. A. Simon, *Planning, Production Inventories, and Work Force*. Englewood Cliffs, New Jersey, Prentice-Hall, Inc., 1960.

13. Howard, R. A., "Systems Analysis of Linear Models," in H. E. Scarf, D. M. Gilford, and M. W. Shelly, eds., *Multistage Inventory Models and Techniques* (Stanford, Stanford University Press, 1963), Chap. 6.

14. Howard, R. A., *Dynamic Programming and Markov Processes*. New York, John Wiley & Sons, Inc., 1960.

15. Huggins, W. H., "Flowgraph Representation of Systems," in C. D. Flagle, W. H. Huggins, and R. H. Roy, eds., *Operations Research and System Engineering*. Baltimore, Maryland, John Hopkins University Press, 1960.

16. Huggins, W. H., "Signal Flowgraphs and Random Signals," *Proc. IRE*, Vol. 9, No. 9 (1957), 74–86.

17. Lorens, C. S., *Flowgraphs for the Modeling and Analysis of Linear Systems*. New York, McGraw-Hill Book Company, 1964.

18. Lynch, W. A. and J. G. Truxall, *Principles of Electronic Instrumentation*. New York, McGraw-Hill Book Company, 1962.

19. Mason, S. J., "Feedback Theory-Some Properties of Signal Flowgraphs," *Proc. IRE*, Vol. 41, No. 9 (1953).

20. Nisbet, T. R. and W. W. Happ, "Visual Engineering Mathematics—A Self-Contained Course," *Electronics Design* (December 9, 1959, December 23, 1959, January 6, 1960, and January 20, 1960).

21. Pritsker, A. A. B., "Applications of Flowgraphs to Queueing Theory," Unpublished paper at Arizona State University, 1963.

22. Robichaud, L. P. A., M. Boisvert, and J. Robert, *Signal Flow Graphs and Applications*. Englewood Cliffs, New Jersey, Prentice-Hall, Inc., 1962.

23. Shannon, G. E., *The Theory and Design of Linear Differential Equation Machines*. CSRD Report 411, January, 1942.

24. Sittler, R. W., "Systems Analysis of Discrete Markov Processes," *IRE Transactions on Circuit Theory*, Vol. CT-3 (1956), 257–266.

25. Whitehouse, G. E., "Model Systems on Paper with Flowgraph Analysis," *Industrial Engineering* (June, 1969), 30–35.

DECISION TREES*

Another form of network analysis, quite different from those discussed so far, is the decision tree *[2, 3]. The decision tree approach, a technique very similar to dynamic programming, was developed as a convenient method for representing and analyzing a series of investment decisions to be made over time. This approach has not, however, been limited to investment decisions. Decision flow diagrams [5] have been developed and used by Raiffa to model and analyze many complicated decision situations. There appears to be very little difference between the decision tree and the decision flow diagram. In this chapter, we will treat these techniques as one and call it* decision tree analysis.

We will discuss such enrichments of decision tree analysis as the inclusion of sampling technology into the decision tree and the effect of uncertainties on the analysis of decision trees. Raiffa [5] covers these subjects in much more detail than we are able to in this chapter. We therefore strongly urge the interested reader to consult Raiffa for further discussion of these subjects. Stochastic decision trees [1] are also discussed in this chapter.

---

*Portions of this Chapter have been adapted from References 1 and 5 and are used with the permission of TIMS, Dr. Hespos, and Addison-Wesley Publishing Co., Inc.

7

7.1 Decision Tree Analysis

The purpose of this section is to discuss decision trees and demonstrate their use by example. The following example, which is an adaption of one presented by Hespos and Strassman [1], is presented to illustrate the concept of decision tree analysis.

A company has decided to introduce a new product, but there has been no determination of whether to introduce the product regionally or nationally. The decision process is modeled by the decision tree shown in Fig. 7.1. Each rectangle is called a *decision node*. These nodes represent places where a decision maker must make a decision. Each branch leading away from a decision node represents one of several possible alternative choices available to the decision maker. For example, node 1 represents the decision to introduce the product regionally or nationally. If the former is chosen, this means that the path leading to node A is followed. Node A is a circled node and thus represents a *chance node*. The chance node represents a point at which the decision maker will discover the response to his decision. Each branch leading away from a chance node represents the outcome of a set of chance factors. For example, 30% of the time a "small regional demand" will be realized at node A, while 70% of the time a "large regional demand" is experienced. If a large regional demand is experienced, the decision tree

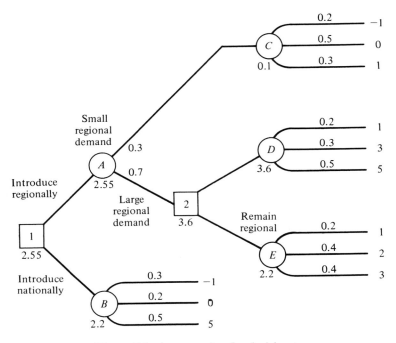

Figure 7.1. An example of a decision tree.

leads us to decision node 2 where another decision must be made: "go national" or "remain regional." If we decide to go national, chance node D is encountered. Node D leads to a set of terminal branches. A terminal branch represents the amount received as a result of a particular sequence of decisions and chance occurrences that lead to the particular branch. Node D will lead to a return of 1, 20% of the time; 3, 30% of the time; and 5, 50% of the time. The remainder of Fig. 7.1 can now be understood. Node B represents the outcome of deciding to go national immediately. Node C represents the results of marketing regionally which yielded a small regional demand, and node E results from a decision to remain regional after a large regional demand, following the initial decision to introduce regionally. The optimal sequence of decisions in a decision tree is found by starting at the right side and "rolling backward." The goal of the rolling backward operation is to maximize the return from the decision situation. At each node, an expected *net present value* (NPV) is calculated, the so-called position value. If the node is a chance node, the position value is calculated as the sum of the products of the probabilities on the branches emanating from the node times their respective position values. If the node is a decision point, the NPV is computed for each of its branches and the highest is selected. This is consistent with the desire to maximize return in the tree. Figure 7.1 shows the position values for our example. For node D the value is found as follows:

$$\text{Expected NPV} = 1(0.2) + 3(0.3) + 5(0.5) = 3.6$$

The value for node 2 is equal to 3.6, which is the maximum of 3.6 and 2.2. Continuing with this analysis, we find that the expected NPV for node 1 is 2.55, which means that the maximum expected return from introducing a new product is 2.55 units. This return will be achieved if we introduce the product regionally and then, if it is successful, go national.

7.1.1 Summary of the decision tree approach. The decision tree approach exemplified in the preceding example can be conveniently summarized in the following manner.

The decision tree approach, a technique very similar to dynamic programming, is a convenient method for representing and analyzing a series of investment decisions to be made over time (Fig. 7.2). Each branch extending from a decision point represents one of the alternatives that can be chosen at this decision point. The chance event represents the various levels of values of a decision parameter. It has a probability associated with each of the branches emanating from it. This probability is the likelihood that the chance event will assume the value assigned to the particular branch.

The optimal sequence of decisions in a decision tree is found by starting

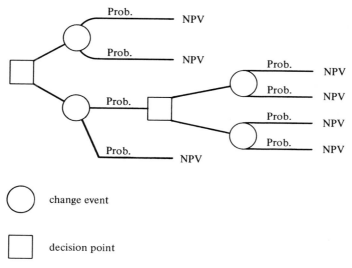

Figure 7.2. A decision tree.

at the right side and rolling backward. At each node, an expected net present value (NPV) must be calculated, the so-called position value. If the node is a chance event node, the expected NPV is calculated as the sum of the products of the probability and the NPV for all paths leaving the node. If the node is a decision point, the expected NPV is computed for each of its branches, and the highest is selected. The procedure continues until the initial node is reached. The position values for this node correspond to the maximum expected return obtainable from the decision sequence.

7.2 Experimental Alternatives

7.2.1 Sample problem statement. A major portion of our discussion in the remainder of this chapter will revolve around the following problem which is adapted from Morris [4].

Suppose a foundry has poured ten lots of castings, one from each of ten heats of iron during a day. Under normal process operating conditions, the product averages 2% defective. After the day's production has been completed, it is discovered that an important additive was omitted from the metal in two of the heats. When this happens, past experience has shown that the castings will average about 20% defective. Due to administrative errors, it is no longer possible to tell which lots of the casting were made of bad metal. Therefore, there are ten lots, two of which average 20% defective and the others which average 2% defective. Suppose it is necessary to ship one lot of castings immediately and that, if a bad lot is shipped, a penalty cost of $1000 is incurred by the company. If a good lot is shipped, no penalty cost arises.

Under these conditions, it is reasonable to say that, if any one of the ten lots is selected at random, the probability that the lot will be bad is 0.20, and the expected penalty cost if it is shipped is $200 = (0.20)($1000).

This problem might be viewed in a different manner. For example, suppose that the decision maker selects a lot at random. The manager can guess that this lot is good and ship it, or he can guess that it is bad and ship any of the other lots. Thus, he has two possible courses of action:

a_1—send the lot selected.
a_2—send any other lot.

Now how about payoffs? The manager's payoff will be determined by the choice that he makes combined with the true state of the lot selected. The true states or possible futures, as they are sometimes called, for this situation are:

S_1—the lot selected is good.
S_2—the lot selected is defective.

The payoffs for this decision are derived as follows:

1. If the manager chooses a_1 and the true state is S_1, then he will incur a cost of $0 because he will send a good lot.
2. If the manager chooses a_1 and the true state is S_2, then he has decided to send a bad lot and will incur a penalty of $1000.
3. If the true state is S_1 and the decision maker chooses a_2, then two of the remaining nine lots will be defective. Since he is selecting one of these lots at random, he will incur the $1000 penalty with probability $\frac{2}{9}$. Therefore, his expected penalty will be $(\$1000)(\frac{2}{9})$.
4. If a_2 is chosen and the true state is S_2, the penalty will be $(\$1000)(\frac{1}{9})$ because, since a bad lot has been selected, the remaining lots will include only one bad lot.

Now the decision maker might like to apply the decision tree concepts discussed in the introduction to decide between a_1 and a_2. The decision tree for this situation is shown in Fig. 7.3. The terminal payoffs for the diagram

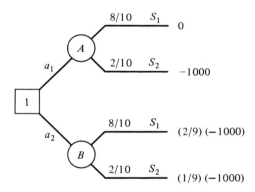

Figure 7.3. Decision tree for the sending of a random lot.

correspond to those just developed. The probabilities of the possible futures are developed from the argument that, since there are two bad lots and since the manager is selecting one lot at random from ten lots, the probability of having selected the bad lot is $\frac{2}{10}$. The probability of selecting a good lot is derived in a similar manner and is equal to $\frac{8}{10}$.

Rolling back the decision tree, we find that the NPV for chance node A is:

$$0(\tfrac{8}{10}) + (-1000)(\tfrac{2}{10}) = -\$200$$

and chance node B has a NPV equal to:

$$(\tfrac{2}{9})(-1000)(\tfrac{8}{10}) + (\tfrac{1}{9})(-1000)(\tfrac{2}{10}) = -\$200.$$

Therefore, the decision tree tells the decision maker that there is no advantage in choosing a_1 over a_2 or a_2 over a_1, and the expected payoff is $-\$200$.

As anticipated, the expected cost of choosing a lot at random is $200. There-fore, if there is no possibility of further information, it makes little difference which lot is shipped.

Now let us assume that it is possible to obtain additional information which might help the manager in selecting between a_1 and a_2. This infor-mation might be available in the form of sampling information and is often called *experimental alternatives*. For the problem at hand, let's assume that the manager has the following experimental alternatives:

e_1—For $10.00, he can sample one unit from the lot selected and determine if the unit is good or bad.
e_2—For $15.00, he can sample two units.
e_3—For $10.00, he can sample one unit from the selected lot and then, after looking at the state of the unit sampled, he can decide whether or not to sample another unit at a cost of another $10.00.

Introducing the idea of experimental alternatives makes this problem much more interesting and, as we will see from the next section, very appro-priate for decision tree analysis.

7.2.2 The development of a decision tree for our sample problem. The manager has four choices in the analysis of this problem:

e_0—no observations.
e_1—one observation.
e_2—two observations.
e_3—sequential observations.

Therefore, we would expect that the initial portion of our tree would appear as shown in Fig. 7.4, where decision node 1 represents a choice of one alter-native from e_0, e_1, e_2, or e_3.

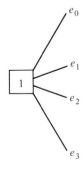

Figure 7.4. First decision point for a value of information decision tree.

Now let us develop in detail the branches of our tree diagram for the e_1 decision as shown in Fig. 7.5. Traveling down the e_1 path, the first item we encounter is a bar (which we will use to designate a fee or toll for choosing a particular experimental alternative) designating a cost of $10.00 to

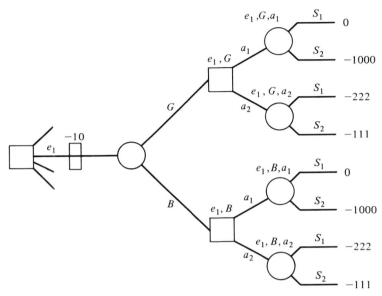

Figure 7.5. The e_1 branch of the value of information decision tree.

sample one unit from the lot. Next we come to a chance node which represents the outcome of the sample taken. The sample might yield a good (G) or a bad (B) unit which is represented by the two paths leaving this node. At this point, the manager is not controlling the outcome of this node—the outcome will be determined by chance. Suppose the part sampled is good; then we follow the path marked G to a node designated by (e_1, G). In complicated trees it has been found advantageous to designate the nodes by the combination of chance occurrences and decisions necessary to reach the node. Thus, node (e_1, G) represents the situation where the manager chooses to sample one unit and finds that it is good. At node (e_1, G) the manager can decide to ship the lot, represented by path a_1, or ship another lot, represented by path a_2. If a_1 is selected, the node reached is (e_1, G, a_1), which is a chance node. This node represents the chance event of sending a good lot (S_1) or a defective lot (S_2). If the route (e_1, G, a_1, S_2) occurs, then besides the toll of $10.00, the manager's terminal payoff would be $-$1000.00. His total cost would be $1010.00. The $1000.00 cost is derived in the same manner as discussed in 7.2.1.

The decision tree for the total problem is presented in Fig. 7.6. We have already discussed the e_0 branch and e_1 branch of Fig. 7.6. The e_2 branch is basically the same as the e_1 branch except that we can have three outcomes from our sample, both good (*GG*), both bad (*BB*), and one good and one bad (*GB*). The e_3 path warrants a little more explanation. After paying a toll of \$10, you encounter your first sampling. Suppose by chance you get

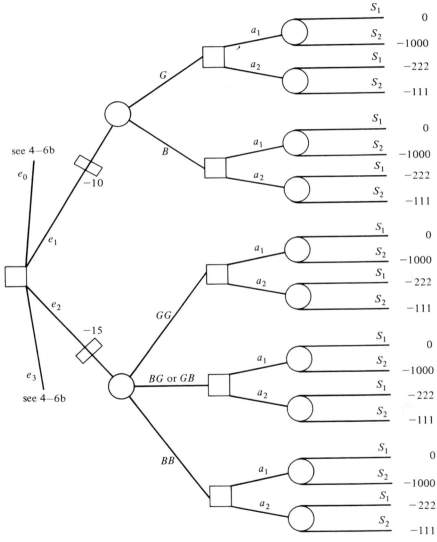

Figure 7.6. Decision tree of a value of information problem.

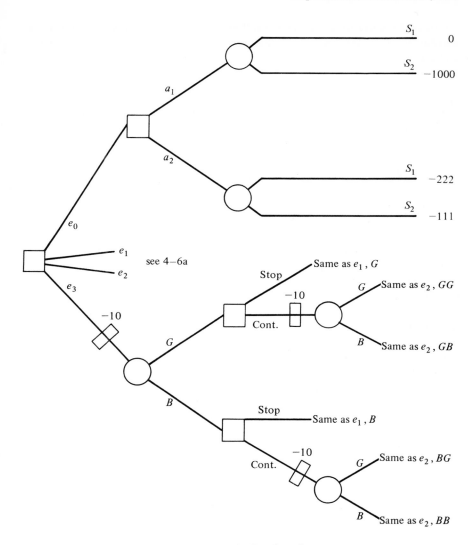

Figure 7.6. Continued.

a good part (take path G); then you will be at the decision node (e_3, G). At this point the manager might decide to stop sampling or continue sampling. If he stops, the tree would continue as it does from node (e_1, G). He will have to pay another toll of $10.00 if he decides to continue sampling. At node (e_3, G, Cont.) it is decided by chance whether the second unit is good or bad. If it was bad, the network would continue as it did from node (e_2, GB). These comments should make this decision tree clear. Before rolling back the tree, we have to determine the probabilities involved in taking the branches emanating from the chance nodes. This will be our concern in the next section.

7.2.3 Adding probability information to the tree. The decision tree in Fig. 7.6 is complete with one exception: no probabilities have been assigned to the branches emanating from the chance nodes. In this section we will discuss the assignment of probabilities to our decision tree.

First, consider the e_0 branch and the chance node (e_0, a_1). The question to be asked is "What is the probability of selecting S_1 by chance?" Since the manager has selected the lot at random from ten lots, two of which are defective and eight of which are good, then the probability of S_1 (having selected a good lot) is 0.8. Similarly, the probability of S_2 is 0.2. These same probabilities can be assigned at node (e_0, a_2).

It is not as easy to make the probability assignment for the remainder of the tree. As an example, let's consider branch e_1 in detail. Suppose that the manager has selected branch e_1, that his sampling of the one unit from the lot selected is good, and that he has decided on action a_1 (to ship the lot). In other words, we are at chance node (e_1, G, a_1) of the tree. What is the probability of S_1? It is no longer 0.8, because the manager knows that a good part has been drawn from the lot. Since a defective lot contains 80% good parts and a nondefective lot contains 98% good parts, the information that a good part has been sampled should increase the probability that the manager has selected a nondefective lot. The question is how to find how much effect the sampling information will have on the probabilities.

For branch e_1, we will need the following information:

1. The conditional probability of S_1 given that a good unit is sampled, or $P(S_1 \mid G)$.
2. The conditional probability of S_2 given that a good unit is sampled, or $P(S_2 \mid G)$.
3. The conditional probability of S_1 given that a bad unit is sampled, or $P(S_1 \mid B)$.
4. The conditional probability of S_2 given that a bad unit is sampled, or $P(S_2 \mid B)$.
5. The probability that e_1 will result in a good unit sampled, or $P(G)$.
6. The probability that e_1 will result in a bad unit sampled, or $P(B)$.

This information will be added to the e_1 branch of the tree as shown in Fig. 7.7. These probabilities can be calculated using the familiar Bayes theorem.

Bayes theorem will allow us to assess $P(S_1 \mid G)$ from the quantities $P(S_1)$, $P(S_2)$, $P(G \mid S_1)$, and $P(G \mid S_2)$. Its development is dependent upon two well-known formulas from probability theory, i.e., if A and B are any two events,

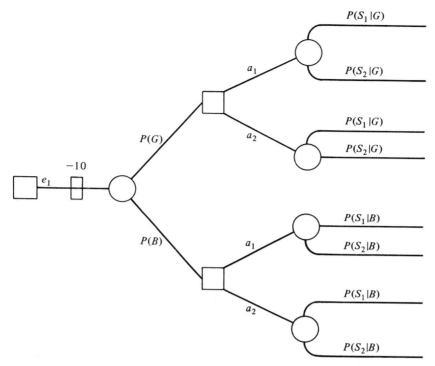

Figure 7.7. Model of the e_1 branch of the value of information decision tree.

then:

$$(1) \qquad P(A \mid B) = \frac{P(A \text{ and } B)}{P(B)}$$

$$(2) \qquad P(A) = P(A \text{ and } B) + P(A \text{ and } (\text{not } B))$$

These relations are shown in Fig. 7.8, and they may now be stated in a more specialized form in terms of S_1, S_2, and G:

$$P(S_1 \mid G) = \frac{P(S_1 \text{ and } G)}{P(G)}$$

$$P(G) = P(S_1 \text{ and } G) + P(S_2 \text{ and } G)$$

$$P(S_1 \text{ and } G) = P(G \mid S_1) P(S_1)$$

$$P(S_2 \text{ and } G) = P(G \mid S_2) P(S_2)$$

From the statement of our problem it is clear that:

$$P(S_1) = 0.8, \qquad P(S_2) = (0.2), \qquad P(G \mid S_1) = 0.98 \qquad P(G \mid S_2) = 0.8$$

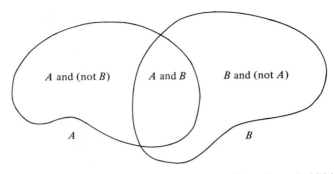

Figure 7.8. Probability space showing conditional probabilities.

Using this information and the formulas just stated, we can find $P(S_1|G)$ as follows:

$$P(S_2 \text{ and } G) = 0.8 \times 0.2 = 0.16$$
$$P(S_1 \text{ and } G) = 0.98 \times 0.8 = 0.784$$
$$P(G) = 0.784 + 0.16 = 0.944$$
$$P(S_1|G) = \frac{0.784}{0.944} = 0.830$$

Thus, the probability that a lot selected at random is nondefective, given that a sample from it yields a good part, is 0.830.

Bayes theorem can be stated in one formula as follows:

$$P(S_1|G) = \frac{P(G|S_1)P(S_1)}{P(G|S_1) \cdot P(S_1) + P(G|S_2) \cdot P(S_2)}$$

Note that all items to the right of the equal sign can be assigned from the data in the problem statement.

Using a more general form of Bayes theorem, we can find the quantities of interest in Fig. 7.7:

$$P(S_j|x) = \frac{P(x|S_j)P(S_j)}{\sum_i P(x|S_i)P(S_i)}$$

$$P(S_2|G) = \frac{P(G|S_2)P(S_2)}{P(G|S_1)P(S_1) + P(G|S_2)P(S_2)} = \frac{(0.8)(0.2)}{(0.98)(0.8) + (0.8)(0.2)} = 0.170$$

$$P(S_1|B) = \frac{P(B|S_1)P(S_1)}{P(B|S_1)P(S_1) + P(B|S_2)P(S_2)} = \frac{(0.02)(0.8)}{(0.2)(0.8) + (0.20)(0.2)} = 0.286$$

$$P(S_2|B) = \frac{P(B|S_2)P(S_2)}{P(B|S_1)P(S_1) + P(B|S_2)P(S_2)} = \frac{(0.20)(0.2)}{(0.02)(0.8) + (0.20)(0.2)} = 0.714$$

$$P(B) = (0.02)(0.8) + (0.20)(0.2) = 0.056$$

The values for the probabilities on all chance branches are calculated in a similar fashion. For example, consider $P(S_1 | BB)$ on path e_2:

$$P(S_1 | BB) = \frac{P(BB | S_1)P(S_1)}{P(BB | S_1)P(S_1) + P(BB | S_2)P(S_2)}$$

$P(BB | s_1)$ can be calculated if we realize that we are sampling from a binomial distribution with a $P(B) = 0.02$; therefore:

$$P(BB | S_1) = \binom{2}{2}(0.98)^0(0.02)^2 = 0.0004$$

Similarly:

$$P(BB | S_2) = \binom{2}{2}(0.80)^0(0.20)^2 = 0.04$$

and

$$P(S_1 | BB) = \frac{0.0004(0.8)}{0.0004(0.8) + 0.04(0.2)} = 0.038$$

The probabilities are shown in Figs. 7.9 and 7.10. Two alternative methods for calculating these probabilities are demonstrated by Raiffa [5]. One involves a tabular calculation and the other the "flipping" of probability trees.

7.2.4 Rolling backward in the tree. The decision tree is now complete. The manager must now use the technique to help him make meaningful deci-

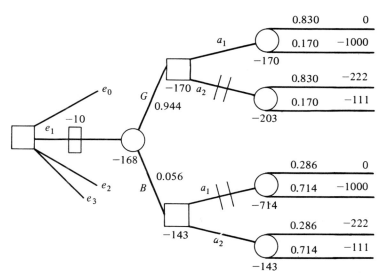

Figure 7.9. Roll-back of the e_1 branch of the value of information decision tree.

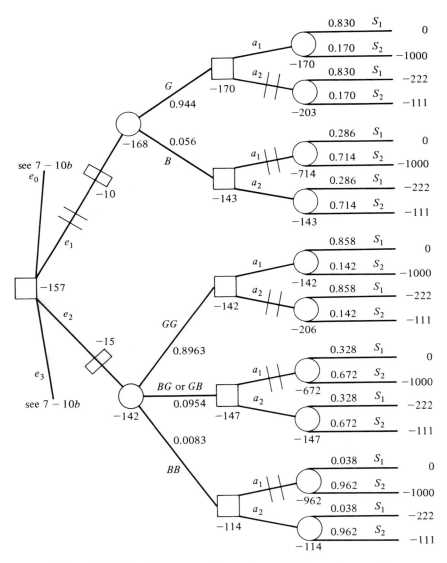

Figure 7.10. Decision tree of a value of information problem including roll-back.

sions at those points where he can exercise his influence. For example, he must decide on e_0, e_1, e_2, or e_3.

The analysis of the problem involves a sequence of calculations which

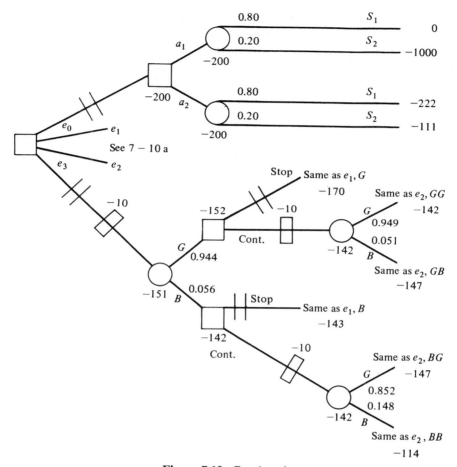

Figure 7.10. Continued.

are summarized in Fig. 7.10. To make sure that this figure is fully understood, let's consider path e_1, which is shown in Fig. 7.9.

The NPV are found by following the rules for analyzing decision trees stated in Section 7.1.

$$\text{NPV}\ (e_1,\ G,\ a_1) = 0(0.830) + (-1000)(0.17) = -170$$
$$\text{NPV}\ (e_1,\ G,\ a_2) = (-222)(0.830) + (-111)(0.17) = -203$$
$$\text{NPV}\ (e_1,\ B,\ a_1) = 0(0.286) + (-1000)(0.714) = -714$$
$$\text{NPV}\ (e_1,\ B,\ a_2) = (-222)(0.286) + (-111)(0.714) = -143$$
$$\text{NPV}\ (e_1,\ G) = \max\ (-170,\ -203) = -170$$

$$\text{NPV } (e_1, B) = \max{(-714, -143)} = -143$$
$$\text{NPV } (e_1) = (-170)(0.944) + (-143)(0.056) = -168$$

Note that those branches not taken at a decision node are marked by double vertical slashes. The analysis of our e_1 branch has shown us that the expected profit incurred when one unit is sampled is $\$-178 = \text{toll} + \text{NPV } (e_1)$ $= -10 - 168$. You should now be in a position to understand the decision tree in Fig. 7.10. From Fig. 7.11, we see that after we pay the initial tolls,

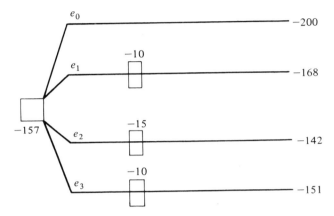

Figure 7.11. Roll-back on the first decision point of the value of information decision tree.

e_2 is better than e_3 which is better than e_1. All the sampling alternatives investigated are better than e_0. The expected value of the game is $-157 = -142 - 15$, or on the average, the company can expect a penalty of $157 caused by the mistake on the shipping dock. To achieve this minimum expected cost, we draw a sample of two from the lot selected; if both units are good, we ship that lot. Otherwise, we ship any other lot.

One additional question might be asked about our problem: "How much can be saved by sampling?" Put another way, "How much is perfect information worth?" In our problem we selected a lot at random and tried to determine if it was a defective lot or not. If we think it is defective, we select another and send it. If we think it is nondefective, we send the lot selected. After selecting a lot, if we are 100% sure that it is good, we will send it and the expected cost will be zero. If we are sure that it is defective, we select another lot and send it. One-ninth of the time the new lot selected will be defective; thus the expected cost will be $\$1000(\frac{1}{9}) = \$111$. Since by chance we will select a good lot 80% of the time and a defective lot 20% of the time, the minimum cost achievable is $0.80(0) + 0.20(111) = \$22.20$. Therefore, perfect information would be worth \$177.80 over just selecting a lot at random.

7.3 Uncertain Payoffs and Measurements

The problem analyzed in Section 7.2 was a "nice" problem in many respects. There was never any doubt about such things as the terminal pay- offs or the cost of sampling. There were no errors in reporting the outcomes of our experiments. In this section, we will briefly discuss how one might handle such things as (1) unknown payoffs, (2) unknown sampling costs, and (3) inaccurate reporting of outcomes.

7.3.1 Uncertain payoffs and sampling costs. For the system discussed in Section 7.2, let us assume that the penalty will not always be \$1000 if a bad part is sent. Assume that one-half the time the penalty will be \$1500 and one-half the time it will be \$500. Then all the terminal payoffs that depend on this penalty will have to be modified. For example, all terminal payoffs of \$1000 in Fig. 7.10 could be replaced with the modification shown in Fig. 7.12(a). Those now ending in \$111 and \$222 would be modified as shown in Fig. 7.12(b) and 7.12(c). In the process of rolling back the tree, the decision maker who wishes to maximize NPV will evaluate these new chance nodes as 1000, 111, and 222; therefore, the rolling back analysis will result in an identical analysis as before. As we shall see in Section 7.5, not all decision makers wish to maximize NPV, and in this case the tree modifications might change their decisions.

Another relatively easy modification to decision trees results if there is uncertainty with respect to the sampling cost. For example, instead of a \$10.00 toll for a single sample from the chosen lot, suppose that the toll was \$8.00, 80% of the time, and \$20, 20% of the time. As we proceed along the e_1 path, we would introduce a chance event, as shown in Fig. 7.13, where the ($-\$168$) is read from the e_1 path in Fig. 7.10(a). The NPV at point A can then be found easily

$$\$-168 - (0.80)(\$-8) + (0.20)(-\$20) = -\$178.40$$

In some problems, the cost of sampling depends on the outcome of the sample. Returning to the problem in Section 7.2, suppose that for e_2 the cost of sampling is not \$15.00, but \$15.00 plus \$3 for each bad part found. One possible procedure for handling this problem is to modify the e_2 path of Fig. 7.10(a) as shown in Fig. 7.14. Another approach is to eliminate all

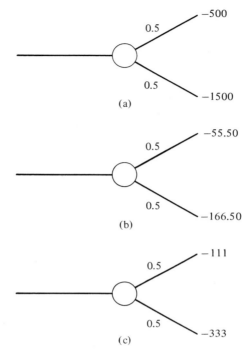

Figure 7.12. Example of uncertain payoffs included in a decision tree.

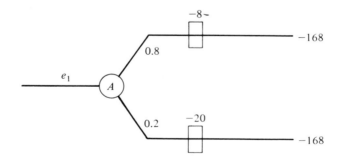

Figure 7.13. Uncertain sampling costs included in a decision tree.

tolls and incorporate them with the terminal payoffs at the tips of the tree. In our example, $15.00 would be subtracted from the tips of the (e_2, GG) branches, $18 from the tips of the (e_2, GB) branches, and $21 from the tips of the (e_2, BB) branches. The rolling backward technique would proceed as shown in the partial decision tree in Fig. 7.15.

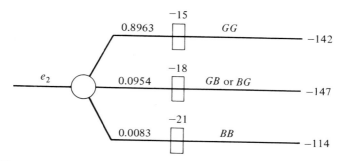

Figure 7.14. An example of sampling costs which change as a function of the number of defects found.

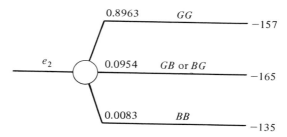

Figure 7.15. Roll-back of the decision branch shown in Fig. 7.14.

7.3.2 Biased measurements. Often you cannot depend totally upon the information you receive (maybe there is noise in your communications channel). In our sample problem, suppose the instrument that we are using to determine if the sample is good or bad is subject to error. Let's assume that we have some statistical data about the instrument's accuracy. The instrument recognizes a good part correctly 90% of the time and bad parts correctly 80% of the time. This information can be tabulated as shown in Table 7.1.

Table 7.1. CONDITIONAL PROBABILITY OF THE MEASUREMENT ACCURACY.

Actual Condition	Instrument Reading	
	Good	Bad
Good	0.90	0.10
Bad	0.20	0.80

We are now interested in seeing how to include this measurement error into the decision tree. Including it in the tree will help answer such questions as:

(1) Should we sample at all? (2) If so, should we adopt strategy e_1, e_2, or e_3? (3) Assuming another instrument is available at a higher cost but is more accurate, would you prefer the cheaper or more expensive instrumentation? We are not interested in the answers to these questions (they are left as student exercises) but in a methodology for attacking such problems. Let us consider Fig. 7.16 which shows a partial representation of the e_1 branch

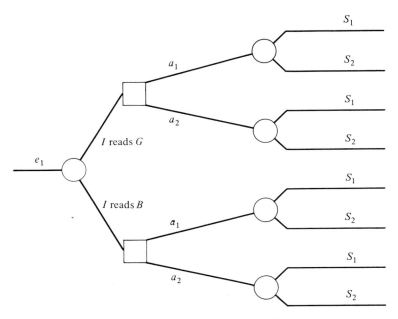

Figure 7.16. Decision tree including biased measurements.

of the tree. "*I* reads *G*" is interpreted as meaning that the testing instrument tells (correctly or incorrectly) that the part is good. We must now be able to find the following probabilities:

$$P\,[I \text{ reads } G] \text{ at } e_1$$
$$P\,[I \text{ reads } B] \text{ at } e_1$$
$$P\,(S_1 \,|\, I \text{ reads } G) \text{ at } (e_1, I \text{ reads } G, a_1)$$
$$P\,(S_2 \,|\, I \text{ reads } G) \text{ at } (e_1, I \text{ reads } G, a_1)$$
$$\text{etc.}$$

But these are relatively easy probability computations which can be made from the data already available.

$$P(I \text{ reads } G) = P(S_1)P(I \text{ reads } G \,|\, S_1) + P(S_2)P(I \text{ reads } G \,|\, S_2)$$
$$= (0.8)((0.98)(0.9) + (0.02)(0.2))$$
$$\quad + (0.2)((0.80)(0.9) + (0.2)(0.2))$$
$$= 0.861$$
$$P(I \text{ reads } B) = P(S_1)P(I \text{ reads } B \,|\, S_1) + P(S_2)P(I \text{ reads } B \,|\, S_2)$$
$$= 0.139$$

From Bayes theorem:

$$P(S_1 \,|\, I \text{ reads } G) = \frac{P(I \text{ reads } G \,|\, S_1)P(S_1)}{P(I \text{ reads } G \,|\, S_1)P(S_1) + P(I \text{ reads } G \,|\, S_2)P(S_2)} = 0.824$$

$$P(S_2 \,|\, I \text{ reads } G) = \frac{P(I \text{ reads } G \,|\, S_2)\,P(S_2)}{P(I \text{ reads } G \,|\, S_1)\,P(S_1) + P(I \text{ reads } G \,|\, S_2)\,P(S_2)} = 0.176$$

The probabilities for the e_1 path are shown in Fig. 7.17. The total tree

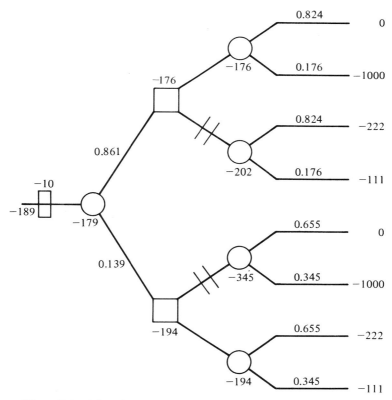

Figure 7.17. Biased measurements included of the e_1 branch of the value of information decision tree.

can be built, and the question mentioned earlier in this section can be answered. For example, rolling backward on the e_1 path, we can see from Fig. 7.17 that the expected cost of sampling one unit with the imperfect testing device is $189. This compares to a cost of $168 with the perfect testing equipment. The cost advantage for the perfect equipment is thus $21.00 for this path.

7.4 Stochastic Decision Trees

One drawback of the decision tree approach is that computations can quickly become unwieldy. The number of end points on the decision tree increases very rapidly as the number of decision points or chance events increases. To make this approach practical, it is necessary to limit the number of branches emanating from chance events to a very small number. This means that the probability distribution of chance events at each node must be represented by a very few point estimates.

A second drawback which sometimes causes the answers obtained from a decision tree analysis to be inadequate is the following. The single answer obtained, i.e., the net present value for each branch of a decision point, is usually close to the expectation of the probability distribution of all possible NPVs. However, it may vary somewhat from the expected NPV, depending on how the point estimates were selected from the underlying distributions and on the sensitivity of the NPV to this selection process. In addition, the decision tree approach gives no information on the range outcomes from the investment or the probability combined with those outcomes. This can be a serious drawback.

As an example, consider the tree in Fig. 7.18. If we compare only the expected NPVs, the outcome of the decision would be to take action A. But with information about the range of the possible outcomes, a decision maker might choose an alternative other than the one which maximizes the NPV. Figure 7.19 shows the distribution of possible NPVs from each decision sequence for the tree in Fig. 7.18.

To overcome these objections to decision tree analysis, Hespos and Strassman [1] introduced the concept of *stochastic decision trees*, which combines concepts of decision trees discussed in this chapter with those of risk analysis.

7.4.1 Risk analysis. *Risk analysis* is a technique for economic evaluation. It provides the analyst with a convenient way to supply management with reliable information about the uncertainties in alternate investment opportunities, corporate strategies, or operational procedures.

The analysis of risks involved in an investment requires an estimation

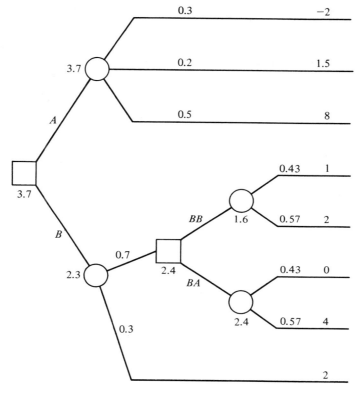

Figure 7.18. A sample decision tree evaluated using expected values.

of the uncertainty associated with each factor affecting an investment decision. Traditional evaluation of a project using only a single "best guess"
for each uncertain factor may produce an incomplete and possibly misleading analysis. With risk analysis, personnel most familiar with the factors
affecting a decision can estimate a range of uncertainty.

The following example, a simplified version of an investment analysis,
illustrates the superiority of the risk analysis approach over the best-guess
analysis:

Factor	Best-Guess Approach	Risk Analysis Approach
Investment	$10,000,000	Optimistic est. $5,000,000
		Pessimistic est. $20,000,000
		Most likely est. $10,000,000
Net annual earning	$2,000,000	Optimistic—$2,500,000
		Pessimistic—$500,000
		Most likely—$2,000,000

Factor	Best-Guess Approach	Risk Analysis Approach
Economic life of new investment before obsolescence	10 years	Optimistic—12 years Pessimistic—6 years Most likely—10 years
Terminal value	$1,000,000	Any value in the range $500,000 to $1,500,000 is as likely as another value in the same range
Project evaluation	Net present value = $2,500,000	Expected net present value = $500,000 Chances of incurring a loss = 25% Chances of incurring a loss of more than $5,000,000 = 5%

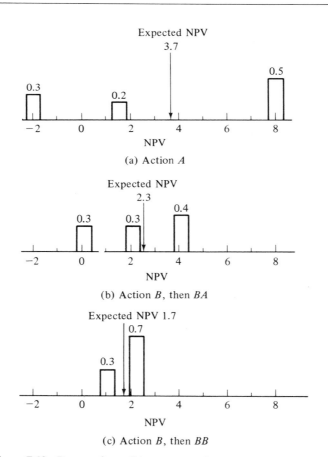

Figure 7.19. Range of possible outcomes for the decision tree shown in Fig. 7.18.

Figure 7.20 shows the risk analysis curve for the problem just discussed. This curve shows the chances of the investment discussed realizing a net present value less than the value on the horizontal axis.

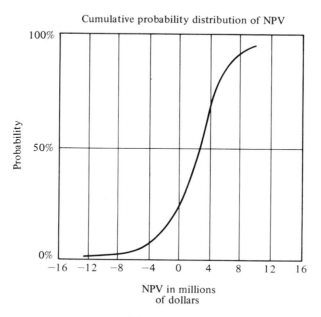

Cumulative probability distribution of NPV

Figure 7.20.

The preceding example shows how much more information is made available for making decisions in the risk analysis approach. Furthermore, the NPV figure of $2,500,000 in the best-guess approach is quite misleading on the favorable side. Cumulative risk factors have been analyzed to show there is a 25% chance that the investment will result in a loss.

The risk analysis approach is basically an application of Monte Carlo simulation. The approach will be demonstrated in the next section when a procedure for the analysis of stochastic decision trees is discussed.

The following are the main areas where risk analysis has been found to be effective:

1. *Investment evaluation.* The analysis of alternate investment opportunities is possibly the most important single risk analysis application. It can, for example, display rate of return versus probability that each level of return will occur in a potential venture where volume, price, investment, or other factors are uncertain.

2. *Evaluation of research and development projects.* Risk analysis can calculate the cumulative profit associated with a proposed research project over a

number of years versus the probability that such a profit will be realized. Uncertainties may include the final outcome of the research, the cost and time to complete it, and the resulting cash flows if the research is eventually a commercial success.

3. *Engineering estimates.* The probable costs of new facilities can be developed using risk analysis. Uncertain estimates of cost items and price uncertainty in labor and materials can be estimated and used to produce a probability distribution of construction costs for use in the evaluation of a new facility.

7.4.2 The stochastic decision tree approach. The first step in performing a stochastic decision tree analysis is to replace all the branches emanating from a chance node, which immediately precede the terminal payoffs, with one branch having a probability distribution of payoffs associated with it. Consider the first example that was presented in this chapter (Fig. 7.1). The output of chance node D could be presented by a histogram which had a 20% chance of yielding a return of 1, a 30% chance of yielding a return of 3, and a 50% chance of yielding a return of 5. Figure 7.1 can thus be modified as shown in Fig. 7.21—consistent with this rule. The output of a chance

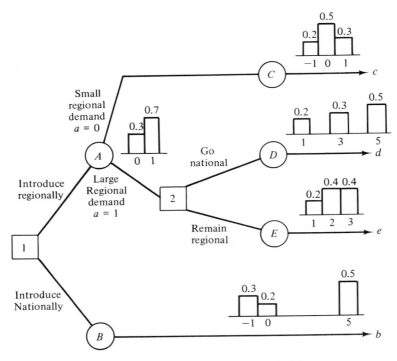

Figure 7.21. A example of a stochastic decision tree.

occurrence at node *B* is represented by observation *b*, shown to the right of the branch. This approach helps to alleviate one of the objections to decision trees in that we have drastically reduced the number of paths through the tree. The terminal payoffs could even be described as continuous functions, which would be impossible with the standard decision tree approach. The continuous function would be analogous to adding an arbitrarily large number of branches at a chance node. This is demonstrated in Fig. 7.22.

The histograms at nodes *B*, *C*, and *E* have similar interpretations rela-

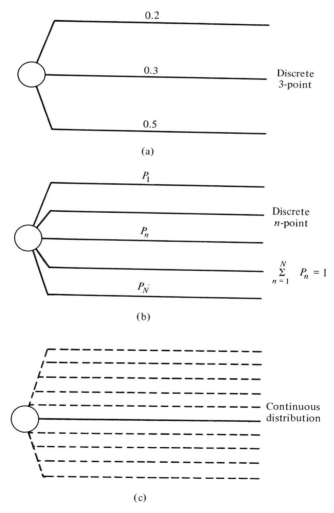

Figure 7.22. Examples of chance node outputs.

tive to the amounts c, d, and e that will be obtained when these nodes are encountered. The histogram at A, however, represents the probability distribution which governs the realization of large and small regional demand. The value 0 shown on the horizontal axis for this histogram is associated with small regional demand. The value 1 is associated with large regional demand. Thus, as already observed, the probability is 0.3 that a zero will occur and 0.7 that a one will occur when node A is encountered.

Now consider each of the following three possible strategies for our example:

1. Introduce nationally.
2. Introduce regionally, and if a large regional demand materializes, then go national.
3. Introduce regionally, and if a large regional demand materializes, then remain regional.

The probability distributions of rewards and penalties for each of these strategies is wanted. In the stochastic decision tree approach—as is also true in risk analysis—this is obtained via a series of simulation runs. In the case of stochastic decision trees, the rules for executing each such simulation are as follows:

1. Each time a decision node. is encountered, take *all* branches leading out from any such node.
2. Each time a chance node is encountered, take only the *one* branch (or value) designed by a chance (or random) drawing which is associated with the histogram at this node.

The point of these rules is to make it possible to obtain probability distributions for *all* relevant combinations of decisions. Evidently, for decision 1, one can hardly do better than simply reproduce the probability distribution for B in Fig. 7.21. This is not the case for the other two decision possibilities, however, since their outcomes are influenced by the event which occurs at A on each trial. Thus a problem arises of combining the probability distribution at A with the distributions at C, D, and E in order to obtain a basis for deciding between the available alternatives.

For the example we are dealing with, observation may be drawn from the distributions at A, B, C, D, and E using one-digit random numbers. For example, the random numbers 0, 1, and 2 might represent a small regional demand, while the numbers 3, 4, 5, 6, 7, 8, and 9 represent large regional demands. Thus, since each random number is equally likely by definition, there will be the desired 30% chance of small regional demand and 70% chance of large regional demand. The relationship between random numbers and outcomes is shown in Table 7.2.

The Monte Carlo simulation technique generates random numbers for each chance node and equates them to the outcome shown in Table 7.2.

Table 7.2. RELATIONSHIP OF RANDOM NUMBERS
TO OUTPUTS AT THE CHANCE NODES OF FIG. 7.21.

Random Number	a	b	c	d	e
0	0	−1	−1	1	1
1	0	−1	−1	1	1
2	0	−1	0	3	2
3	1	0	0	3	2
4	1	0	0	3	2
5	1	5	0	5	2
6	1	5	0	5	3
7	1	5	1	5	3
8	1	5	1	5	3
9	1	5	1	5	3

For example, if a random number of 4 is associated with node *D*, this is equivalent to *d* taking the value 3. After each assignment of random numbers, the tree is evaluated to determine what the outcome would have been if each of the available strategies had been adopted. These results are recorded and the procedure is repeated many times. The result is a distribution of the returns achievable for each strategy.

Let us now consider a couple of simulations of the tree. In Table 7.3, two simulations are represented:

Table 7.3. SIMULATION EXAMPLES.

Node	Sample 1		Sample 2	
	R.N.	Result	R.N.	Result
A	6	1	0	0
B	4	0	9	5
C	3	0	6	0
D	6	5	3	3
E	0	1	4	2

For sample 1, the second random number drawn is 4. Entering Table 7.2 we would find that a decision to market nationally would yield a return of 0.

Similar remarks apply to nodes *C*, *D*, and *E*. Note, however, that the random number 6 which was drawn for *A* is associated with the occurrence of a large regional demand. Hence the value at *C* is really irrelevant and can be ignored. To put the matter differently, the node at *A* serves as a switch which cuts in (or out) part of the diagram of Fig. 7.21 in accordance with the instruction assigned to the number obtained in any such random drawing. Thus, in particular, the random drawing of a 6 for *A* produces the situation shown in Fig. 7.23, that is, branches leading from node A to node C and beyond in Fig. 7.21 are eliminated.

Thus, there are three outcomes of sample 1:

1. $b = 0$ which is associated with a decision to market nationally.
2. $d = 5$ which is associated with a decision to introduce regionally, and if a large regional demand materializes, then go national.
3. $e = 1$ which is associated with a decision to introduce regionally, and if a large regional demand materializes, to stay regional.

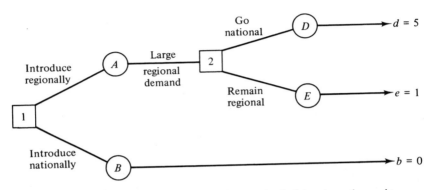

Figure 7.23. A trial result for the stochastic decision tree shown in Fig. 7.21.

In sample 2, the random number for node A implies a small regional demand; so the only meaningful outputs are:

1. $b = 5$ which is associated with a decision to market nationally.
2 and 3. $c = 0$ which is associated with both alternatives involving a trial regional campaign.

After many simulations, it is possible to construct histograms similar to those shown in Fig. 7.24 which are the results of 1000 simulations. The next problem is how to make meaningful decisions based upon results similar to those in Fig. 7.24. If the decision maker wishes to maximize NPV, then this can be done using standard decision trees; but he might be interested in other criteria. Alternate decision criteria will be discussed in the next section.

7.5 Decision Making for Those Not Interested in Maximizing NPV

Much has been written about decision making [4, 5, 6]. Therefore, I do not intend to give a detailed discussion of decision theory but rather to give a flavor of what is involved. Two topics will be addressed. The first

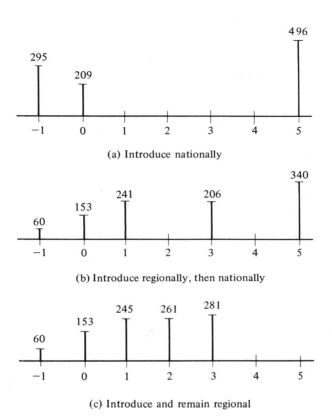

(a) Introduce nationally

(b) Introduce regionally, then nationally

(c) Introduce and remain regional

Figure 7.24. Simulation results of the stochastic decision tree shown in Fig. 7.21.

concerns ways in which the reader can analyze outputs similar to that shown in Fig. 7.24. The second introduces the concept of *utility theory*.

7.5.1 Analysis of the output of stochastic decision trees. In this section we will discuss three common criteria of decision making which could be applied to the output of stochastic decision trees. First, however, we will summarize the results shown in Fig. 7.24 and Table 7.4:

Table 7.4. ESTIMATED PROBABILITIES OF VARIOUS RETURNS FOR THE MARKETING PROBLEM.

Alternatives/Return	−1	0	1	2	3	4	5
1. National	0.295	0.209	0	0	0	0	0.496
2. Regional-National	0.060	0.153	0.241	0	0.206	0	0.340
3. Regional-Regional	0.060	0.153	0.245	0.261	0.281	0	0

These results will be used to illustrate the methods of decision making.

MAXIMIZE EXPECTED RETURN

In this criterion, the decision maker chooses the alternative which maximizes expected return. This would describe the decision maker that wants to maximize NPV in the decision tree. The use of this criterion should yield the same results as the standard decision tree approach did in Section 7.1.

Alternative 1:

$$\text{Expected return} = -1(0.295) + 0(0.209) + 5(0.496) = 2.185$$

Alternative 2:

$$\begin{aligned}\text{Expected return} &= -1(0.060) + 0(0.153) + 1(0.241) + 3(0.206) \\ &\quad + 5(0.340) \\ &= 2.499\end{aligned}$$

Alternative 3:

$$\begin{aligned}\text{Expected return} &= -1(0.060) + 0(0.153) + 1(0.247) + 2(0.261) \\ &\quad + 3(0.281) \\ &= 1.552\end{aligned}$$

Thus, the maximum expected return criterion tells the decision maker to choose alternative 2.

THE MOST-PROBABLE-FUTURE PRINCIPLE

Another popular criterion, often called *assumed certainty*, overlooks all but the most likely return (the one with the highest probability of occurrence) and chooses the alternative which has the maximum *most probable future*.

The most probable futures for our example are:

Alternative	M.P.F.	Probability
1	5	0.496
2	5	0.340
3	3	0.281

Thus, the most-probable-future decision maker views alternatives 1 and 2 as equally attractive. If he wanted to break the tie, he probably would choose alternative 1 because of its higher probability of occurrence.

THE ASPIRATION-LEVEL PRINCIPLE

Often a manager will base his decision on thoughts such as, "I don't care which alternative, as long as it doesn't lose money," or "Let's choose the alternative which has the best chance of having a respectable return."

A manager of this type is exercising a criterion called the *aspiration level principle*. He wants to choose the alternative which maximizes the probability of obtaining some level return, say, A. For example;

If $A = 4$

> Prob (Alt. $1 \geq 4$) = 0.496
>
> Prob (Alt. $2 \geq 4$) = 0.340
>
> Prob (Alt. $3 \geq 4$) =   0
>
> Choose alternative 1

If $A = 2$

> Prob (Alt. $1 \geq 2$) = 0.496
>
> Prob (Alt. $2 \geq 2$) = 0.546
>
> Prob (Alt. $3 \geq 2$) = 0.542
>
> Choose alternative 2

The obvious question to ask at this point is, "Which criteria should be used?" Unfortunately, there is no easy answer to this question. These techniques are merely attempts to model individual decision-maker's behavior. They can be viewed as attempts at approximating a decision maker's utility curve. The concept of utility curves will now be investigated.

7.5.2 Utility theory. Which would you prefer: a tax-free gift of $10,000 or a 5% chance of getting $250,000 and 95% chance of getting nothing? Most people would take the first alternative even though its expected payoff is $10,000 compared to $12,500 for the second alternative. This apparent paradox is explained by the *utility theory*. Basically, it states that every dollar is not equally valuable to an individual. For example, one dollar is worth much less to an individual who has lots of money than it is to a poor man. The utility theory approach attempts to determine a utility curve for a decision maker. This curve converts dollars to an arbitrary utility measure. The choice between alternatives is then made based upon maximizing expected utilities instead of maximizing expected return.

Some typical forms of utility curves are shown in Fig. 7.25. Figure 7.25(a) shows a typical utility curve where the slope decreases as money values increase. Figure 7.25(b) shows the utility curve for a decision maker

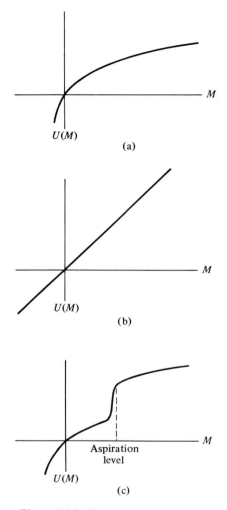

Figure 7.25. Examples of utility curves.

who says that every dollar amount is equally valuable. In this case, expected utility would give the same result as expected return analysis. This may be a realistic form of the curve for large companies and the government. This is why large companies practice self-insurance and individuals do not. An individual with an aspiration level would have a curve similar to that in Fig. 7.25(c). The utility curve takes a sharp jump at the aspiration level.

Let's assume that the company's utility values for the dollar amounts in the marketing example are as shown in Table 7.5:

Table 7.5. UTILITY VALUES FOR THE
MARKETING EXAMPLE.

Dollars	−1	0	1	2	3	4	5
Utility values		0	4	7	10	12	14	15

To choose between alternatives, the expected utility should be calculated and the alternative with maximum expected utility selected.

Alternative 1:

   Expected utility $= 0(0.295) + 4(0.209) + 15(0.496) = 8.276$

Alternative 2:

   Expected utility $= 0(0.060) + 4(0.153) + 7(0.241) + 12(0.206)$
   $+ 15(0.340)$
   $= 9.871$

Alternative 3:

   Expected utility $= 0(0.060) + 4(0.153) + 7(0.245) + 10(0.261)$
   $+ 12(0.281)$
   $= 8.310$

Therefore, alternative 2 would be chosen. This analysis is not meaningful unless the decision maker's utility curve is known. Consider the following approach to the construction of a utility curve:

1. Formulate an alternative (a_1) which promises two rewards with equal probability (say $0, $10,000).
2. Assign arbitrary utility values to the two dollar values (maybe $U($0) = 0$, $U($10,000) = 1$).

$$E(U(a_1)) = \tfrac{1}{2}U(0) + \tfrac{1}{2}U(10,000)$$
$$= \tfrac{1}{2}(0) + \tfrac{1}{2}(1) = 0.5$$

3. Try to find an alternative (a_2) to which the decision maker is indifferent when compared to a_1 (suppose a guarantee of $3000 would be equivalent to a_1).
$$U($3000) = U(a_1) = 0.5$$

4. Find additional points by constructing alternatives composed of points with known utilities; then find alternatives to these for which the decision maker is indifferent.

$$(U(\$5500) = \tfrac{1}{2}U(\$3000) + \tfrac{1}{2}U(\$10,000)$$
$$= \tfrac{1}{2}(0.5) + \tfrac{1}{2}(1) = 0.75$$

5. What if we need utilities for dollar values outside the range considered in step 1? We might ask:

$$U(5500) = \tfrac{1}{2}U(0) + \tfrac{1}{2}U(?)$$

Maybe

$$U(5500) = \tfrac{1}{2}U(0) + \tfrac{1}{2}U(100,000)$$

and

$$U(100,000) = 1.50$$

Or we might ask

$$U(3000) = pU(-1000) + (1 - p)U(10,000)$$

Maybe

$$U(3000) = \tfrac{1}{3}U(-1000) + \tfrac{2}{3}U(10,000)$$
$$\therefore \; U(-1000) = -0.5$$

6. Construct the utility curve, e.g.,

Money	Utility Index
$100,000	1.50
10,000	1.00
5,500	0.75
3,000	0.50
900	0.25
0	0.00
−1,000	−0.50
−2,000	−1.50

7. Base the decisions on expected utility. For example, should the decision maker bet $1000 to get a 5% chance of winning $100,000?

$$U(\text{bet}) = 0.95(U(-1000)) + 0.05(U(100,000))$$
$$= 0.95(-0.50) + 0.05(1.50) = -0.40$$
$$U(\text{don't bet}) = 0$$

DON'T BET

Raiffa [5] and Schlaifer [6] give detailed discussions on the relationship of utility theory to decision trees.

7.6 Applications of Decision Trees

Decision trees can be applied to many sequential decision situations. In this section, we will briefly view some applications suggested by Raiffa [5] and Magee [2, 3].

Raiffa concentrated on the use of decision trees for analyzing the value of additional information in suggesting the following areas for consideration.

An Oil Drilling Problem

An oil company must decide whether or not to drill at a given location before its option expires. There are many uncertainties: cost of drilling, the value of the deposit, etc. There are records available for similar and not-so-similar drillings in the area. They could obtain more information about the geophysical structure at this site by conducting a sounding. Soundings are very expensive. *The problem*—Should the company collect more information before making their final decision to drill or not to drill?

Introduction of a New Product

A chemical firm must decide whether or not to market a new, long-lasting house paint. The decision must be whether or not to manufacture the product themselves and if they do, what size plant to build or whether they should sell or lease their patents and technical know-how to a firm that deals exclusively in house paints. The uncertainty lies in the proportion of the market they will get at a given price and advertising expenditure if they manufacture the product themselves, and the time before a competitor introduces a similar product. Expensive market surveys could be run, but they might be misleading. *The problem*—How should the company proceed?

Treatment of Illness

A doctor does not know whether his patient's sore throat is caused by streptococci or by a virus. If he knew it was a strep throat, he would prescribe penicillin, whereas if it was caused by a virus, he would prescribe rest, gargle, and aspirin. Failure to treat strep throat might result in serious disease; however, penicillin cannot be used indiscriminately since it may cause a penicillin reaction.

The physician can take a throat culture, which will indicate the presence of streptococci. The test is not 100% accurate. *The problem*—What should the doctor do? (a) Take no culture, treat the infection as viral; (b) Take no culture, prescribe penicillin injection; (c) Take no culture, prescribe penicillin pills for ten days; (d) Take a culture, prescribe penicillin if positive; (e) Take a culture and prescribe pills, and then continue pills if positive and stop if negative.

Magee [2, 3] discusses the following two decisions in detail.

PLANT MODERNIZATION

A company management is faced with a decision on a proposal by its engineering staff which wants to install a computer-based system in the company's major plant. The cost of the control system is $30 million. The benefits of this program depend upon the level of product throughput during the next decade. The project, according to the engineers, will yield 20% return on investment. There are many questions that had to be considered in the construction of the decision tree for this example. Will the process work? Will it achieve the economies expected? Will your competitors follow your lead if you are successful? Will new products or processes make the new facility obsolete? Will the controls last ten years? The alternatives considered were (1) to install the new control system, (2) postpone action until trends in the market and/or competition become clearer, or (3) initiate more investigation or an independent evaluation. Each alternative was followed by various other decisions depending upon the outcomes of the other alternatives, e.g., in the case of the decision to install, the consequent decisions to continue or abandon operation.

NEW FACILITY

The choice of alternatives in building a new plant depends upon market forecasts, and the alternative chosen will affect the market outcome. A company has won a contract to produce a new type of military engine suitable for army transport vehicles. The company has a contract to build for its productive capacity and to produce at a specified level over a period of three years. The company is not sure whether the contract will be continued at a relatively high rate after the third year. There is a possibility that a large commercial market might develop for the engine. This possibility is largely dependent upon the cost of the item. Three alternative approaches are evaluated with the tree: (1) It might subcontract all fabrication and set up a simple assembly line, (2) It might undertake the major part of the fabrication but use general purpose machine tools, or (3) the company could build a highly mechanized plant with specialized fabrication and assembly equipment. Either of the first two alternatives would be better adapted to low-volume production than would the third. The uncertainties considered were the cost-volume relationships under the alternative manufacturing methods, the size and structure of the market, and the possibilities of competitive developments which render the product competitively or technologically obsolete.

Magee also suggests the use of decision trees in the following situations: (1) A chemical company faced the need to expand capacity to meet near term demand, (2) A study at the U.S. Mint regarding plant sizes and multiple-shift operation, (3) A proposed field order—and sales—recording system being considered by the management of a company, and (4) A company that needs to evaluate the ten-year market study for a new labor-saving device.

EXERCISES

1. Model the physician's problem described in Section 7.6. What information does the doctor need before making his decision?

2. Model and analyze the oil company problem from Section 7.6. The following information is available. Drilling which costs $1,000,000 might lead to payloads which can be classed as

Big	$3,000,000
Moderate	1,000,000
Small	250,000

Before soundings, it is guessed that the probability of a big payoff is 20%, a moderate payoff is 50%, and a small payoff is 30%. The sounding costs $150,000 and yields the following:

	Actual Result		
Sounding Result	Big	Moderate	Small
Big	0.60	0.30	0.10
Moderate	0.20	0.70	0.10
Small	0.10	0.30	0.60

The probabilities of the various sounding results are big 0.25, moderate 0.50, and small 0.25.

3. Interpret Exercise 2 as a stochastic decision tree. Determine by simulation the distribution of payoffs for each decision alternative.

4. Develop decision trees for three different decision situations. Develop the appropriate probabilities and payoffs. Analyze your trees.

5. Two batches of a certain product are available for shipment. The product can be tested only by destructive means. Suppose that it is known in advance that one lot is approximately 10% defective and the other 5% defective. Suppose

that you are allowed to sample up to two parts. The cost of testing a part is $100. You receive an order for ten items under the following agreement:

No. of Defectives	Revenue
0	$10,000
1	8,000
2	6,000
3 or more	2,000

Use standard decision-tree analysis for this problem.

6. Use stochastic decision tree analysis on the problem described in Exercise 5.

7. In Exercise 5, assume that the test is not 100% accurate. The test will say that a good part is bad 10% of the time and that a bad part is good 25% of the time. What strategy would you adopt in this case?

8. In Exercise 7, how much would it be worth to have a test device that will say that a good part is bad 5% of the time and that a bad part is good 0% of the time?

9. Consider the following marketing situation: The company has two alternatives:

a_1—market the product
a_2—do not market the product

Assume that there are three possible futures:

	Profit	A Priori Probability
S_1	8,000,000	0.60
S_2	3,000,000	0.20
S_3	−2,000,000	0.20

If a test marketing is made, there are two results—z_1, high sales potential, and z_2, low sales potential. The cost of the test marketing is $500,000.
The conditional probabilities of the test results are:

$$P(z_1 | S_1) = 0.75; \quad P(z_1 | S_2) = 0.50; \quad P(z_1 | S_3) = 0.20$$
$$P(z_2 | S_1) = 0.25; \quad P(z_2 | S_2) = 0.50; \quad P(z_2 | S_3) = 0.80$$

(a) Develop a decision tree for this problem.
(b) Analyze it to maximize NPV.
(c) Analyze this problem as a stochastic decision tree problem.

(d) Evaluate the results in (c) based upon the following utility curve:

Dollars	Utility
10,000,000	10.0
8,000,000	9.7
6,000,000	9.2
4,000,000	8.5
2,000,000	7.5
1,000,000	6.5
0	5.0
−500,000	4.0
−1,000,000	3.0
−1,500,000	2.0
−2,000,000	1.0
−3,000,000	0

10. Using the procedure outlined in Section 7.5.2, construct a utility curve for someone. Try to predict how he would behave in the following decision situation: The person can pay $100 to get one play in the following game or $300 to get two plays. The game yields a return of $0, 50% of the time, and $400, 50% of the time.

11. Develop a procedure for introducing the time value of money concept into decision tree analysis.

12. Complete the analysis started in Section 7.3.2. Should the decision maker choose alternative e_0, e_1, e_2, or e_3?

13. A company makes a two-stage decision regarding the installation of pollution control equipment. The decision situation is shown in Fig. 7.26. Assume the company's rate of return is 10%. What is the best initial decision?

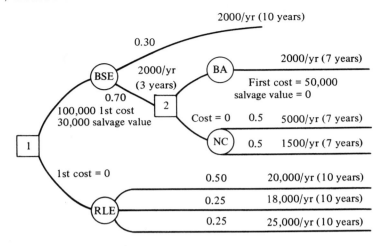

Where

BSE ⟶ Buy small equipment
RLE ⟶ Rent large equipment
BA ⟶ Buy additional equipment
NC ⟶ No change
Interest = 10%/yr

Figure 7.26. Pollution control equipment decision tree.

14. A city must decide between buying a Dodge riot car or a smaller Jeep riot car. The Dodge costs $30,000 and will last six years and be worthless after six years. The Jeep costs $10,000 and lasts three years with no salvage value. After three years we estimate that a Jeep will cost either $12,000 or $14,000 with equal probability. The annual costs are $1500 for a Dodge and $1000 for a Jeep. Assuming that the city's rate of return is 10%, what should the city do to cover its riot needs?

15. Use a decision tree to describe a company's alternatives regarding an applied research project. Include such decisions as (a) start applied research, (b) start development, and (c) start commercialization.

16. One of the decisions we all will be making over the years to come will be the purchase of automobiles. Develop a personal decision tree which will help you in your decision-making over the next ten years or so? Is utility theory appropriate for your decision?

REFERENCES

1. Hespos, R. F. and P. A. Strassman, "Stochastic Decision Trees for the Analysis of Investment Decisions." *Management Science*, Vol. 11, No. 10 (August, 1965).

2. Magee, J. F., "Decision Trees for Decision Making." *Harvard Business Review* (July–August, 1964).

3. Magee, J. F., "How to Use Decision Trees in Capital Investment." *Harvard Business Review* (September–October, 1964).

4. Morris, W. T., *The Analysis of Management Decisions.* Homewood, Illinois, Richard D. Irwin, Inc., 1964.

5. Raiffa, H., *Decision Analysis—Introductory Lectures on Choices under Uncertainty.* Reading, Mass., Addison-Wesley Publishing Co., Inc., 1968.

6. Schlaifer, R., *Analysis of Decisions under Uncertainty.* New York, McGraw-Hill Book Co., 1969.

GERT: AN ANALYTIC APPROACH TO
THE SOLUTION OF STOCHASTIC NETWORKS

Recently much interest has developed in the area of stochastic networks [3, 4, 5, 14]. The properties of a stochastic network are:

1. *Each network consists of nodes denoting logical operations and transmittances.*

2. *A transmittance has associated with it a probability that the activity represented by the network will be performed.*

3. *Other parameters describe the activities which the transmittances represent. These parameters are usually additive, such as time or cost.*

4. *A realization of a network is a particular set of transmittances and nodes which describes the network for one experiment.*

5. *If the time associated with a transmittance is a random variable, then a realization also implies that a fixed time has been selected for each transmittance.*

Eisner [3] suggested the use of logical elements in the PERT-type networks, and Elmaghraby [4, 5] developed a notation for a multiparameter branch network and the logical elements previously presented. Elmaghraby also developed an algebra and coined the phrase generalized activity

8*

networks *to describe such networks. Elmaghraby's algebra is limited to branches that have constant times associated with them.*

Huggins [9] and Howard [7, 8] have employed flowgraphs to represent and analyze probabilistic systems.

Pritsker, Happ, and Whitehouse [14, 15, 18] introduced a new graphical technique which they called GERT (graphical evaluation and review technique). GERT is a procedure which combines the disciplines of flowgraph theory, moment generating functions, and PERT to obtain a solution to stochastic problems. The authors claim that their procedure makes it possible to analyze complex systems and problems in a less inductive manner than ever before. In the remainder of this text we will explore GERT in detail. In this chapter we consider the analytic background of GERT. In the following chapters a simulation approach to the solution of GERT will be considered, and in Chapters 10 and 11 we will explore applications of GERT in detail.

*The material in this chapter has been adapted from References 10, 11, 14, 15, 16, and 18 and is used with the permission of Drs. Pritsker and Ishmael and the American Institute of Industrial Engineers.

8.1 Elements of the GERT Network

Stochastic networks are characterized by logical nodes, probabilistic activity realization, and additive stochastic parameters on the transmittances. In this section we view these characteristics in detail.

8.1.1 Logical nodes. A node in a stochastic network consists of an input (receiving, contributive) side and an output (emitting, distributive) side. In this chapter we will consider three logical relations on the input side and two types of relations on the output side. The three logical relations on the input side are:

Name	Symbol	Characteristic
EXCLUSIVE-OR	◁|	The realization of any branch leading into the node causes the node to be realized; however, one and only one of the branches leading into this node can be realized at a given time.
INCLUSIVE-OR	◁	The realization of any branch leading into the node causes the node to be realized. The time of realization is the smallest of the completion times of the activities leading into the INCLUSIVE-OR node.
AND	(	The node will be realized only if all the branches leading into the node are realized. The time of realization is therefore the largest of the completion times of the activities leading into the AND node.

On the output side, the two relations are defined as:

Name	Symbol	Characteristic
DETERMINISTIC	D	All branches emanating from the node are taken if the node is realized, i.e., all branches emanating from this node have a p parameter equal to 1.
PROBABILISTIC	▷	Exactly one branch emanating from the node is taken if the node is realized.

For notational convenience, the input and output symbols are combined below to show that there are six possible types of nodes:

Before proceeding to the mathematical analysis of these different node types, an example of the use of stochastic networks in modeling is given to assist in understanding stochastic networks. This example was originally suggested by Pritsker [16]. The example is for illustrative purposes only, and no analysis will be performed on the derived networks.

Consider a space mission involving the rendezvous of two vehicles. In order for the mission to have a chance for success, both vehicles must be successfully launched. The stochastic network for this problem is shown in Fig. 8.1.

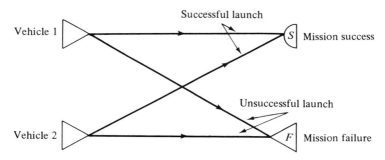

Figure 8.1. Stochastic network model of the rendezvous of two vehicles.

For the node S to be realized, both branches leading into it must be realized (a characteristic of the AND node). Node F will be realized if either branch incident to it is realized (INCLUSIVE-OR node). Obviously, this model is simple, but it does illustrate the modeling and communication aspects of stochastic networks. To extend the model somewhat, assume that if both vehicles are successfully launched, at least one of the vehicles must be capable of maneuvering for the mission to be a success. The network for this situation is shown in Fig. 8.2. In this case nodes 1 and 2 are added to specify the event that both vehicles are successfully launched. The S node now will be realized if either branch incident to it is realized since the assumption was that a maneuverability capability was necessary for only one vehicle to obtain mission success.

The above networks represent highly aggregated models of complex operations. One of the beauties of a stochastic network is its usefulness at many levels within a problem area. For example, the branch "successful launch" can be divided into many branches and nodes. Figure 8.3 illustrates this concept. In this network the AND node plays a predominant role in the activities up to and including the terminal countdown. This is due to the fact that all activities must be performed prior to lift-off. This, of course, is a simplified view of the system; however, it serves the purpose of illustrating

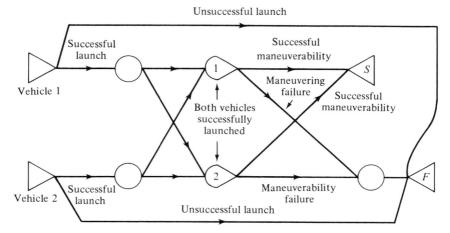

Figure 8.2. Figure 8.1 modified to include maneuvering success of the rocket.

that part of a stochastic network can be a PERT-type network. After the terminal countdown, either-or possibilities are presented and the probabilistic output node is shown. The event represented by the node labeled "successful orbit" is an EXCLUSIVE-OR node since a successful orbit can occur in two mutually exclusive ways: (1) proper operation during boost phase and (2) unsuccessful orbit after boost phase with orbit correction achieved. The BROKEN lines represent activities that do not contribute to the successful launch but are branches associated with the system modeled. In this case they would lead to the node "unsuccessful launch," which is an INCLUSIVE-OR node, because any of the branches leading into the node can be realized and any of them causes the node to be realized.

Continuing the example, consider the branch "terminal countdown," a segment of which can be represented as shown in Fig. 8.4. The network shows three preparatory actions such as power-on, stimuli calibrated, and recorder-on, which are required before the test can begin. The test is performed and, based on the results of the test, the countdown is continued, diagnosis is initiated, or the test is performed over. This last action illustrates the concept of feedback in a stochastic network.

Obviously, the above are not complete descriptions; but they illustrate the communication capabilities of GERT. Also, by decomposing the problem into segments, we can compute the parameters of interest for an aggregate model. Thus the probability of a successful launch could be computed by evaluation of the more detailed networks.

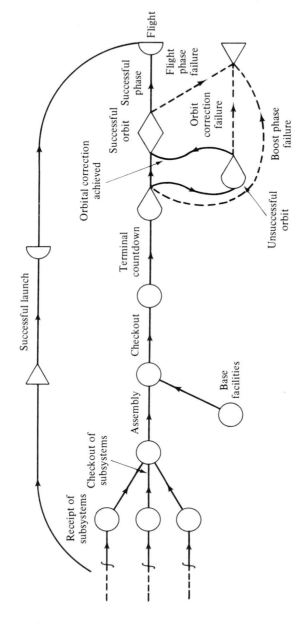

Figure 8.3. Detailed model of the successful launch branch of Fig. 8.2.

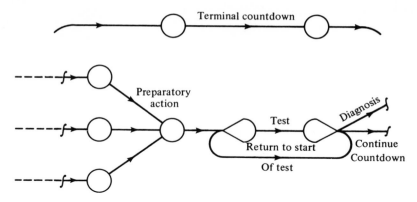

Figure 8.4. Detail model of the terminal countdown branch of Fig. 8.3.

8.1.2 Transmittance parameters. To illustrate the transmittance para-meters of interest in a GERT model, consider the following example sug-gested by Elmaghraby [4]:

> Consider a shop which produces electromechanical equipment characterized by the following sequence. There is a series of operations that terminates at an inspection station. Inspected units are dispatched to one of two areas: a further testing area or an adjustment operation. Units in the test area are either accepted and sent to the adjustment operation or rejected and sent to the repair area. From the repair area, material flows back to be tested. After the adjustment, the units are packed and delivered to stock.

The network for this project can now be drawn as shown in Fig. 8.5. Table 8.1 shows the meaning of the activities along with their pertinent parameters. Thus we can see that the transmittances on this graph will be

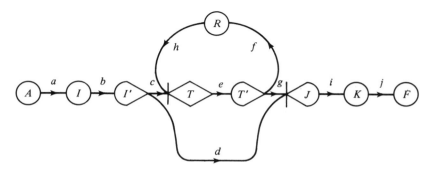

Figure 8.5. GERT model of a production shop.

Table 8.1. DESCRIPTION OF ACTIVITIES FOR FIG. 8.5.

Activity	Meaning	Parameters
a	Manufacturing sequence	Production time
b	Inspection	Inspection time
c	Movement from inspection to test	Transit time; probability item fails inspection
d	Movement from inspection to adjustment	Transit time; probability item passes inspection
e	Test	Test time
f	Movement from test to repair	Transit time; probability item fails the test
g	Movement from test to adjustment	Transit time; probability an item passes the test
h	Repair	Repair time
i	Adjustment	Adjustment time
j	Packing	Packing time

multidimensional: (1) the probability of taking a given path and (2) the distribution of the time to traverse a path. The solution of this problem will be deferred until the GERT approach has been developed. It should be noted that the time parameter is additive, while the probability factor is multiplicative.

8.1.3 Relationship of GERT, PERT, and flowgraphs. We can now state the relationship between PERT-type networks and flowgraphs and stochastic networks:

1. PERT-type networks are stochastic (GERT-type) networks with all AND-DETERMINISTIC nodes.
2. Flowgraphs are stochastic networks with a single multiplicative parameter (all additive parameters such as time are set to zero). The probabilistic interpretation for the multiplicative parameter is removed.

8.2 Evaluation of EXCLUSIVE-OR GERT Networks

The analytic developments in GERT concentrate primarily on networks involving the EXCLUSIVE-OR type of nodes. The development in this section concerns itself with the EXCLUSIVE-OR nodes. Discussion of the evaluation of other nodes will be deferred until later sections.

8.2.1 Use of the topological equation as a means of solution. Each transmittance of the GERT network has two parameters associated with it:

(1) the probability that the path will be taken, given that the node from which it originates is realized, (p_{ij}) and (2) a function of the time required to complete the activity which the transmittance represents $(f_{ij}(t))$. The transmittance in the GERT network is always a directed path as shown in Fig. 8.6. The

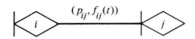

Figure 8.6. A directed branch.

conditional probability to take a given path will be assumed to be constant throughout the development of GERT. t will be allowed to be a random variable. The sum of all the conditional probabilities of the branches emanating from a node must equal one. This implies that there is no absorption in any of the nodes of the network. Anything that enters a node must leave along one of the branches which leave the node.

In the theory of flowgraphs, three basic types of combinations of elements were discussed: (1) elements in series, (2) elements in parallel, and (3) loops. It is possible to reduce any flowgraph by stepwise reducing it from these basic elements to a single-element graph. For large graphs this is not the approach taken, but the more powerful topological equation is employed. For a topological equation to be employed in a graph, the three basic elements in that graph must behave as they do for the flowgraph.

Now consider the desired results for equivalent probability (p_e) and expected time (t_e) for the three basic elements of the EXCLUSIVE-OR GERT network in which all times are assumed to be constants. For the series system shown in Fig. 8.7(a), $p_{1,3} = p_a p_b$, and $t_{1,3} = t_a + t_b$. $p_{1,2} = p_a p_b$ and $t_{1,2} = (p_a t_a + p_b t_b)/(p_a + p_b)$ are the results for the parallel system shown in Fig. 8.7(b). For the loop system shown in Fig. 8.7(c) $p_{1,2} = p_a /(1 - p_b)$ and $t_{1,2} = t_a + ((p_b/(1 - p_b))t_b)$.

We noted earlier that the probability elements of the GERT network were multiplicative, while the time variable was additive in nature. The flowgraph system is a multiplicative system; so it would appear that the probability elements could be handled by the normal flowgraph approach but that trouble could be anticipated with the time elements. There are a number of transforms that will turn an additive operation into a multiplicative operation. The one chosen by the originators of GERT was the moment generating function. There is a well-known theorem in statistics which states that the moment generating function of the sum of *independent* variables is equal to the product of the moment generating function of the individual variables. Pritsker and Happ [14] had other reasons for choosing the MGF of the time as their transmittance for the graph. This allows them to incorporate random

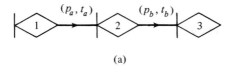

(a)

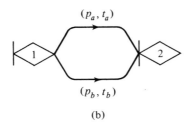

(b)

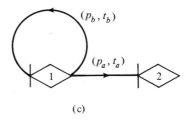

(c)

Figure 8.7. Basic graph configurations for the "EXCLUSIVE-OR" node.

variables for the time. The distribution of the time can be specified by the form of the MGF chosen. Most frequency distributions can be represented by the MGF transform which is equal to $E(e^{st})$. The characteristic function might be considered as a substitute for the MGF if a frequency function is encountered that cannot be represented by the MGF function. No change would be noted in the logic suggested by Pritsker and Happ. The transmittance $(W_{ij}(s))$ suggested was the product of the conditional probability (p_{ij}) and the MGF of the time to traverse the activity $(M_{ij}(s))$. The basic element of the GERT network is illustrated in Fig. 8.8. $W_{ij}(s)$ is referred to as the W function.

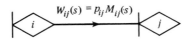

Figure 8.8. GERT element.

REDUCTION OF THE BASIC CONFIGURATION

Pritsker and Happ [14] showed that their proposed transmittance could be treated as a flowgraph transmittance and hence can be treated with the usual technology of flowgraphs. Below we show that flowgraph theory applied to the basic series, parallel, and loop systems yields the desired results.

Basic series system. For the series system shown in Fig. 8.9(a), standard flowgraph operations would reduce it to the graph shown in Fig. 8.9(b), which expresses the relationship:

$$W_{1,3}(s) = p_a p_b M_a(s) M_b(s)$$

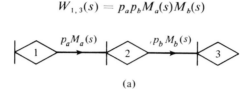

(a)

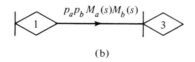

(b)

Figure 8.9. Basic series GERT system.

The probability of realizing two successive independent elements is the intersection of their individual probabilities. Thus;

$$P_a \cap P_b = P_a P_b$$

The time to traverse the network would equal the sum of the two times, and the MGF of the sum of the times will be equal to the product of their MGF. Thus:

$$M_{1,3}(s) = p_{1,3} M_{1,3}(s) = p_a p_b M_a(s) M_b(s)$$

which is the result obtained by standard flowgraph methods.

Basic parallel system. For the parallel system shown in Fig. 8.10(a), standard flowgraph operations would reduce it to the graph shown in Fig. 8.10(b) which expresses the relationship

$$W_{1,2}(s) = p_a M_a(s) + p_b M_b(s)$$

The probability of realizing two parallel elements is equal to the union of two

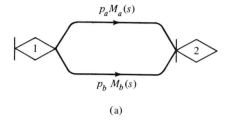

(a)

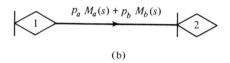

(b)

Figure 8.10. Basic parallel GERT system.

individual probabilities. Thus:

$$P_{1,2} = P_a \cup P_b = P_a + P_b - P_a \cap P_b$$

But the two paths cannot both be realized for a given realization of the parallel elements, so $P_a \cap P_b = 0$. The time to traverse the network is derived as follows. The a path will be taken with probability p_a, and b will be taken with probability p_b. When a is taken, the MGF will be $M_a(s)$, and likewise, for the b path the MGF will be $M_b(s)$. Since only one of the paths can be taken at any one given passage through the network, the result is $p_a M_a(s) + p_b M_b(s)$. But, by definition, the time portion of the GERT network assumed that the path 1–2 is taken. Therefore, the $p_a M_a(s)$ must be modified by the probability of taking 1–2 path which is $p_a + p_b$. Consequently:

$$M_{1,2}(s) = \frac{p_a M_a(s) + p_b M_b(s)}{p_a + p_b}$$

When the equivalent probability is multiplied by the equivalent MGF, $p_a M_a(s) + p_b M_b(s)$ is obtained. This is consistent with the results obtained by applying flowgraph techniques.

Basic loop system. For the loop system shown in Fig. 8.11(a), standard flowgraph operations would reduce it to the graph shown in Fig. 8.11(b), which expresses the relationship:

$$W_{1,2}(s) = \frac{p_a M_a(s)}{(1 - p_b M_b(s))}$$

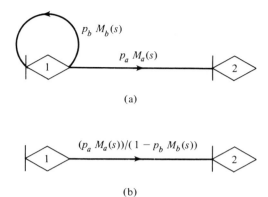

Figure 8.11. Basic loop GERT system.

Enumerating the possible paths through this system shown in Fig. 8.11(a), we have:

Path A	Path B	P	M(s)	pM(s)
1	0	p_a	$M_a(s)$	$p_a M_a(s)$
1	1	$p_a p_b$	$M_a(s)M_b(s)$	$p_a p_b M_a(a)M_b(s)$
1	2	$p_a p_b^2$	$M_a(s)M_b(s)^2$	$p_a p_b^2 M_a(s)M_b(s)^2$
.	.	.		
.	.			
.	.			
1	i	$p_a p_b^i$	$M_a(s)M_b(s)^i$	$p_a p_b^i M_a(s)M_b(s)^i$
.	.			
.	.			
.	.			

Summing the last column, we obtain $p_a M_a(s)/(1 - p_b M_b(s))$. If we modify this by the probability of going between 1 and 2, $p_a/(1 - p_b)$, the conditional MGF, given that the path from 1 to 2 is realized, will be obtained. This conditional MGF multiplied by the probability of going from 1 to 2 yields the result obtained by flowgraph methods. The basic loop system is also consistent with the transmittance suggested by Pritsker and Happ.

Since the basic series, parallel, and loop systems behave according to the flowgraph laws, it would appear that the GERT system involving EXCLUSIVE-OR nodes reduces to a flowgraph system. All the flowgraph theorems will apply to this specialized GERT network.

The output, $W_e(s)$, of the GERT system represents the product of the probability of realizing the portion of the graph being investigated and the MGF of the time to traverse this portion, given that it has been traversed. Since all MGFs become equal to one, when s is equated to 0, $p_e = W_e(s)|_{s=0}$.

The MGF will be the quotient of the output of the graph ($W_e(s)$) divided by p_e, i.e., $M_e(s) = W_e(s)/p_e$. The nth derivative of $M_e(s)$, evaluated at s equal to 0, will yield the expected value of the nth power of t.

Now consider the solution of the system suggested by Elmaghraby [4] which was shown in Fig. 8.5. The quantities in Table 8.2 define numerically the characteristics of the system to be analyzed. Applying these transmittances to the network, we obtain Fig. 8.12. And when we apply Mason's rule for solving flowgraphs, we get $W_e(s)$.

Table 8.2. LIST OF TRANSMITTANCE PARAMETERS FOR FIG. 8.5.

Activity	P	M(s)	W(s)
a	1	e^{25s}	e^{25s}
b	1	e^{3s}	e^{3s}
c	0.7	e^{3s}	$0.7e^{3s}$
d	0.3	e^{3s}	$0.3e^{3s}$
e	1	e^{2s}	e^{2s}
f	0.3	e^{s}	$0.3e^{s}$
g	0.7	e^{2s+3s^2}	$0.7e^{2s+3s^2}$
h	1	e^{4s}	e^{4s}
i	1	e^{s}	e^{s}
j	1	e^{s}	e^{s}

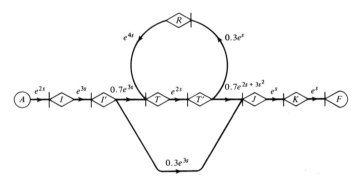

Figure 8.12. GERT representation of Elmaghraby's production line example.

$$W_e(s) = \frac{0.49e^{37s+3s^2} + 0.3e^{33s}(1 - 0.3e^{7s})}{(1 - 0.3e^{7s})}$$

$$p_e = W_e(s)|_{s=0} = \frac{(0.49 - 0.3(0.7))}{0.7} = 1$$

$$M_e(s) = \frac{W_e(s)}{W_e(s)|_{s=0}} = W_e(s)$$

$$E(t) = \left.\frac{dM_e(s)}{ds}\right|_{s=0} = 37.9$$

Higher moments can also be obtained from $M_e(s)$. We should note that the element g was assumed to be a random variable having normal distribution for illustrative purposes.

8.2.2 Steps in the solution of problems by means of GERT. The foregoing material described the principles and procedures of GERT with EXCLUSIVE-OR nodes. A review of the steps employed in applying GERT are listed by Pritsker and Happ as follows [14]:

1. Convert a qualitative description of a system or problem to a model in network form.
2. Collect necessary data to describe the transmittances of the network.
3. Apply the topology equation to determine the equivalent function or functions of the network.
4. Convert the equivalent function into the following two performance measures of the network.
   4.1 The probability that a specific node is realized.
   4.2 The moment generating function of the time associated with a node if it is realized.
5. Make inferences concerning the system under study from the information obtained in 4.

A number of examples of the applications of GERT suggested by Pritsker and Whitehouse [15] will now be given to illustrate the potentials of this technique.

8.2.3 Application of GERT to probabilistic problems.

In this section GERT will be applied to:

1. The development of moment generating functions of probability laws.
2. The solution of complex probability problems.

8.2.3.1 Development of the moment generating function.

DEVELOPMENT OF THE MGF FOR THE NEGATIVE BINOMIAL
PROBABILITY LAW

The problem is the determination of the number of failures encountered before the rth success in a sequence of independent Bernoulli trials. The GERT network is shown in Fig. 8.13 for this problem. Note that since only failures are counted, the number of trials for a success is set equal to zero. Applying Mason's rule, we have:

$$W_e(s) = \frac{p^r}{1 - \sum_{j=1}^{r} \binom{r}{j}(-1)^j (qe^s)j}$$

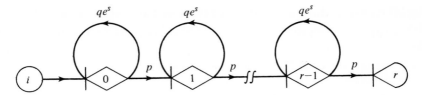

Figure 8.13. GERT representation of the development of the MGF of the negative binomial distribution.

Then we employ the binomial expansion:

$$W_e(s) = \left(\frac{p}{1 - qe^s}\right)^r = M_e(s)$$

since $W_e(s) = 1$.

An alternate approach for solving this stochastic network is to reduce the network in segments. Consider the basic elements as shown in Fig. 8.14.

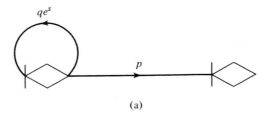

(a)

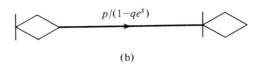

(b)

Figure 8.14. Alternative GERT representation of the development of the negative binomial MGF.

From Fig. 8.14 it is seen that there are r of these equivalent elements in series, and

$$W_e(s) = M_e(s) = \left(\frac{p}{1 - qe^s}\right)^r$$

This example demonstrates the importance, in some cases, of employing the reduction procedure first for part of the network and then for the entire network.

An MGF for a Modified Negative Binomial Probability Law

As an extension of the previous example, consider the distribution of the number of trials required before the rth success. In this case, the network is a series of r subnetworks of the form shown in Fig. 8.15, and:

$$W_e(s) = M_e(s) = \left(\frac{pe^s}{1 - qe^s}\right)^r$$

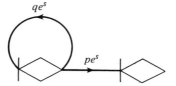

Figure 8.15. GERT representation of the development of the MGF for a modified negative binomial distribution.

The Binomial Probability Law

We next consider the distribution of the number of successes in n independent Bernoulli trials. The network is shown in Fig. 8.16 and it contains no feedback loops. After trial n there have been $0, 1, 2, \ldots, n$ successes

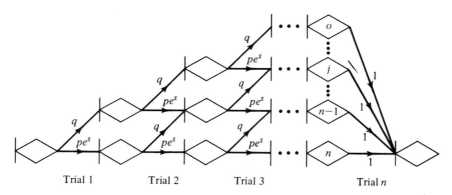

Figure 8.16. GERT model of the development of the MGF of the binomial distribution.

represented by the $(n + 1)$ nodes preceding the terminal node. Since these outcomes are mutually exclusive, the $(n + 1)$ nodes can be connected to a single output node. This permits the distribution of the number of successes in n trials to be obtained. The topological equation for this network is:

$$W_e(s) = \sum_{j=0}^{n} \binom{n}{j}(pe^s)^j q^{n-j} = 0$$

and, as expected:

$$W_e(s) = M_e(s) = (pe^s + q)^n$$

8.2.3.2 Solution of probability problems. The application of GERT to selected probabilistic problems will be discussed below, including the drawing of the network and the derivation of the equivalent network equations.

DICE THROWING

Consider the problem of determining the number of throws of a pair of dice required to obtain three consecutive sevens if the probability of obtaining a seven is p and the probability of not obtaining a seven is q. The network for this problem is shown in Fig. 8.17. For this network:

$$W_e(s) = \frac{(pe^s)^3}{1 - qe^s - pq(e^s)^2 - p^2q(e^s)^3}$$

$$= \frac{(pe^s)^3(1 - pe^s)}{1 - e^s + (pe^s)^3(1 - pe^s)}$$

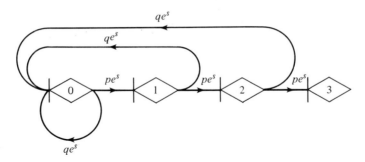

Figure 8.17. GERT model of the die tossing example.

Since $W_e(0) = 1$, $M_e(s) = W_e(s)$ and the MGF for this specific problem is obtained. Extension to the general case of n consecutive values of seven (or any other possible number) is straightforward, with the result that:

$$W_e(s) = M_e(s) = \frac{(pe^s)^n(1 - pe^s)}{1 - e^s + (pe^s)^n(1 - pe^s)}$$

For example, the MGF of the number of throws to obtain ten sevens, where the probability of obtaining a seven is $\frac{1}{6}$, is:

$$M_e(s) = \frac{(\frac{1}{6}e^s)^{10}(1 - \frac{1}{6}e^s)}{1 - e^s + (\frac{1}{6}e^s)^{10}(1 - \frac{1}{6}e^s)}$$

THE THIEF OF BAGDAD PROBLEM

The following problem has been abstracted from Parzen [13]. The thief of Bagdad has been placed in a dungeon (node D) with three doors. One door leads to freedom (node F), one door leads to a long tunnel, and a third door leads to a short tunnel. The tunnels return the thief to the dungeon. If the thief returns to the dungeon, he attempts to gain freedom again. But his past experiences do not help him in selecting the door which leads to freedom, i.e., the probabilities associated with the thief's selection of doors remain constant. The network for this problem is shown in Fig. 8.18, and:

$$W_e(s) = M_e(s) = \frac{(p_f M_f(s))}{(1 - p_L M_L(s) - p_s M_s(s))}$$

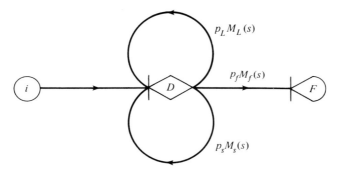

Figure 8.18. GERT representation of the "Thief of Bagdad" problem.

From this equation for $M_e(s)$, the moments with regard to the time it takes for the thief to reach freedom are completely characterized. For the example given in Parzen, $p_f = p_s = p_L = \frac{1}{3}$ and $M_f(s) = e^0 = 1$, $M_L(s) = e^{3s}$, and $M_s(s) = e^s$. The expected exit time is:

$$\mu_1 = \frac{dM_e(s)}{ds}\bigg|_{s=0} = t_f + \frac{1}{p_f}(p_L t_L + p_s t_s) = 4 \text{ time units}$$

This problem would not be changed conceptually if random variables were considered for the time on each branch. The algebraic manipulations may, however, be increased. Higher moments of the time to traverse the network could be calculated in the standard way.

A THREE-PLAYER GAME

Another problem from Parzen [13] will now be examined to give further insight into GERT analysis. Three players (denoted by A, B, and C) take turns playing a game according to the following rules. At the start, A and B play while C sits out. The winner of the match between A and B plays C. The winner of the second match then plays the loser of the preceding match until a player wins twice in succession, in which case he is declared the winner of the game. The network for this game is given in Fig. 8.19, where p_{ij}

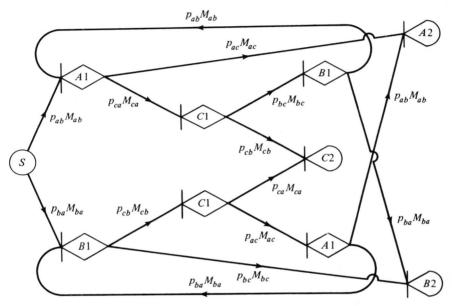

Figure 8.19. GERT representation of a three-player game.

denotes the probability that i defeats j, and M_{ij} denotes the time required for i to defeat j. The reduced network of Fig. 8.19 would appear as shown in Fig. 8.20. The three branches of the resulting network shown in Fig. 8.20 can be examined separately. Consider the branch from S to A_2. The graph could be redrawn as shown in Fig. 8.21. Applying Mason's rule, we get:

$$W_{S,A_2}(s) = \frac{W_{ab}W_{ac}(1 - W_{cb}W_{ac}W_{ba}) + W_{ba}W_{cb}W_{ac}W_{ab}(1 - W_{ca}W_{bc}W_{ab})}{1 - W_{ca}W_{bc}W_{ab} - W_{cb}W_{ac}W_{ba} + W_{ca}W_{bc}W_{ab}W_{cb}W_{ac}W_{ba}}$$

The expressions for $W_{S,B_2}(s)$ and $W_{S,C_2}(s)$ can be found in a like manner. The procedure for finding the probability that A, B, or C wins can be applied.

$$p_{S,A_2} = W_{S,A_2}(s)|_{s=0}$$

and

$$M_{S,A_2}(s) = \frac{W_{S,A_2}(s)}{p_{S,A_2}}$$

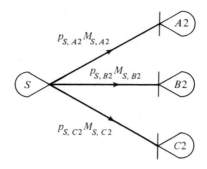

Figure 8.20. Simplified GERT representation of the three-player game.

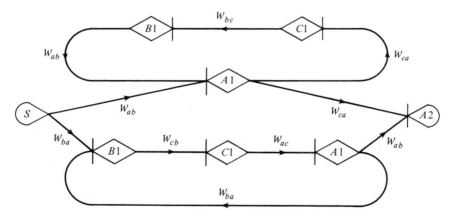

Figure 8.21. GERT representation of the branch from S to $A2$ in a three-player game.

If it is desired to obtain the MGF of the time to the end of the game, then three branches can be added to the resulting network and Fig. 8.22 is obtained. Then,

$$W_{S,\text{end}}(s) = W_{S,A_2}(s) + W_{S,B_2}(s) + W_{S,C_2}(s) = M_{S,\text{end}}(s)$$

since the probability that the game ends is 1.

War Gaming

In this example, a simple air-duel model will be structured in stochastic network form and GERT applied for evaluation purposes.

An interceptor is alerted and is assigned a specific bomber as its target. The time for the interceptor to climb to altitude, approach the bomber, and make a pass is a random variable with MGF, $M_{GBK}(s)$, if the interceptor

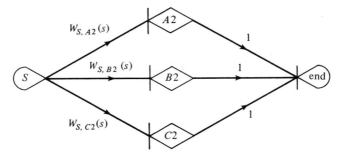

Figure 8.22. GERT representation of a three-player game adapted to calculate the duration of the game.

shoots down (kills) the bomber. If the interceptor misses, then it will be assumed that the time taken is from the distribution whose MGF is $M_{gbm}(s)$. There is a third possibility, viz., the bomber will shoot down the interceptor on the first pass. The MGF for this case is symbolized by $M_{gik}(s)$. If the interceptor misses, then there are successive passes made at the bomber; however, after each pass there is a probability that the interceptor's mission will have to be aborted. First an infinite number of passes will be considered, then a restriction on the number of passes will be imposed. The stochastic network is shown in Fig. 8.23, and

$$W_{gb}(s) = \frac{p_{gbk}M_{gbk}(s)[1 - p_{bm}M_{bm}(s)] + p_{gbm}M_{gbm}(s)p_{bk}M_{bk}(s)}{[1 - p_{bm}M_{bm}(s)]}$$

The expressions for $W_{gi}(s)$ and $W_{ga}(s)$ are computed in a similar manner.

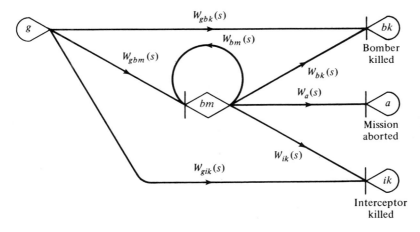

Figure 8.23. GERT representation of a simple air duel with unlimited passes.

If the number of passes is restricted and the probabilities and distributions of times change for each pass, then the network would be as shown in Fig. 8.24. For this network we have

$$W_{gk}(s) = p_{gbk}M_{gbk}(s) + \sum_{j=1}^{n} \left\{ \left[\prod_{i=1}^{n} p_{mi}M_{mi}(s) \right] p_{bj}M_{bj}(s) \right\}$$

$$W_{gi}(s) = p_{gik}M_{gik}(s) + \sum_{j=1}^{n} \left\{ \left[\prod_{i=1}^{j} p_{mi}M_{mi}(s) \right] p_{ij}M_{ij}(s) \right\}$$

and

$$W_{ga}(s) = \sum_{j=1}^{n} \left\{ \left[\prod_{i=1}^{j} p_{mi}M_{mi}(s) \right] p_{aj}M_{aj}(s) \right\}$$

More complex air-duel situations can be modeled along the lines presented in this example.

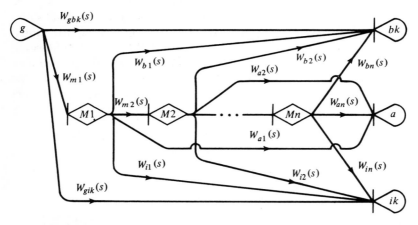

Figure 8.24. GERT representation of a simple air duel with the number of passes limited.

8.2.4 Evaluation of complex networks.
In this section we will use GERT to evaluate (1) a network with multiple loops and (2) a network with multiple input and output nodes. Specific applications were discussed in the preceding sections; thus the emphasis here is entirely on the mechanics of the evaluation portion of GERT with a minimum of descriptive material.

MULTIPLE FEEDBACK LOOPS

Consider the network given in Fig. 8.25. In Table 8.3 a list of the loops of order 1 and 2 (nontouching pairs) is presented. From Table 8.3 and Mason's rule, we have:

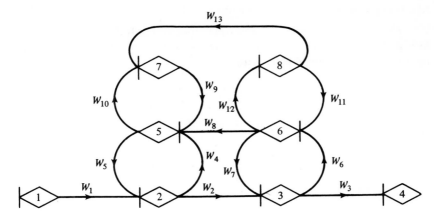

Figure 8.25. A complex GERT network.

Table 8.3. LIST OF LOOPS FOR THE COMPLEX NETWORK OF
FIG. 8.25.

Loop	Elements of Loop	Nontouching Associated Loops
L_1	$W_4 W_5$	L_2, L_5
L_2	$W_6 W_7$	L_1, L_4
L_3	$W_2 W_6 W_8 W_5$	
L_4	$W_9 W_{10}$	L_2, L_5
L_5	$W_{11} W_{12}$	L_1, L_4
L_6	$W_2 W_6 W_{12} W_{13} W_9 W_5$	

$$W_e = \frac{W_1 W_2 W_3 (1 - (W_9 W_{10} + W_{11} W_{12}) + W_9 W_{10} W_{11} W_{12})}{H}$$

where

$$\begin{aligned}
H = 1 &- (W_4 W_5 + W_6 W_7 + W_2 W_6 W_8 W_5 + W_9 W_{10} + W_{11} W_{12} \\
&+ W_2 W_6 W_{12} W_{13} W_9 W_5) + W_4 W_5 (W_6 W_7 + W_{11} W_{12}) \\
&+ W_6 W_7 W_9 W_{10} + W_9 W_{10} W_{11} W_{12}
\end{aligned}$$

A digital computer program has been written that computes the probability that an output node is realized and the first two moments of the time
to realize the output node, given that it is realized. The program is discussed
in Section 8.2.5. This program makes the analysis for larger problems purely
mechanical.

MULTIPLE INPUT AND OUTPUT NODES

The next example involves multiple input and output nodes, as illustrated in Fig. 8.26. From the figure, the following equivalent branch equations can be obtained by using Mason's rule:

$$W_{e1} = \frac{W_3(W_1 + W_4 W_6)}{1 - W_2}$$

$$W_{e2} = W_4 W_7$$

$$W_{e3} = \frac{W_3 W_5 W_6}{1 - W_2}$$

$$W_{e4} = W_5 W_7$$

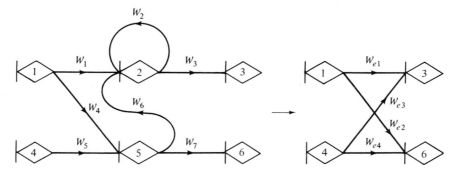

Figure 8.26. A GERT network with multiple input and output nodes.

The probabilities of the branch being part of a realization and the moments of the times associated with an equivalent branch, if it is part of the realization, are computed from W_{ej} as previously discussed.

Suppose it is given that p_a proportion of the time node 1 is the starting node and $(1 - p_a)$ proportion of the time node 4 is the starting node. Given this information, we can write directly from Fig. 8.27 that:

$$W_{s3} = p_a W_{e1} + (1 - p_a) W_{e3}$$

and

$$W_{s6} = p_a W_{e2} + (1 - p_a) W_{e4}$$

The relationship between this network (including nodes 2 and 5) and Markov chains is seen in the transition probability matrix given below, where a blank

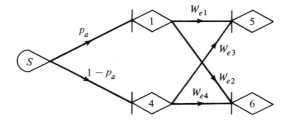

Figure 8.27. Figure 8.26 modified to have a single input node.

indicates a zero entry:

	S	1	2	3	4	5	6
S		p_a			$1 - p_a$		
1			p_1			p_4	
2			p_2	p_3			
3				1			
4						p_5	
5			p_6				p_7
6							1

A network represents a sparse transition probability matrix. The stochastic network also includes the concept of transition time. For the above matrix it was assumed that if either node 3 (state 3) or node 6 (state 6) was realized, then the network would be realized. Thus nodes 3 and 6 represent absorbing states, and once reached, the process never leaves these nodes. This concept corresponds to a self-loop about nodes 3 and 6 or a 1 in diagonal of the above matrix in rows 3 and 6.

A stochastic network corresponds closely to processes of the semi-Markov variety. The main theorems of semi-Markov processes pertain to processes whose underlying Markov chain is *ergodic*. In network terminology, an ergodic chain is one in which every node can be reached from every other node in a finite number of branch transitions. Quantities of interest in semi-Markov process theory are the steady-state probability of being in a particular state and the steady-state percentage of time spent in a particular state, where a state includes all activities represented by the branches leaving a node. From these quantities other pertinent information can be obtained, such as mean recurrence time of a state. For stochastic networks, the states are normally transient and the basic theorems of semi-Markov processes are not of interest.

8.2.5 A computer program for solving GERT EXCLUSIVE-OR systems.
Ishmael and Pritsker [10, 11] have developed a digital computer program written in FORTRAN for analyzing GERT networks that contain nodes of the EXCLUSIVE-OR, PROBABILISTIC type, and branches which have both a probability and a time associated with them. The time associated with a branch can be a random variable. The FORTRAN code for the program is available. See the preface for details on how to obtain the coding.

Input to the program includes appropriate problem identification information and the branches of the network. Information concerning each branch includes the start node and end node for the branch, the probability of realizing the branch, and data about the moment generating function (MGF) of the random variable associated with the branch. The MGF is described by a three-letter code and up to two parameters of the MGF.

Given the above inputs, the program determines the source nodes, the sink nodes, the paths connecting the source nodes to the sink nodes, and the loops of the network. The standard output from the program includes (1) appropriate problem identification headings, (2) the paths and loops of a network, (3) the probability of realizing a sink node from any source node, and (4) the mean and variance of the time to realize a sink node, given that the sink node is realized and given an initial source node. The option exists to print the loops and/or paths of a network.

8.2.5.1 Calculation of network statistics.
The program accepts the input information and determines the source and sink nodes and all paths and loops of the network. In addition, the program determines the following three values associated with a path or loop:

1. Probability of traversal.
2. The mean time of traversal.
3. The second moment of the time of traversal.

These values are determined by the following methods. The W function associated with a path or loop is the product of the W function of the branches which make up the path or loop. Letting L represent a path or loop, we have

$$W_L(s) = \prod_{i \in L} W_i(s)$$

Now the probability associated with L is:

$$p_L = W_L(0) = \prod_{i \in L} W_i(0) = \prod_{i \in L} p_i$$

The expected time to traverse L is given by:

$$\mu_{1L} = \frac{1}{W_L(0)} \frac{\partial W_L(s)}{\partial s}\bigg|_{s=0} = \frac{1}{W_L(0)} \left(\prod_{i \in L} W_i(s) \right) \left(\sum_{i \in L} \frac{1}{W_i(s)} \frac{\partial W_i(s)}{\partial s} \right)\bigg|_{s=0}$$

$$\mu_{1L} = \frac{1}{W_L(0)} \left(\prod_{i \in L} W_i(0) \right) \left(\sum_{i \in L} \frac{\partial M_i(s)}{\partial s}\bigg|_{s=0} \right) = \sum_{i \in L} \mu_{1i}$$

This says that the expected time to traverse a path or loop is the sum of the expected times of the branches of the path or loop. The complex analysis is given to lay the foundation for obtaining an equation for the second moment. From the W function we have:

$$\mu_{2L} = \frac{1}{W_L(0)} \frac{\partial^2 W_L(s)}{\partial s^2}\bigg|_{s=0}$$

$$= \frac{1}{W_L(0)} \left(\prod_{i \in L} W_i(s) \right) \left\{ \left[\sum_{i \in L} \frac{1}{W_i(s)} \frac{\partial W_i(s)}{\partial s} \right]^2 - \sum_{i \in L} \left(\frac{1}{W_i(s)} \frac{\partial W_i(s)}{\partial s} \right)^2 \right. $$

$$\left. + \sum_{i \in L} \frac{1}{W_i(s)} \frac{\partial^2 W_i(s)}{\partial s^2} \right\}\bigg|_{s=0}$$

$$= \mu_{1L}^2 - \sum_{i \in L} \mu_{1i}^2 + \sum_{i \in L} \mu_{2i}$$

The computer program computes p_L, μ_{1L}, and μ_{2L} for all paths and loops of the network, including loops which are products of nontouching loops. We then combine these values through the topological equation to obtain the output statistics desired. The equivalent W function for one path, A, between the two nodes of interest is given by:

$$W_E(s) = \frac{A(s) \left[1 + \sum_{i=1}^{\infty} (-1)^i \sum_{k=1}^{n_i} W_{L_k}^{(i)}(s) \right]}{\left[1 + \sum_{j=1}^{\infty} (-1)^j \sum_{v=1}^{n_j} W_{L_v}^{(j)}(s) \right]} = \frac{A(s)B(s)}{D(s)} = \frac{N(s)}{D(s)}$$

where: $A(s)$ = product of the values of all branches in the path considered,
$W_{L_k}^{(i)}(s)$ = product of the values of i disjoint loops having no nodes in common with path A,
n_i = the number of loops composed of i disjoint loops,
$W_{L_v}^{(j)}(s)$ = product of the values of any j disjoint loops, and
n_j = the number of loops composed of j disjoint loops.

Also, $B(s)$, $D(s)$, and $N(s)$ are direct substitutions. If there is more than one path, then the W functions associated with each path would be summed. For convenience, consider the one-path case. The output statistics can be computed from the following equations:

$$p_E = W_E(0)$$

$$\mu_{1E} = \frac{1}{W_E(0)} \frac{\partial W_E(s)}{\partial s}\bigg|_{s=0} = \frac{1}{W_E(0)} \left[\frac{D(s)\dfrac{\partial N(s)}{\partial s} - N(s)\dfrac{\partial D(s)}{\partial s}}{D^2(s)} \right]\bigg|_{s=0}$$

$$\mu_{2E} = \frac{1}{W_E(0)} \frac{\partial W_E^2(s)}{\partial s^2}\bigg|_{s=0}$$

$$= \frac{1}{W_E(0)} \left\{ \frac{D(s)\left[D(s)\dfrac{\partial^2 N(s)}{\partial s} - N(s)\dfrac{\partial^2 D(s)}{\partial s} \right] - 2\dfrac{\partial D(s)}{\partial s}\left[D(s)\dfrac{\partial N(s)}{\partial s} - N(s)\dfrac{\partial D(s)}{\partial s} \right]}{D^3(s)} \right\}$$

$$\sigma_E^2 = \mu_{2E} - \mu_{1E}^2$$

In these equations, the values of $\partial N(s)/\partial s$, $\partial^2 N(s)/\partial s^2$, etc., evaluated at $s = 0$, are obtained from the previously compiled values of μ_{1L}, μ_{2L}, etc.

8.2.5.2 Program operating procedure. The input specifications for the first data card to the program and the first data card for each network are given in Table 8.4. Each branch of a network must be described by a separate card as shown in Table 8.5.

The equations and moments of the distributions that have been programmed are shown in Table 8.6. Since only the mean and variance of each branch are used in the calculation of the mean and variance of the system, other distributions can be accommodated by the program, if we specify the normal distribution with mean and standard deviation values of the distribution of interest. Further, it can be shown that nth central moments for the network depend only on j central moments of the branches, $j = 1, 2, \ldots, n$. This should aid in developing programs for computing higher moments.

For the computer program, only one branch can join a given start and end node. There can, however, be another branch between the two nodes if the roles of the start node and end node are reversed. If the network of interest has more than one branch between the same two nodes, then a dummy node should be inserted for one of the branches.

An option to delete higher-order loops which have a low probability of occurrence can be exercised through field 9 of data card 1. If this option field is left blank, all loops will be considered. If, however, loops should be deleted that have a probability of realization less than or equal to δ, then δ can be read in as the deletion probability in field 9, and no loops with lower probabilities will be included. The effect upon the accuracy of the results for the equivalent network will depend upon the size, the complexity of the network, and the magnitude of δ. Since the deletion of higher-order loops having low probability of occurrence introduces some error into the final output, the probabilities for all equivalent branches of the network emanating from a given source node may not sum to one. The user can cause the

Table 8.4. INPUT TO GERT PROGRAM

FIRST CARD OF THE DATA INPUT DECK
  The first card of the input deck is the means by which the distribution codes are placed in the machine. These codes are then used to check the distribution codes in the input network
  Field 1 (cc. 1–9) = *ABDEGNOPU*

FIRST CARD OF EACH INPUT NETWORK IN THE INPUT DECK (HEADER CARD)
  The first card of each input network contains the user name and the problem identification. Also contained in this card are the three program control options available to the user.

Field	Card Columns	Definition
1	1–4	Blank.
2	5–16	User name.
3	17–20	Problem designation (right-justified) may be any combination of alphabetic or numeric characters.
4	21–22	Month (right-justified).
5	23–24	Day (right-justified).
6	25–28	Year.
7	29–30	Loop printout control option: If blank, do not print loops; if positive, print loops.
8	31–32	Path printout control option: If blank, do not print paths; if positive, print paths.
9	33–42	Loop deletion probability (F10.8); if this is left blank, no loops will be deleted.
10	43–44	Option for adjustment of the equivalent branch printout when the loop deletion option is exercised; if this is left blank, the actual calculated branch values are printed. If this field contains a positive number, the equivalent branch printout will be adjusted so that the probabilities for all equivalent branches emanating from a given source node sum to one.

Table 8.5. INPUT TO GERT PROGRAM FOR EACH BRANCH OF THE NETWORK

Each network is specified by defining its branches as follows:
1. Node beginning the branch.
2. Node terminating the branch.
3. Type of distribution of time associated with the branch.
4. Probability of realizing the branch if its beginning node is realized.
5. Coefficients defining the time distribution.

One data card is required for each branch of the network. The order of the data cards is not significant. The fields on each data card are:

Field	Card Columns	Definition
1	1–4	Node beginning branch (right-justified).
2	6–9	Node terminating branch (right-justified).
3	11–13	Type of distribution [$B, D, E, GA, GE, NB, NO, P, U$] (left justified).
4	14–20	
5	21–27	
6	28–34	
7	35–41	The definition of these fields depends on the type
8	42–48	of distribution (see below).
9	49–55	The format for all fields is F7.3.
10	56–62	
11	63–69	

Type of Distribution*	Field							
	4	5	6	7	8	9	10	11
B (Binomial)	Prob.	n	p	—	—	—	—	—
D (Discrete)	Prob. 1	T_1	Prob. 2	T_2	Prob. 3	T_3	Prob. 4	T_4
E (Exponential)	Prob.	$1/a$	—	—	—	—	—	—
GA (GAmma)	Prob.	$1/a$	b	—	—	—	—	—
GE (GEometric)	Prob.	p	—	—	—	—	—	—
NB (Neg. Binomial)	Prob.	r	p	—	—	—	—	—
NO (NOrmal)	Prob.	m	σ	—	—	—	—	—
P (Poisson)	Prob.	λ	—	—	—	—	—	—
U (Uniform)	Prob.	a	b	—	—	—	—	—

*See Table 8.6 for definition of the parameters.

program to adjust the probabilities so that they sum to one by specifying a positive value in field 10 of the first data card. The mean and variance of the time to traverse the equivalent branch are adjusted by dividing by the nonadjusted probabilities.

Multiple networks can be analyzed if the data decks are separated with a blank card. The last data deck should be followed by a blank card and a

Table 8.6. DISTRIBUTIONS ACCEPTABLE TO GERT PROGRAM

Type of Distribution	$M_E(s)$	Mean	Second Moment	Input Variables
Binomial (B)	$(pe^s + 1 - p)^n$	np	$np(np + 1 - p)$	$W_E(0); n, p$
Discrete (D)	$\dfrac{p_1 e^{sT_1} + p_2 e^{sT_2} + \cdots}{p_1 + p_2 + \cdots}$	$\dfrac{p_1 T_1 + p_2 T_2 + \cdots}{p_1 + p_2 + \cdots}$	$\dfrac{p_1 T_1^2 + p_2 T_2^2 + \cdots}{p_1 + p_2 + \cdots}$	$W_E(0); p_1, T_1, p_2, T_2$
Exponential (E)	$\left(1 - \dfrac{s}{a}\right)^{-1}$	$\dfrac{1}{a}$	$\dfrac{2}{a^2}$	$W_E(0); \dfrac{1}{a}$
Gamma (GA)	$\left(1 - \dfrac{s}{a}\right)^{-b}$	$\dfrac{b}{a}$	$\dfrac{b(b+1)}{a^2}$	$W_E(0); \dfrac{1}{a}, b$
Geometric (GE)	$\dfrac{pe^s}{1 - e^s + pe^s}$	$\dfrac{1}{p}$	$\dfrac{2-p}{p^2}$	$W_E(0); p$
Negative Binomial (NB)	$\left(\dfrac{p}{1 - e^s + pe^s}\right)^r$	$\dfrac{r(1-p)}{p}$	$\dfrac{r(1-p)(1+r-rp)}{p^2}$	$W_E(0); r, p$
Normal (NO)	$e^{(sm + (1/2)s^2\sigma^2)}$	m	$m^2 + \sigma^2$	$W_E(0); m, \sigma$
Poisson (P)	$e^{\lambda(e^s - 1)}$	λ	$\lambda(1 + \lambda)$	$W_E(0); \lambda$
Uniform (U)	$\dfrac{e^{sa} - e^{sb}}{(a - b)s}$	$\dfrac{a + b}{2}$	$\dfrac{a^2 + ab + b^2}{3}$	$W_E(0); a, b$

Table 8.7. RESULTS OF A THREE-PLAYER GAME WITH $p_{AB} = p_{BC} = p_{CA}$

p_{AB}	A Wins			B Wins			C Wins		
	Prob.	Expected Number of Games	Variance	Prob.	Expected Number of Games	Variance	Prob.	Expected Number of Games	Variance
0.10	0.3591	9.5445	85.9734	0.3330	10.0537	89.1947	0.3079	10.8341	88.5718
0.20	0.3711	4.9210	17.4220	0.3343	5.1259	19.0015	0.2946	5.8055	18.1899
0.30	0.3724	3.5479	6.1122	0.3391	3.5958	6.7711	0.2885	4.2334	5.9870
0.35	0.3702	3.2087	4.0221	0.3427	3.2206	4.4116	0.2871	3.8456	3.6645
0.40	0.3666	2.9931	2.8553	0.3471	2.9895	3.0647	0.2862	3.6039	2.3538
0.45	0.3621	2.8707	2.2638	0.3521	2.8653	2.3554	0.2858	3.4710	1.6783
0.50	0.3571	2.8286	2.1094	0.3571	2.8286	2.1094	0.2857	3.4286	1.4694

card with a negative value in columns 1–4. The GERT EXCLUSIVE-OR program has been tested on many computers. Numerous examples are given in reference 1, one of which is given in the next section.

8.2.5.3 An example of the use of the GERT EXCLUSIVE-OR computer package.

The GERT program was used to analyze the three-player game given in Section 8.2.3.2, in which the game is won by the first player to win two consecutive games. The network is shown in Fig. 8.19. It will be assumed in this analysis that $p_{AB} = p_{BC} = p_{CA}$, that is, the probabilities of A beating B, of B beating C, and of C beating A are equal. Also, the time element will be assumed to be 1 for each game played. The GERT program was run to analyze the game when the independent variable was $p_{AB} = p_{BC} = p_{CA}$. The results of the program are given in Table 8.7. Since the game is symmetrical, only values of $p_{AB} \leq 0.5$ are tabulated.

The results presented in Table 8.7 show that player A has the highest probability of winning the game for values of p_{AB} in the range of 0.10 to 0.50. This is somewhat contrary to intuition. It is also evident that, as p_{AB} moves away from the value 0.50, the expected number and the variance of the games played increase until a winner is declared.

8.3 Counters and Conditional Moment Generating Functions for GERT Networks

As we will see in Chapters 10 and 11, the study of inventory, reliability, queueing, and maintenance problems showed that it would be advantageous to develop a device to relate the number of executions of a given portion of the graph to the time of execution of the entire graph. The development of a method of determining the distribution of the number of times an element is executed in the GERT network is also of great interest.

8.3.1 Development of the MGF of the number of times an element is traversed.

BRANCHES

The transmittance e^c is placed on the branch being investigated. It is noted that e^c is the equivalent to the moment generating function of a constant equal to 1. If there were no MGF on the branches with respect to the time of execution of the branch, the graph would merely be a GERT network with respect to the number of executions of a given element. It should be remembered that the GERT approach is not restricted to the time parameter. Investigation of other parameters such as cost, counts, and manpower are also of interest. Since all MGF become equal to 1 when s is equated to 0, the time parameter on a branch could be suppressed for this counter analysis if s is set to zero.

Consider the problem illustrated in Fig. 8.28(a) where there is an interest in determining the number of passages through the top loop. First the tag, e^c, is placed on the path as shown in Fig. 8.28(b). The graph is now solved by Mason's rule:

$$W(s, c) = M(s, c) = \frac{\frac{2}{3}e^{4s}(1 - \frac{1}{2}e^c e^{2s})}{1 - \frac{1}{6}e^{6s} - \frac{1}{2}e^c e^{2s}}$$

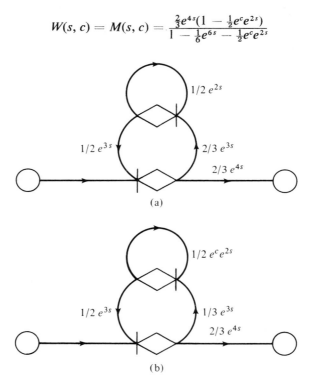

(a)

(b)

Figure 8.28. GERT network adapted to find the distribution of the number of realizations of a branch.

When we set s equal to zero, the moment generating functions with respect to time will be reduced to unity, and the MGF of the count will remain:

$$M(c, s)|_{s=0} = M(c) = \frac{\frac{2}{3} - \frac{1}{3}e^c}{\frac{5}{6} - \frac{1}{2}e^c}$$

The moments of the distribution of the counts can now be found in the normal manner. The first moment is:

$$E(\text{count}) = \frac{dM(c)}{dc}\bigg|_{c=0} = -\frac{1}{3}e^c\left(\frac{5}{6} - \frac{1}{2}e^c\right)^{-1}$$

$$+ \frac{1}{2}e^c\left(\frac{5}{6} - \frac{1}{2}e^c\right)^{-2}\left(\frac{2}{3} - \frac{1}{3}e^c\right)\bigg|_{c=0} = \frac{1}{2}$$

For higher moments about the origin:

$$E(\text{count}^n) = \frac{d^n M(c)}{dc^n}\bigg|_{c=0}$$

In terms of the $M(s, c)$ function, this relationship is expressed as follows:

$$E(\text{count}^n) = \frac{\partial^n M(s, c)}{\partial c^n}\bigg|_{\substack{s=0 \\ c=0}}$$

NODES

The approach for finding the MGF for the number of passages through a given node is exactly the same as that for the branch approach, except that each branch entering the given node is tagged with e^c. The reason for this is that the sum of the transitions of all the paths leading into a given node is equivalent to the number of times this node is realized.

To determine the MGF of the random variable, the number of times the A node in Fig. 8.29 has been accomplished, both paths entering the node are

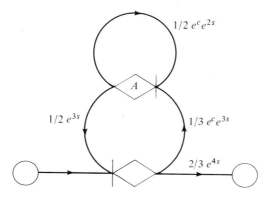

Figure 8.29. GERT network adapted to find the distribution of the number of transitions of a node.

tagged; this yields:

$$W(s, c) = M(s, c) = \frac{[\frac{2}{3}e^{4s}(1 - \frac{1}{2}e^c e^{2s})]}{[1 - \frac{1}{6}e^c e^{6s} - \frac{1}{2}e^c e^{2s}]}$$

$$M(c) = M(s, c)|_{s=0} = \frac{(\frac{2}{3} - \frac{1}{3}e^c)}{(1 - \frac{2}{3}e^c)}$$

$$E(\text{count}) = \frac{dM(c)}{dc}\Big|_{c=0} = \left(\frac{2}{3} - \frac{1}{3}e^c\right)\left(\frac{2}{3}e^c\right)\left(1 - \frac{2}{3}e^c\right)^{-2}$$

$$+ \left(-\frac{1}{3}e^c\right)\left(1 - \frac{2}{3}e^c\right)^{-1}\Big|_{c=0} = 1$$

SETS OF ELEMENTS

In addition to the number of times a given element is traversed, there is interest in the number of transitions that are made in a group of elements. One example is the number of transitions that occur within the system before the system goes from one point to another. The case of major interest here is the MGF of the number of transitions in going from the source to the sink of the network. To accomplish this, the e^c tag is placed on all paths of the network.

Consider the expected number of transitions to traverse the network shown in Fig. 8.30:

$$W(s, c) = M(s, c) = \frac{[\frac{2}{3}e^{4s}e^c(1 - \frac{1}{2}e^c e^{3s})]}{[1 - \frac{1}{6}e^{2c}e^{6s} - \frac{1}{2}e^c e^{2s}]}$$

$$M(c) = M(s, c)|_{s=0} = \frac{(\frac{2}{3}e^c - \frac{1}{3}e^{2c})}{(1 - \frac{1}{6}e^{2c} - \frac{1}{2}e^c)}$$

$$E(\text{count}) = \frac{dM(c)}{dc}\Big|_{c=0} = \left(\frac{2}{3}e^c - \frac{1}{3}e^{2c}\right)\left(\frac{1}{3}e^{2c} + \frac{1}{2}e^c\right)$$

$$\times \left(1 - \frac{1}{6}e^{2c} - \frac{1}{2}e^c\right)^{-2} + \left(\frac{2}{3}e^c - \frac{2}{3}e^{2c}\right)$$

$$\times \left(1 - \frac{1}{6}e^{2c} - \frac{1}{2}e^c\right)^{-1}\Big|_{c=0} = \frac{5}{2}.$$

8.3.2 Development of conditional moment generating functions.

CONDITIONAL MGF INVOLVING ONE PARAMETER

The basis for the development of the conditional moment generating function is an adaptation of the idea of generating functions as discussed by Feller [6].

A generating function is a power series in terms of an undefined variable, z. When used in conjunction with networks, the variable z is used to

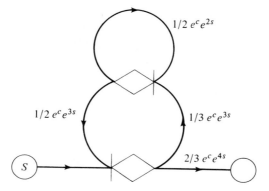

Figure 8.30. GERT network adapted to find the distribution of the number of transitions in the entire network.

multiply the W function associated with a branch; therefore, the power of z specifies the number of times branches were traversed whose values were multiplied by a z. When a branch value is multiplied by z, we say the branch is *tagged*. If we define $W(s|j)$ as the W function associated with a network when the branches tagged with a z are taken j times, then the equivalent function associated with the network can be written as a generating function (hereafter referred to as the W generating function) as shown in Eq. (8-1):

$$W(s, z) = W(s|0) + W(s|1)z + W(s|2)z^2 + \cdots + W(s|j)z^j + \cdots$$

$$= \sum_{j=0}^{\infty} W(s|j)z^j \qquad (8\text{-}1)$$

where $W(s|j)$ is the conditional W function associated with the network, given that branches tagged with a z are traversed j times.

The relationship between the conditional W function and the conditional moment generating function (MGF) is shown in Eq. (8-2):

$$W(s|j) = p(j)M(s|j) \qquad (8\text{-}2)$$

with

$$p(j) = W(0|j) \qquad (8\text{-}3)$$

where $p(j)$ is the probability that the network is realized when branches tagged with a z are traversed j times, and $M(s|j)$ is the conditional MGF associated with the network, given that branches tagged with a z are traversed j times.

The conditional W function can be obtained from the W generating function as shown in Eq. (8-4):

$$W(s \mid j) = \frac{1}{j!} \frac{\partial^j W(s, z)}{\partial z^j}\bigg|_{z=0} \tag{8-4}$$

In some cases, the W generating function can be written as a power series, as in Eq. (8-1), in which case $W(s \mid j)$ can be obtained directly as the coefficient of z^j.

Normally, it is the conditional MGF that is desired. We obtain it by using Eq. (8-4) to derive the conditional W function from which $p(j)$ can be obtained using Eq. (8-3). Then we use Eq. (8-2) to get the conditional MGF as:

$$M(s \mid j) = \frac{W(s \mid j)}{p(j)} \tag{8-5}$$

Two expressions which can be obtained directly from the W generating function are:

$$W(s \mid 0) = W(s, z)|_{z=0} \tag{8-6}$$

where $W(s \mid 0)$ is the W function of the network with all tagged branches deleted, and

$$W(s, 1) = W(s, z)|_{z=1} \tag{8-7}$$

where $W(s, 1)$ is the W function of the network with no branches tagged.

An example of the computational procedure to obtain the statistics described in the preceding sections will now be given. The network for the example is shown in Fig. 8.31.

Consider first tagging the upper loop with a z, that is, multiplying $W_f(s)$

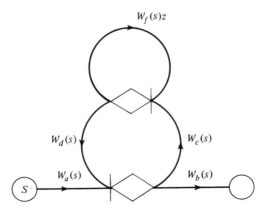

Figure 8.31. A z-tagged GERT network.

by z. Using Mason's rule we have the W generating function as:

$$W(s, z) = \frac{W_a(s)W_b(s)[1 - W_f(s)z]}{[1 - W_c(s)W_d(s) - W_f(s)z]}$$

To obtain the conditional W function, the partial derivatives of $W(s, z)$ are required. Taking the derivative with respect to z and combining terms yields:

$$\frac{\partial W(s, z)}{\partial z} = \frac{W_a(s)W_b(s)W_c(s)W_d(s)W_f(s)}{[1 - W_c(s)W_d(s) - W_f(s)z]^2}$$

and

$$\frac{\partial^2 W(s, z)}{dz^2} = 2\frac{W_a(s)W_b(s)W_c(s)W_d(s)W_f^2(s)}{[1 - W_c(s)W_d(s) - W_f(s)z]^3}$$

By induction it can be shown that:

$$\frac{\partial^j W(s, z)}{\partial z^j} = j! \cdot \frac{W_a(s)W_b(s)W_c(s)W_d(s)W_f^j(s)}{[1 - W_c(s)W_d(s) - W_f(s)z]^{j+1}}, \qquad j \geq 1 \qquad (8\text{-}8)$$

Substituting Eq. (8-8) into Eq. (8-4) yields:

$$W(s\,|\,j) = \frac{1}{j!}\frac{\partial^j W(s, z)}{\partial z^j}\bigg|_{z=0} = \frac{W_a(s)W_b(s)W_c(s)W_d(s)W_f^j(s)}{[1 - W_c(s)W_d(s)]^{j+1}}, \qquad j \geq 1$$

and

$$W(s\,|\,0) = W(s, 0) = \frac{W_a(s)W_b(s)}{1 - W_c(s)W_d(s)}$$

To make the example numerical, let:

$$W_a(s) = 1$$
$$W_b(s) = \tfrac{2}{3}e^{4s}$$
$$W_c(s) = \tfrac{1}{3}e^{3s}$$
$$W_d(s) = \tfrac{1}{2}e^{3s}$$
$$W_f(s) = \tfrac{1}{2}e^{2s}$$

(For convenience, all branch times are taken as constant.) with these values, the W function for the network when the top loop is taken exactly j times is:

$$W(s\,|\,j) = \frac{(\tfrac{2}{18})(e^{10s})(\tfrac{1}{2}e^{2s})^j}{(1 - \tfrac{1}{6}e^{6s})^{j+1}}$$

Thus,

$$W(s\,|\,2) = \frac{\tfrac{1}{36}e^{14s}}{(1 - \tfrac{1}{6}e^{6s})^3}$$

The probability that the top loop is taken exactly two times is:

$$p(2) = W(0 \,|\, 2) = \frac{\frac{1}{36}}{(\frac{5}{6})^3} = \frac{6}{125}$$

and the conditional MGF, $M(s \,|\, z)$, is:

$$M(s \,|\, 2) = \frac{1}{(\frac{6}{125})}\left|\frac{\frac{1}{36}e^{14s}}{(1 - \frac{1}{6}e^{6s})^3}\right|$$

This quantity is the MGF of the time to realize the network, given that the top loop is realized exactly two times.

THE W GENERATING FUNCTION WITH MULTIPLE PARAMETERS

Consider the W generating function in the two parameters z_1 and z_2:

$$W(s, z_1, z_2) = \sum_{i=0}^{\infty} \sum_{j=0}^{\infty} W(s \,|\, i, j) z_1^i z_2^j$$

where $W(s \,|\, i, j)$ is the W function of the network, given that branches tagged with a z_1 are traversed i times and branches tagged with a z_2 are traversed j times.

In general, if there are k parameters, the W generating function is:

$$W(s, z_1, \ldots, z_k) = \sum_{i_1, i_2, \ldots, i_k = 0}^{\infty} W(s \,|\, i_1, i_2, \ldots, i_k) z_1^{i_1} z_2^{i_2} \ldots z_k^{i_k}$$

The techniques for obtaining the probabilities, MGFs, and conditioned W functions are similar to those presented for the one-parameter case; they are:

$$p(i_1, i_2, \ldots, i_k) = W(0 \,|\, i_1, i_2, \ldots, i_k)$$

$$M(s \,|\, i_1, i_2, \ldots, i_k) = \frac{W(s \,|\, i_1, i_2, \ldots, i_k)}{p(i_1, i_2, \ldots, i_k)} \tag{8-9}$$

and

$$W(s \,|\, i_1, i_2, \ldots, i_k) = \frac{\partial^{\left(\sum_{j=1}^{k} i_j\right)} W(s, z_1, \ldots, z_k)}{(\partial^{i_1} z_1)(\partial^{i_2} z_2) \ldots (\partial^{i_k} z_k)}\Bigg|_{z_j = 0 \text{ all } j}$$

Consider the two-parameter case shown in Fig. 8.32. The W generating function for the network of Fig. 8.32 is:

$$W(s, z_1, z_2) = \frac{W_a(s)W_b(s)(1 - W_f(s)z_2)}{1 - W_c(s)W_d(s)z_1 - W_f(s)z_2}$$

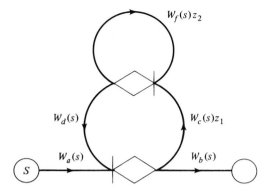

Figure 8.32. GERT network with multiple z-parameters.

By series expansion, we have:

$$W(s, z_1, z_2) = W_a(s)W_b(s) \sum_{i_1=0}^{\infty} \cdot \left\{ \sum_{i_2=0}^{\infty} [W_f(s)z_2]^{i_2} [W_c(s)W_d(s)z_1] \right\}^{i_1}$$

Thus, for given values of i_1 and i_2, we have

$$W(s \mid i_1, i_2) = W_a(s)W_b(s)[W_c(s)W_d(s)]^{i_1}[W_f(s)]^{i_2} \cdot \binom{i_1 + i_2 - 1}{i_1 - 1}$$

and

$$p(i_1, i_2) = p_a p_b (p_c p_d)^{i_1} p_f^{i_2} \binom{i_1 + i_2 - 1}{i_1 - 1}$$

$M(s \mid i_1, i_2)$ is obtained if we substitute the above into Eq. (8-9).

8.3.3 Development of MGF of passage times.

THE MGF OF THE FIRST PASSAGE TIME TO A NODE, GIVEN THAT THE NODE IS REALIZED

The MGF of the first passage time to the node can easily be calculated if the activities entering the node are tagged with z's. The graph is next evaluated from the source to the node of interest. Since this function can be expressed as the W generating function, $W(s \mid 1)$ will represent the path to the node for the first passage. Thus:

$$M(s)_{1\text{st passage time}} = \frac{\dfrac{\partial W(s, z)}{\partial z}\Big|_{z=0}}{\dfrac{\partial W(s, z)}{\partial z}\Big|_{\substack{s=0 \\ z=0}}}$$

Consider the example in Fig. 8.33, and the first passage to A is:

$$W_{S,A}(s, z) = \frac{(\frac{1}{3}ze^{3s})}{(1 - \frac{1}{2}ze^{2s} - \frac{1}{6}ze^{6s})}$$

$$\frac{\partial W_{S,A}(s, z)}{\partial z} = \left(\frac{1}{3}e^{3s}\right)\left(1 - \frac{1}{2}ze^{2s} - \frac{1}{6}ze^{6s}\right)^{-1} + \left(\frac{1}{3}e^{3s}\right)$$

$$\times \left(\frac{1}{2}e^{2s} + \frac{1}{6}e^{6s}\right)\left(1 - \frac{1}{2}ze^{2s} - \frac{1}{6}ze^{6s}\right)^{-2}$$

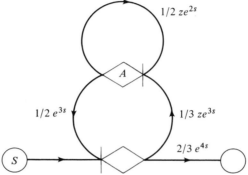

Figure 8.33. GERT network adapted to find the MGF of the passage time to a node.

and

$$M(s)_{1\text{st passage to } A} = \frac{\dfrac{\partial W_{S,A}(s, z)}{\partial z}\Big|_{z=0}}{\dfrac{\partial W_{S,A}(s, z)}{\partial z}\Big|_{\substack{s=0 \\ z=0}}} = e^{3s}$$

THE MGF OF THE nTH PASSAGE TIME TO A NODE, GIVEN THAT THE NODE IS REALIZED

The first passage time is a special case of the more general nth passage time problem. The coefficients of the various powers of z in the W generating function described in the previous section represent the MGF of the passage time from the source to the point of interest, which has already passed this point $n - 1$ times in the case of the z^n term. This MGF is modified by the probability of occurrence.

$$\begin{matrix} \text{MGF of } n\text{th passage time, given} \\ n \text{ realizations of the node} \end{matrix} = \frac{\dfrac{\partial^n W(s, z) \text{ to point}}{\partial z^n}\Big|_{z=0}}{\dfrac{\partial^n W(s, z) \text{ to point}}{\partial z^n}\Big|_{\substack{s=0 \\ z=0}}}.$$

In the previous example, the MGF of the second passage time is found as follows:

$$\frac{\partial^2 M_{S,A}(s,z)}{\partial z^2} = \left(\frac{1}{3}e^{3s}\right)\left(\frac{1}{2}e^{2s} + \frac{1}{6}e^{6s}\right)\left(1 - \frac{1}{2}ze^{2s} - \frac{1}{6}ze^{6s}\right)^{-2}$$

$$+ \left(\frac{1}{3}e^{3s}\right)\left(\frac{1}{2}e^{2s} + \frac{1}{6}e^{6s}\right)\left(1 - \frac{1}{2}ze^{2s} - \frac{1}{6}ze^{6s}\right)^{-2}$$

$$+ \left(\frac{2}{3}ze^{3s}\right)\left(\frac{1}{2}e^{2s} + \frac{1}{6}e^{6s}\right)^2\left(1 - \frac{1}{2}ze^{2s} - \frac{1}{6}ze^{6s}\right)^{-3}$$

$$\frac{\left.\frac{\partial^2 M_{S,A}(s,z)}{\partial z^2}\right|_{z=0}}{\left.\frac{\partial^2 M_{S,A}(s,z)}{\partial z^2}\right|_{\substack{s=0\\z=0}}} = \frac{\left(\frac{2}{3}e^{3s}\right)\left(\frac{1}{2}e^{2s} + \frac{1}{6}e^{6s}\right)}{\left(\frac{4}{9}\right)}$$

$$= \frac{3}{4}e^{5s} + \frac{1}{4}e^{9s}$$

nTH PASSAGE TIME TO AN ACTIVITY

The only difference between this and the node case is that it is suggested that the path of interest be broken into two parts: the first includes the probability of taking the path and the z tag, while the second portion contains the MGF of the time to traverse the path. This split is necessary to assure that the path is taken and that the time of the path will not be counted in the nth passage time. The graph is solved from the source to the dummy node located between the two sections of the path that have been created. The procedure is exactly the same as before.

Consider the graph in Fig. 8.28 in which the MGF of the second passage time through the top loop is to be derived. Redrawing the graph as shown in Fig. 8.34, we can derive the MGF of interest.

$$W_{S,A'}(s,z) = \frac{\left(\frac{1}{6}ze^{3s}\right)}{\left(1 - \frac{1}{2}ze^{2s} - \frac{1}{6}e^{6s}\right)}$$

$$\frac{\partial W_{S,A'}(s,z)}{\partial z} = \left(\frac{1}{6}e^{3s}\right)\left(1 - \frac{1}{2}ze^{2s} - \frac{1}{6}e^{6s}\right)^{-1} + \left(\frac{1}{6}e^{3s}\right)$$

$$\times \left(\frac{1}{2}e^{2s}\right)\left(1 - \frac{1}{2}ze^{2s} - \frac{1}{6}e^{6s}\right)^{-2}$$

$$\frac{\partial^2 W_{S,A'}(s,z)}{\partial z^2} = \left(\frac{1}{6}e^{3s}\right)\left(\frac{1}{2}e^{2s}\right)\left(1 - \frac{1}{2}ze^{2s} - \frac{1}{6}e^{6s}\right)^{-2}$$

$$+ \left(\frac{1}{12}e^{5s}\right)\left(1 - \frac{1}{2}ze^{2s} - \frac{1}{6}e^{6s}\right)^{-2} + \left(\frac{1}{3}ze^{3s}\right)$$

$$\times \left(\frac{1}{2}e^{2s}\right)^2\left(1 - \frac{1}{2}ze^{2s} - \frac{1}{6}e^{6s}\right)^{-3}$$

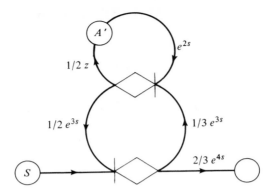

Figure 8.34. GERT network adapted to find the passage time to an activity.

and

$$
\frac{\left.\dfrac{\partial^2 W_{S,A'}(s,z)}{\partial z^2}\right|_{z=0}}{\left.\dfrac{\partial^2 W_{S,A'}(s,z)}{\partial z^2}\right|_{\substack{s=0 \\ z=0}}} = \frac{(\frac{1}{6}e^{5s}(1 - \frac{1}{6}e^{6s})^{-2}}{(\frac{6}{25})}
$$

8.4 The Analytic Approach to the AND and INCLUSIVE-OR Node

In the previous section we saw that there is a close relationship between EXCLUSIVE-OR GERT networks and flowgraphs. The relationship allowed the analytic solution of GERT EXCLUSIVE-OR systems. Unfortunately, the analytic solution of networks involving AND and INCLUSIVE-OR nodes is more difficult. In fact, to date, no one has developed satisfactory procedures for the analytic solution of GERT systems involving AND and INCLUSIVE-OR nodes. Bell [1] and Pritsker [16] have suggested approaches, but these approaches have not proved computationally feasible. Networks involving AND and INCLUSIVE-OR nodes are usually analyzed using the simulation procedures discussed in Chapter 9. In this section we will view some of the problems involved in the analytic solution of AND and INCLUSIVE-OR networks.

8.4.1 The AND logic element. The AND node requires that all branches entering that node be realized prior to the realization of the branch emanating from the AND node. A simple example of AND nodes is shown in Fig. 8.35. The network in Fig. 8.35(a) is equivalent to the network shown

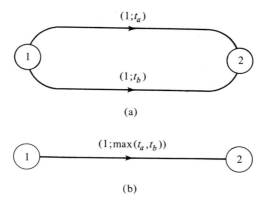

(a)

(b)

Figure 8.35. Reduction of a simple parallel AND system.

in Fig. 8.35(b). The AND node only creates a problem when the graph we are dealing with has parallel paths. The series path reduction is the same as the series path reduction for the EXCLUSIVE-OR approach. It is impossible for a loop to return to an AND node, because this would imply that elements which follow the given node must also precede it.

Pritsker and Happ [14] suggested the following three characteristics which make the AND node networks difficult to analyze.

1. If the probability of taking a branch is less than one and that branch eventually enters an AND node, then there is a probability that the entire network will not be realized. For example, if, in the simple network given in Fig. 8.35, one of the parallel branches had a probability of only 0.3 associated with it, then 70% of the time a branch emanating from node 2 could not be realized.
2. The use of expected values leads to incorrect results. Consider the network given in Fig. 8.36(a). Reducing the branch between nodes 2 and 3 to the equivalent branch with only an expected value for its time parameter results in the graph shown in Fig. 8.36(b). This would reduce according to the maximum operator to the graph in Fig. 8.36(c). This result is obtained even though we know that 30% of the time it will take ten time units to go from 1 to 4. The fallacy is in the use of expected values.
3. The introduction of random variables for the time elements requires the computation of the distribution of the maximum of random variables, which is a complex computational problem.

Solutions for the first two of these problems were proposed by Pritsker and Happ [14], but the third problem remains an unanswered question. Consequently, networks involving AND nodes can only be handled analytically with constant elements.

According to the originators of GERT, the problem presented in item

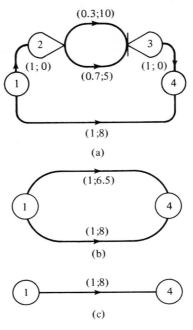

Figure 8.36. Demonstration of an error in logic in dealing with an AND node.

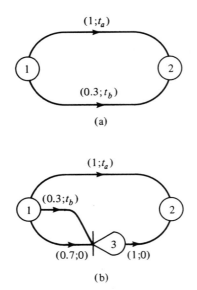

Figure 8.37. Example of the reduction of an AND node parallel system.

1 appears to be mainly a semantic one. If two activities emanate from an AND node, then the sum of the probabilities associated with these activities should be two. If this is not the case, an activity has probably been omitted. This activity is usually of zero time occurrence. Consider as an example the network shown in Fig. 8.37(a). This network should appear as in Fig. 8.37(b). In conclusion, the sum of the probabilities of the branches emanating from an AND node should be a positive integer.

As previously stated, the only problem in the analysis of the AND node occurs with parallel branches. The approach we take to this problem is to convert parallel branches leading to an AND node to an equivalent set of branches leading to an EXCLUSIVE-OR node. For example, the graph in Fig. 8.38(a) is equivalent to the graph in Fig. 8.38(b). Also, the graph in Fig. 8.39(a) can be written (when zero time elements are included) as shown

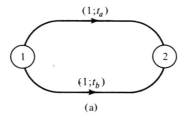

$(1;t_a)$

$(1;t_b)$

(a)

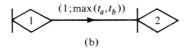

$(1;\max(t_a,t_b))$

(b)

Figure 8.38. Example of the reduction of an AND node parallel system.

in Fig. 8.39(b). There are four possible outcomes regarding the length of time to go from 1 to 2 in Fig. 8.39. These are:

Outcome	Probability	Time
1	$(1 - p_a)(1 - p_b)$	0
2	$(1 - p_b)p_a$	t_a
3	$(1 - p_a)p_b$	t_b
4	$p_a p_b$	Max (t_a, t_b)

Therefore, the two parallel paths between two AND nodes can be redivided into four parallel paths between two EXCLUSIVE-OR nodes as shown in

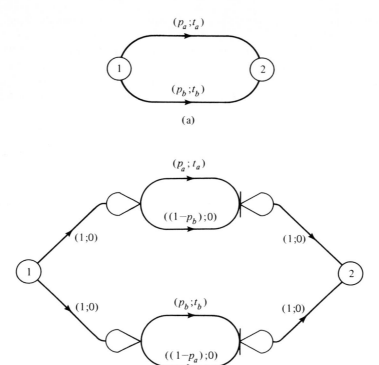

(a)

(b)

Figure 8.39. Example of the reduction of an AND node parallel system.

Fig. 8.40. The W function for the network in Fig. 8.40 is:

$$W_{1,2}(s) = (1 - p_a)(1 - p_b) + (p_a(1 - p_b)e^{st_a})$$
$$+ (p_b(1 - p_a)e^{st_b}) + (p_a p_b e^{s\max(t_a, t_b)})$$

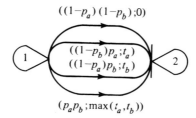

Figure 8.40. Representation of parallel AND paths as an EXCLU-SIVE-OR network.

To add further insight into the reduction procedure, another illustration is given in Fig. 8.41(a), where $p_a + p_b = 1$ and $p_d + p_c = 1$. The equivalent network is shown in Fig. 8.41(b).

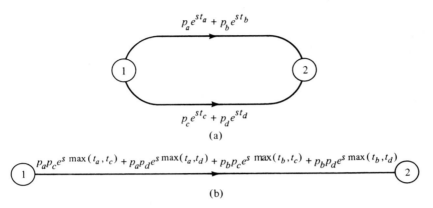

$$p_a e^{st_a} + p_b e^{st_b}$$

$$p_c e^{st_c} + p_d e^{st_d}$$

(a)

$$p_a p_c e^{s\,\max(t_a,\,t_c)} + p_a p_d e^{s\,\max(t_a,\,t_d)} + p_b p_c e^{s\,\max(t_b,\,t_c)} + p_b p_d e^{s\,\max(t_b,\,t_d)}$$

(b)

Figure 8.41. Reduction of a GERT network composed of AND nodes.

A procedure has now been developed for combining parallel branches. The next step taken by Pritsker and Happ was to provide a procedure for reducing combinations of parallel and series branches which make segments of the network independent of other segments. In this manner, segments which are either parallel or series will result. They are independent of the other segments and, hence, can be reduced by the previously discussed procedure. This procedure was explained in the following example taken from the original GERT paper.

REDUCTION OF A NETWORK WITH AND NODES

Assume that there is a nominal schedule of activities for a project as shown in Fig. 8.42 in network form. From the network, it is seen that all

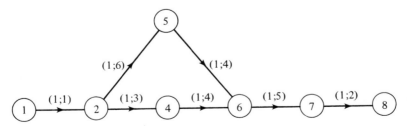

Figure 8.42. GERT representation of a network work with AND nodes.

activities (branches) must be performed and that each activity requires a constant amount of time. Suppose that two branches are superimposed on this network between nodes 2 and 6 and between 5 and 7. These new branches represent activities which do not have to be performed all the time and, indeed, are only performed 2% and 30% of the time, respectively. Examples of such activities are the need of a repair action, a spare part, a demand for a specific serivce, etc. According to the previous discussion, two branches must be added for each new activity because, if the new activity is not performed, it must be shown in the network in order to make it complete. This is illustrated in Fig. 8.43. The addition of the branches between nodes 2 and

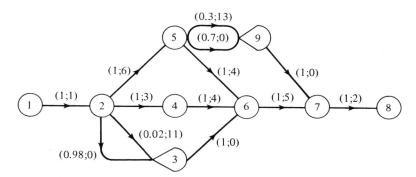

Figure 8.43. Representation of a network composed of AND nodes as one with EXCLUSIVE-OR nodes.

3 does not cause any difficulties since they are in parallel with the series combination between nodes 2 and 6. Thus, an independent segment exists and reduction would proceed as previously discussed. This is not the case for the branches added between nodes 5 and 9. To circumvent this interdependence, an extra node, 5′, is added to the network, and node 2 will be connected to node 5′ by a branch with the same characteristics as the branch between nodes 2 and 5. The addition of this branch does not affect the network (nor in this case the distribution of the times to reach any node). The reduction procedure can now proceed in steps as shown in Fig. 8.44. This last network yields:

$$W_e(s) = M_e(s) = 0.30e^{22s} + 0.014e^{19s} + 0.686e^{18s}$$

The moments can now be obtained as discussed in the previous section.

As we can see, this procedure is much more cumbersome than the procedure discussed in the previous section.

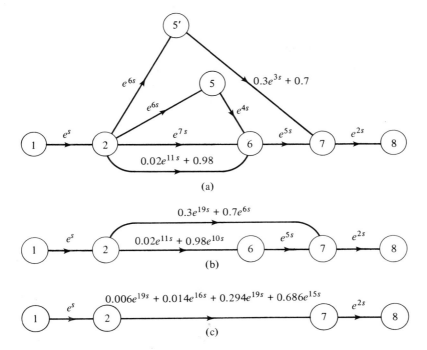

(a)

(b)

(c)

Figure 8.44. Reduction of a network composed of AND nodes.

8.4.2 The INCLUSIVE-OR logic element. The INCLUSIVE-OR node specifies that any branch entering the node causes the node to be realized. It differs from the EXCLUSIVE-OR node in that more than one branch leading to the node can be realized. This necessitates a minimum operator which transforms the network as shown in Fig. 8.45. The graph in Fig. 8.45(a) is equivalent to the one in Fig. 8.45(b).

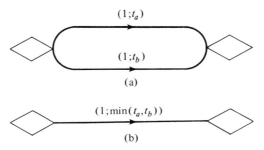

(a)

(b)

Figure 8.45. Parallel INCLUSIVE-OR network reduction.

The INCLUSIVE-OR node is handled essentially the same as the AND node. There are many characteristics of the INCLUSIVE-OR node which are the same as for the AND node. For example:

1. Only parallel branches are of interest since an INCLUSIVE-OR with a feedback branch in an input is not meaningful, and a series network with INCLUSIVE-OR nodes can be replaced by a series network with EXCLUSIVE-OR nodes.
2. At the current time only branches with constant time elements are computationally tractable.

Under these limiting conditions, consider the simple network shown in Fig. 8.46, where $0 \leq p_a \leq 1$ and $0 \leq p_b \leq 1$. This network can be trans-

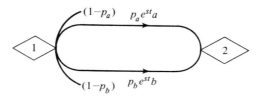

Figure 8.46. Complex parallel INCLUSIVE-OR network reduction.

formed into a network of EXCLUSIVE-OR nodes as shown by Pritsker and Happ if we enumerate all mutually exclusive paths from node 1 to node 2. This is shown in Fig. 8.47. The flowgraph approach to solving this network can then be applied directly.

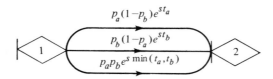

Figure 8.47. Parallel INCLUSIVE-OR network representation as a parallel EXCLUSIVE-OR network.

8.5 Other Topic Areas

8.5.1 Confidence statements. The primary outputs of a GERT analysis are the probability of realizing a node and the MGF of the time to realize a node, given that it is realized. A typical question which might be asked is: "What is the probability of realizing a node in T time units?" The answer to this question involves the joint occurrence of realizing the node and the time to realize the node in less than T time units. Symbolically, this can be written as

$$P(AB) = P(A)P(B \mid A)$$

where A denotes the event "node is realized" and B denotes the event "time of realization of node in less than T time units."

The quantity $P(A)$ is obtained directly from the GERT analysis. The quantity $P(B \mid A)$ must be derived from the MGF obtained from the GERT analysis. A clarification in terminology is required here: the MGF obtained from the GERT analysis is really a conditional MGF, since it is conditioned on the realization of a specific node. Therefore, to obtain $P(B \mid A)$, it is necessary to obtain the distribution function associated with the derived MGF. This problem is referred to in the literature as the *inverse transform problem*.

The most common inversion method is a table look-up operation. For GERT problems this does not appear practical due to the complexity of the MGF derived.

A second inversion method is to use an inversion formula. For systems involving only constant times, the inversion formula can be applied by inspection, since all terms are of the form pe^{st}, and the density function is described by all pairs of p and t. For more complex expressions it may be more appropriate to employ the Laplace transform or the characteristic function in the GERT analysis. For these transforms, complex inversion formulas exist.

A third inversion method is to calculate the first n moments of the distribution function from the MGF. These moments (the number of moments used depends on the technique employed and the accuracy desired) can then be used to approximate the distribution function. The two most widely used techniques for this approximation are Pearson's curves and Gram–Charlier series. These techniques are discussed in the literature [2, 12] and will not be presented here.

As an alternative approach to making confidence statements, the form of the distribution might be assumed. Then confidence statements can be made using the assumed distribution function. For example, the distribution of the time to reach a node could be assumed to be normal, and with a knowledge of the mean and variance, confidence statements can be made. The appropriateness of assuming a distribution form depends on the specific problem under study.

8.5.2 Multiplicative transmittance parameters. In this chapter we assumed that all varying transmittance parameters, such as time, were additive. Some parameters, such as reliability, are multiplicative. Pritsker [16] suggested Mellin transforms as a possible method of introducing multiplicative random variables into the GERT system.

EXERCISES

1. In Chapter 2, calculation of expected quantities using generating functions was discussed. Using these procedures, develop methods for counting using the W generating function.

2. For the network shown in Fig. 8.28, find the expected number of loops realized. Also find the variance.

3. For the Thief of Bagdad problem, find:
   (a) The number of times the dungeon is reentered.
   (b) The number of times the top loop is taken.
   (c) The conditional MGF, given that the long tunnel is taken twice.
   (d) The conditional MGF, given that either tunnel is taken twice.
   (e) The MGF of the first passage time to D.
   (f) The MGF of the third passage time to D.

4. How many times must you flip a coin before you get four straight tails?
   (a) Model in GERT.
   (b) Find the mean and variance of the number of throws.
   (c) Find the MGF of the first passage time until you get two straight heads.
   (d) How many heads will you toss in this game?

5. Model the following system using GERT: When a federal income tax return is filed at the regional IRS installation, an initial perusal of the return is made and the decision to audit or not to audit is also made. The problem concerns the processing of a return selected for an audit. After the return is received in the local IRS Audit Division, it is assigned to and examined by an auditor. This examination will result in one of two actions: either the auditor will request additional information about items on the return from the taxpayer, or the auditor can make a final decision about the case on the basis of the information already there. The items needing additional information are sent to the correspondence area in the department. Those cases on which a final decision has been made go to a reviewer in the department. Cases approved by the reviewer are completed cases and are sent to have the computations verified. From this area, the return can go to one of two places: if computation is correct, the return goes to the assessment/refund area; if computation is incorrect, the return goes to the adjustment area for computation correction and subsequently to the assessment/refund area. We are interested in the expected amount of time needed for a case to go through the auditing procedure.

6. Consider the following problem: Two animals are mated. From their offspring, two animals are selected by some method, and they are mated. The procedure is then repeated. Consider the case where we classify parents according to the pair of genes they carry. Assuming the gene is either type a or b, the individual animals must be of type aa, ab, or bb. The method of selection is as follows: One mate is selected at random and this offspring selects a mate. In its selection it is k times as likely to select a given animal unlike itself than a given

like animal. ($k = 1$ implies random selection, $k = 0$ implies selection like itself, *aa* and *ab* are alike, *bb* is different.) Mates must be one of the following types:

1	(*aa, aa*)	4	(*ab, bb*)
2	(*bb, bb*)	5	(*aa, bb*)
3	(*aa, ab*)	6	(*ab, ab*)

This is a stochastic process. The transition probabilities are as follows:

State	1	2	3	4	5	6
1	1	0	0	0	0	0
2	0	1	0	0	0	0
3	$\frac{1}{4}$	0	$\frac{1}{2}$	0	0	$\frac{1}{4}$
4	0	$\frac{1}{2(k+1)}$	0	$\frac{k}{k+1}$	0	$\frac{1}{2(k+1)}$
5	0	0	0	0	0	1
6	$\frac{1}{4(k+3)}$	$\frac{1}{4(3k+1)}$	$\frac{1}{k+3}$	$\frac{2k(k+1)}{(k+3)(3k+1)}$	$\frac{k(k+1)}{(k+3)(3k+1)}$	$\frac{1}{k+3}$

Model in GERT and find the expected number of matings and the probability of getting to state 1, given that the first pair were (*ab, bb*).

7. Consider the following hypothetical situation: There are a series of five equally spaced traffic lights. The probability of a given traffic light being green when the car reaches it is 0.1 less than the probability of the previous light being green, if the car did not stop for the previous light. If the car did stop for the previous light, the probability of making the present light green is 0.9. Thus, there is an effect of the traffic lights being timed. The time to drive between lights is taken as being normally distributed with a mean of 1 minute and a standard deviation of 0.1 minutes. If the car must stop for a light, the driving time becomes 1.6 minutes with a standard deviation of 0.2 minutes. Find the mean time to traverse the lights.

8. At a picnic, it is expected that when the food is first set out, 40% will eat one hamburger, 50% will eat two, and 10% will eat three. After a reasonable rest period, 80% will play volleyball. After the game it is estimated that, of those who did play, 30% will not want anything to eat, 50% will want one hamburger, and 20% will want two. Of those who didn't play, 20% will want one hamburger. How many hamburgers per person should be prepared?

9. A contractor is interested in building a piece of heavy equipment for the government. He has obtained the specifications for the equipment. Under the government's "total package procurement" policy, the contractor realizes that there are many pitfalls on the road to successful completion of the contract. From his previous experience on other contracts he estimates his chances and activity distributions as follows:
   (a) Once he submits his bid, he has a 50:50 chance of getting the contract.

(b) If he gets the contract, he has to complete the development and build a prototype model for government testing. The time to complete this activity is normally distributed with a mean of 36 weeks and a standard deviation of 4 weeks.

(c) Prototype testing time is normally distributed with a mean of 6 weeks and standard deviation of 1 week.

(d) At the end of prototype testing, four things could happen:

1–The project could be cancelled with probability 0.1.

2–Minor redesign could be called for with probability 0.2. If this happens, the time for redesign and building of replacement parts is distributed Poisson with a mean of 6 weeks. Return to activity 3.

3–With probability of 0.2, the government could call for a major redesign. In this case, return to activity 2.

4–With probability 0.5, the prototype model will be accepted and a go-ahead will be given for production.

(e) Production time is normally distributed with a mean of 20 weeks, and a standard deviation of 2 weeks.

(f) The first production model has to be tested. Testing time is a constant 2 weeks.

(g) At the end of 2 weeks, the government has the option to order additional modification (probability 0.1), cancel the project (probability 0.1), or order full production (probability 0.8). Any additional modification will require 4 weeks and the entire model has to be retested for 2 weeks.

What is the probability that the contractor's project will be cancelled or rejected? What is the mean time until he knows whether he can go ahead with production? What is the probability that he receives an order for full production?

10. In the world of gambling, GERT may be used to model the dice game commonly known as "craps." The game requires the player to throw two dice one or more times until he wins or loses. The player wins if on the first pass the sum of the two dice is either seven or eleven or, alternatively, if the first sum is 4, 5, 6, 8, 9, or 10 and the same sum appears before either 7 or 11 or subsequent passes. The player loses if on the first pass either of the sums 2, 3, or 12 appears, or if the first sum was 4, 5, 6, 8, 9, or 10, or 7 or 11 appears on subsequent passes before the first number shows again. Model the game and find the length of game and probability of winning, given that the first toss is a 6.

11. A destroyer has been assigned two regions in which to search for an enemy submarine. The captain has decided that the submarine will more likely be in region A; so he orders 60% of the search runs to be made in region A and 40% in region B. He also orders that the region be chosen randomly before each search. The submarine captain has decided to employ an evasive procedure which rules that after each search, if he is in region A, he will remain there with probability 0.6 and move to region B with probability 0.4; if he is in B, he will remain there with probability 0.5 and move to region A with probability 0.5. If the submarine is in region A and the destroyer searches region A,

the destroyer has a 0.8 chance of finding it. If the submarine is in region B and the destroyer searches there, the destroyer has a 0.7 chance of finding the submarine. What is the mean number of searches the destroyer will make before it finds the submarine?

12. Three contestants have weapons which they can use against one of their opponents. Man A has a probability of p_A of killing an opponent when he uses his weapon; man R has a probability of p_R of killing an opponent when he uses his weapon; and man C has a probability of p_C of killing his opponent when using his weapon. The object of A, R, and C is to defeat the other opponents. Suppose $p_A > p_R > p_C$ and R fires at A first. Then A would return R's fire. C being rational decides he would be better off not to fire until only one opponent is left. Two questions of interest for A are: (a) for what values of p_A, p_R, and p_C should A fire first, and (b) if $p_A = 0.90$, $p_R = 0.75$, and $p_C = 0.40$, what is the probability under the above firing sequence of A winning?

13. A child is instructed to wash four forks and four spoons. He decides to wash all the forks before starting on the spoons. Unfortunately, he is not tall enough to look into the sink, and he must reach in the sink and pull out a utensil. If he pulls out a fork, he will wash it; but if he pulls out a spoon, he will put it back into the sink until all the forks have been washed. What is the expected time to wash the forks if it takes two minutes to wash a fork and $\frac{1}{2}$ a minute to reach into the sink?

14. John and Henry are going to play handball. John is a much better player and the probability of him winning a point or service is 0.7, while the probability of Henry winning a point or service is 0.3. In handball, only the server can win points. To improve Henry's chances of winning a two-point game, John must score two consecutive points, and Henry's points do not have to be earned consecutively. Henry gets first service. What is the probability that John wins the game, and what is the expected number of services for John to win?

15. Three persons alternately flip a coin. The one who gets a head first is the winner. Find the probability of each player winning, and find the expected number of trials for each.

16. A player has three dollars. At each play of a game he loses one dollar with probability of 0.75, but he wins two dollars with a probability of 0.25. He stops playing if he has lost his three dollars or if he has won at least three dollars. What is his probability of going broke? How long will the game last?

17. I. Hunter, a big game hunter, is lost in the jungle and he has to overcome three obstacles to get out safely. The first obstacle is a fork in the road with three possible paths leading out. The first path takes him into a tribe of cannibals where he is eaten. The second path returns him to the fork, and the third path takes him to the next obstacle which is a treacherous river. At the river he has a choice of four boats. This first boat has a bad leak, which he does not know, and it causes him to be eaten by crocodiles. The second boat has a slow leak, which allows him time enough to return to shore. The third

boat causes him to return to the first obstacle, and the fourth boat allows him to proceed on to the next obstacle. The last obstacle is a cave with three tunnels. The first tunnel causes him to return to the first obstacle. The second tunnel causes him to fall off a hidden cliff. The third tunnel gets him to safety. Find the probability of the hunter getting out alive and the expected time for him to get out alive. This hunter has no memory and it takes one day to complete each path.

REFERENCES

1. Bell, W. A., "An Investigation of the Extension of Analytic GERT to Generalized Logic Circuits." Lehigh M. S. Thesis, June, 1971.

2. Cramer, H., *Mathematical Methods of Statistics*. Princeton, N.J., Princeton University Press, 1945.

3. Eisner, H., "A Generalized Network Approach to the Planning and Scheduling of a Research Program." *Operations Research*, Vol. 10, No. 1 (1962), pp. 115–122.

4. Elmaghraby, S. F., "An Algebra for the Analysis of Generalized Activity Networks." *Management Science*, Vol. 10, No. 3 (1964), pp. 494–514.

5. Elmaghraby, S. F., "On Generalized Activity Networks." *Journal of Industrial Engineering*, Vol. 18, No. 11 (November, 1966), pp. 621–631.

6. Feller, W., *An Introduction to Probability Theory and Its Applications*. New York, John Wiley & Sons, Inc., 1950.

7. Howard, R. A., *Dynamic Programming and Markov Processes*. Technology Press, Massachusetts Institute of Technology, Cambridge, Mass., and John Wiley & Sons, Inc., London, 1960.

8. Howard R. A., "Systems Analysis of Semi-Markov Processes." *IEEF Transactions on Military Electronics* (April, 1964), pp. 114–124.

9. Huggins, W. H., "Signal Flow Graphs." *Proc. IRE*, Vol. 9, No. 9 (1957), pp. 74–86.

10. Ishmael, P. C. and A. A. B. Pritsker, *User Manual for GERT EXCLUSIVE-OR Program*. NASA/ERC, NGR 03-011-034 (July, 1968).

11. Ishmael, P. C. and A. A. B. Pritsker, *Definitions and Procedures Employed in the GERT EXCLUSIVE-OR Program*. NASA/ERC, NGR 03-001-034 (July, 1968).

12. Kendall, M. G. and A. Stuart, *The Advanced Theory of Statistics*, Vol. 1. New York, Hanger Publishing Company, 1958.

13. Parzen, E., *Stochastic Processes*. San Francisco, Holden-Day, Inc., 1962.

14. Pritsker, A. A. B. and Happ, W. W., "GERT: Graphical Evaluation and Review Technique, Part I, Fundamentals." *Journal of Industrial Engineering*, Vol. 17, No. 5 (May, 1966) 267.

15. Pritsker, A. A. B. and Whitehouse, G. E., "GERT: Graphical Evaluation and Review Technique, Part II, Probabilistic and Industrial Engineering Applications." *Journal of Industrial Engineering*, Vol. 17, No. 6 (June, 1966).

16. Pritsker, A. A. B., *GERT: Graphical Evaluation and Review Technique.* Rand Memorandum RM-4973-NASA (April, 1966).

17. Whitehouse, G. E., *Extensions, New Developments, and Applications of GERT.* PhD. Dissertation, Arizona State University, August, 1965.

18. Whitehouse, G. E. and A. A. B. Pritsker, "GERT-Generating Functions, Conditional Distributions, Counters, Renewal Times, and Correlations." *AIIE Transactions* (March, 1969).

GERTS III: A SIMULATION APPROACH TO THE ANALYSIS OF STOCHASTIC NETWORKS

In Chapter 8 we saw that the analytic approach to solving GERT networks was inadequate in some circumstances. These inadequacies lead to the development of a simulation approach to the analysis of GERT networks. Pritsker [1, 2, 3, 4] has developed a series of GERT simulation programs. In this chapter we concentrate on the last of these programs, GERTS III. The material in this chapter is an adaptation of the program documentation for this system which was authored by Pritsker and Burgess [1].

The GERT simulation program is a general purpose program for simulating networks. The program is written in FORTRAN IV. The input to the program is a description of the network in terms of its nodes and branches, along with control information for setting up the simulation conditions.

The following list describes the features in GERTS III.

1. Branches that are characterized by:
(a) A probability of being included in the network.
(b) A time required to complete the activity represented by the branch. The time is specified by defining a parameter set number and a distribution type.
(c) A counter type to identify the branch as belonging to a particular group of branches.
(d) An activity number.

9 *

2. *Nodes that are characterized by:*
   (a) *The number of releases required to realize the node*
      for the first time.
   (b) *The number of releases required to realize the node*
      after the first time.
   (c) *The removal of events that are scheduled to release the node.*
   (d) *The method for scheduling the activities emanating from the node*
      (DETERMINISTIC or PROBABILISTIC).
   (e) *The statistical quantities to be estimated for the node.*
3. *Modification of the network based on the occurrence of end-of-activity*
   events during the simulation of the network.
4. *A method for tracing a set of simulation runs.*
5. *Automatic printout of the description of the network*
   and the final results.

During the research leading to GERTS III, the authors explored
the following concepts:

1. *Nodes that provided a storage or queue capability—a Q-node.*
2. *Costs associated with the performance of activities.*
3. *Activities that required resources.*

---

*The material in this chapter has been adapted from Reference 1 and is used with the permission of Dr. Pritsker.

Pritsker and Burgess were able to incorporate to some degree each of these features in GERTS III. They developed three separate programs: (1) GERTS IIIQ, a GERT network simulation program that includes Q-nodes; (2) GERTS IIIC, a GERT network simulation program that collects cost statistics; and (3) GERTS IIIR, a GERT network simulation program that involves resource allocation decisions. Each of these programs contained the basic GERTS III program and could be used to solve the GERTS III system.

In this chapter we will concentrate on GERTS III and GERTS IIIQ. The GERTS IIIQ system used to solve the examples in this and succeeding chapters is available as described in the preface.

9.1 GERTS III Network Characteristics

GERT networks consist of nodes and directed branches. We will first consider the characteristics that describe a node. The number of releases associated with a node specifies the number of times activities incident to the node must be realized before the node can be realized. So when the number of releases is 1, we can think of the input side of the node being an OR operation. If the number of releases equals the number of activities incident to the node, the node is thought of as an AND operator. However, it is permissible to specify the number of releases to be less than or greater than the number of activities incident to the node. For example, the number of releases can be 2, whereas the number of activities incident to the node could be 3. This would represent the case where, if 2 of the 3 activities were realized, the node is realized. Alternatively, the number of releases can be 2, and the number of activities incident to the node could be 1. This would represent the case where the activity must be realized twice before the node is realized.

In Fig. 9.1 we illustrate the node symbolism for GERTS III. As in GERT, we see that the semicircle, $\supset$, on the output side of a node, is used to represent a DETERMINISTIC output, and a lazy V, $>$, is used for for a

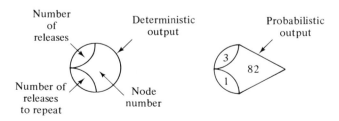

Figure 9.1. Node system for GERTS III.

PROBABILISTIC output. Nodes are also characterized by their function in the network. A GERT analyst can specify a node as:

1. A source node.
2. A sink node.
3. A statistics node.
4. A mark node.

Activities emanating from a source node are started at time zero. A sink node is a node that indicates that the network may be realized when it is realized. A statistics node is one on which statistics are maintained. All sink nodes are automatically made statistics nodes. A mark node establishes a reference time and permits the calculation of the time it takes to go between two nodes of the network.

For statistics nodes, GERTS III obtains statistical estimates associated with the time a node is realized. Five types of time statistics are possible:

F. The time of *first* realization of a node.
A. The time of *all* realizations of a node.
B. The time *between* realizations of a node.
I. The time *interval* required to go between two nodes in the network.
D. The time *delay* from first activity completion on the node until the node is realized.

The nodes on which statistics are to be collected and the type of statistics desired are part of the description given to a node by the input to GERTS III. They are not part of the graphical representation.

The branches of GERT networks represent activities and/or information transfers. We will use the term *activity* to identify both. Associated with activities are a probability that the activity will be realized, given that its start node is realized, and a time to perform the activity, given that the activity is realized. For GERTS III the time variable is specified by a parameter set number and a distribution type. We can use the following nine distribution types:

1. Constant.
2. Normal.
3. Uniform.
4. Erlang.
5. Lognormal.
6. Poisson.
7. Beta.
8. Gamma.
9. Beta fitted to three parameters as in PERT.

The parameter set number along with the distribution type completely describe the time variable associated with an activity. Each distribution type specifies the arrangement of the parameters in a parameter set. With GERTS III, two additional characteristics can be associated with an activity. These are a counter type and an activity number.

The counter type number specifies the counter to be increased by 1 every time the activity is realized. The number of counter types permitted is limited to 4, but changes in the dimension of two arrays can increase this value. Any number of activities may be associated with a counter type. However, activities incident to nodes on which delay statistics are collected cannot have counter types associated with them.

Statistics are automatically kept on the counter types. At the end of all simulation runs, the average and standard deviation of the number of times a counter type was realized prior to the realization of each node for which statistics are collected is determined and printed. In addition, the minimum and maximum numbers of times activities having the specified counter type were realized during a simulation are printed. Since the number of counts is always referenced to the realization of a node, the number of counts occurring prior to the realization of a node may be different in different simulation runs due to the sequence in which the nodes are realized.

Activity numbers are placed on activities to permit network modifications based on the realization of the activity. Specification of an activity number does not automatically indicate that the network will be modified. However, only activities with activity numbers can cause the network to be modified. Network modification involves the replacing of a node by another node on the output side only. Thus when a node is realized, the activities to be started depend on the modifications that have taken place. For example if node 8 replaces node 5, then when node 5 is realized, the activities emanating from node 8 are scheduled to start. A node may be changed many times before it is actually realized.

The activity number causing the network modification along with all the nodes to be replaced, and the nodes to be inserted, are specified by the user. Figure 9.2 illustrates the branch and node modification notation that will be used throughout this book. Modifications will be shown by a dashed branch with the activity number attached in a square. The modification in Fig. 9.2 is read "the output of node 2 is replaced by node 4 when activity 1 is realized."

As an illustration of the use of activities, we will consider the network shown in Fig. 9.3. This network represents the changing of the network structure when the self-loop about node 2 is taken three times. The change is accomplished in the following manner. The output of node 3 is DETERMINISTIC; therefore, every time node 3 is realized, both branches emanating from node 3 are taken. The branch from node 3 to node 4 is used to count the number of times node 3 is realized. Node 4 is realized only when the branch incident to it is realized three times (of course, this corresponds

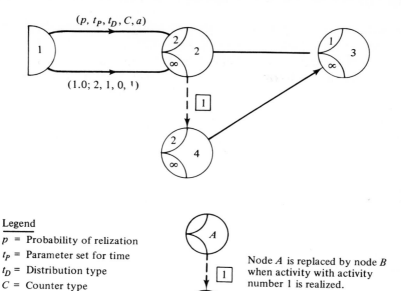

Legend

p = Probability of relization

t_P = Parameter set for time

t_D = Distribution type

C = Counter type

a = Activity number

Node A is replaced by node B when activity with activity number 1 is realized.

Figure 9.2. Illustration of branch descriptors and network modification symbolism.

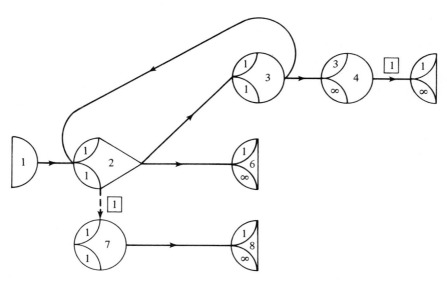

Figure 9.3. Network containing information branches in addition to activities.

to three traversals of the self-loop). When node 4 is realized, the activity labeled activity number 1 (a 1 on the network in this case) causes node 2 to be replaced by node 7, and the objective of changing the network is achieved. Care is required here to ensure that the branch from node 4 to node 5 is realized prior to the realization of the branch from node 3 to node 2. If a zero time is associated with both branches, normal operation would have the branch from node 3 to node 2 realized first since it was scheduled first. By assigning a small negative time (-0.000001) to branch from node 4 to node 5, the desired ordering can be obtained. Of significance in the foregoing network is the incorporation into the network of branches representing activities and branches representing information transfers. The inclusion of different types of branches within a GERTS III network expands the network modeling capability within the GERT framework.

9.2 Program Operating Procedure

The GERTS III program performs a simulation of a network by advancing time from event to event. In simulation parlance this is termed a *next-event simulation*. The events associated with a simulation of a GERT network are: (1) start of the simulation, (2) end of an activity, and (3) completion of a simulation run of the network. Since GERTS III is a FORTRAN IV program, the operating procedure is the standard FORTRAN operating procedure. Many concepts of GERTS III were adopted from GASP IIA [5].

The start event causes all source nodes to be realized and schedules the activities emanating from the source nodes according to the output type of the source node. The output type for all nodes is either DETER-MINISTIC or PROBABILISTIC. In the former case, all activities emanating from the node are scheduled, and in the latter case, only one of the activities emanating from the node is scheduled. By scheduling an activity is meant that an event "end of activity" is caused to occur at some future point in time. The simulation proceeds from event to event until the conditions which indicate that the simulation of the network is completed are obtained. The process is then repeated for a specified number of simulations of the network.

As part of the input data, the number of releases required to realize a node is specified. Each time an end-of-activity event occurs, the number of releases for the end node of that activity is decreased by one. When the number of releases remaining is zero, the node is realized. At this time the number of releases is set equal to the number of releases required to realize the node after the first time, and the activities emanating from the node are scheduled.

Again, the number of activities scheduled depends on the type for the node.

For each activity scheduled, an end-of-activity event is put in a file containing all events in chronological order. The end-of-activity events are removed from the event file one at a time, and at each removal instant, a test is performed to determine if a node is realized. If a node is not realized, the next event is removed from the event file. If a node is realized, activities from that node are scheduled and the simulation is continued. The simulation ends when a prescribed number of sink nodes have been realized. As part of the input data, the number of source nodes, sink nodes, nodes on which statistics are collected as well as their node numbers, and the number of nodes required to realize the network are defined.

The above process describes one simulation of a network. The program is written to allow multiple simulations to be performed. The number of simulation runs to be performed is part of the input data. The GERT simulation program automatically initializes the pertinent variables in order that consecutive simulations of the same network can be performed and, if desired, permits simulations of different networks to be performed consecutively.

9.3 Input to GERTS III and Limitations

The input requirements for GERTS III consist of at most seven different types of data cards. These seven cards describe the network and the control information for performing the simulation. A general description of each card is given as follows:

Data Card Type	General Description
1	Identification information, number of times simulation is to be performed, and an initial random number seed (one card).
2	General node, counter, and network modification data (one card).
3	Description of each node (one card for each node).
4	Parameters of time variables associated with activities (one card for each parameter set).
5	Description of each activity (one card for each activity).
6	Network modifications desired (one card for each activity that modifies network. If none, no data card type 6 is required.).
7	Run numbers to be traced (one card only if tracing is requested by using a negative project number).

The following represents a detailed description of the input format for GERTS III:

Data Card 1

Field 1	The analyst's name (6A2).
Field 2	The project number (I4). (If negative, data card 7 is required to indicate the runs to be traced.)
Field 3	The month number (I2).
Field 4	The day number (I2).
Field 5	The year (I4).
Field 6	The number of times the network is to be simulated (I4).
Field 7	The number of activities with different time characteristics (I4).
Field 8	The number of branches in the network plus an estimate of the maximum number of activities which can occur simultaneously (I4).
Field 9	An integer random number seed (I8).
Field 10	A floating point random number seed (F10.4).

Data Card 2

Field 1	The largest node number of the network including all possible modifications to the network (I3). *The smallest node number permitted is 2. Node 1 is illegal in the GERTS III System.*
Field 2	Number of source nodes (I3).
Field 3	Number of sink nodes (I3).
Field 4	Number of sink nodes that must be realized before the network is realized (I3).
Field 5	Number of nodes which statistics are to be collected on, including all sink nodes (I3).
Field 6	Number of types of counts (I3).
Field 7	1 if network modifications exist; 0 otherwise (I3).

Data Card 3

Field 1	The node number (descriptor) associated with the node characteristics given on this card (I3).
Field 2	Special characteristic of the node. Codes for special characteristics are: 1. Source node 2. Sink node 3. Node on which statistics are collected 4. Mark node If field 2 is left blank, no special characteristic is associated with the node (I3).

Field 3	The number of releases required to realize the node for the first time (I3).
Field 4	The number of releases required to realize the node after the first realization (I3).
Field 5	Output characteristic of the node. Codes for input are: *P* for PROBABILISTIC; and *D* for DETERMINISTIC (A1).
Field 6	If events that have been scheduled to end on this node are to be removed (cancelled) when this node is realized, an *R* should be put in this field. If removal is not desired, leave blank (A1).

Fields 7, 8, and 9 are used only if:

The node is a sink node or a statistics node (code 2 or 3 in field 2).

Field 7	The lower limit of the second cell for the histogram to be obtained for this node. The first cell of the histogram will contain the number of times the node was realized in a time less than the value given in this field (F6.2).
Field 8	The width of each cell of the histogram. Each histogram contains 32 cells. The last cell will contain the number of times the node was realized in a time greater than or equal to the lower limit (specified in field 7) + $30 \times$ [cell width (specified by field 8)] (F6.2).
Field 9	Statistical quantities to be collected (A1):

    F. The time of *first* realization of the node.
    A. The time of *all* realizations of the node.
    B. The time *between* realizations of the node.
    I. The time *interval* required to go between two nodes.
    D. The time *delay* from first activity completion on the node until the node is realized.

The last card of this type must have a zero in field 1.

DATA CARD 4

The parameters associated with the distribution of the time to perform each activity are contained on data card 4. One card is required for each activity with a different time characterization. The number of cards is specified by data card 1, field 7. A maximum of 300 is permitted. The cards must be arranged by ascending parameter number, and the parameters must be numbered consecutively or blank cards appropriately placed. Nine distribution types are available:

1. Constant.
2. Normal.
3. Uniform.
4. Erlang.
5. Lognormal.
6. Poisson.
7. Beta.
8. Gamma.
9. Beta fitted to three parameters as in PERT.

The fields required are dependent on the distribution type of the activity and are as follows:

For Distribution Type 1 (Constant)

Field 1   The constant time (F10.4).

For Distribution Type 2 (Normal); 5 (Lognormal); 7 (Beta); and 8 (Gamma)

Field 1	The mean value (F10.4).
Field 2	The minimum value (F10.4).
Field 3	The maximum value (F10.4).
Field 4	The standard deviation (F10.4).

For Distribution Type 3 (Uniform)

Field 1	Not used (F10.4).
Field 2	The minimum value (F10.4).
Field 3	The maximum value (F10.4).
Field 4	Not used (F10.4).

For Distribution Type 4 (Erlang)

Field 1	The mean time for the Erlang variable divided by the value given to field 4 (F10.4).
Field 2	The minimum value (F10.4).
Field 3	The maximum value (F10.4).
Field 4	The number of exponential deviates to be included in the sample obtained from the Erlang distribution (F10.4).

If field 4 is set equal to 1, an exponential deviate will be obtained from distribution type 4.

For Distribution Type 6 (Poisson)

Field 1	The mean minus the minimum value (F10.4).
Field 2	The minimum value (F10.4).
Field 3	The maximum value (F10.4).
Field 4	Not used (F10.4).

Care is required when using the Poisson since it is not usually used to represent an interval of time. The interpretation of the mean should be the mean number of time units per time period.

For Distribution Type 9 (Beta Fitted to Three Values as in PERT)

Field 1	The most likely value, m (F10.4).
Field 2	The optimistic value, a (F10.4).
Field 3	The pessimistic value, b (F10.4).
Field 4	Not used (F10.4).

Samples are obtained from the distributions such that, if a sample is less than the minimum value, the sample value is given the minimum value. Similarly, if the sample is greater than the maximum value, the sample value is assigned the maximum value. This is not sampling from a truncated distribution, but sampling from a distribution with a given probability of obtaining the minimum and maximum values.

DATA CARD 5

One data card for each activity associated with the network:

Field 1	Probability of realization (F8.3).
Field 2	Start node (I3).
Field 3	End node (I3).
Field 4	Parameter number (I3).
Field 5	The distribution type (I3).
Field 6	Count type (I3).
Field 7	Activity number (I3).

The last data card of this type *must* have a zero (or blank) in field 2.

DATA CARD 6

Required only if number of nodes modified is greater than zero (field 7, data card 2).

Field 1	An activity number (I3).
Field 2	The number of the node to be replaced if the activity given in field 1 is realized (I3).
Field 3	The number of the node to be inserted into the network in place of the node specified in field 2 when the activity in field 1 is realized (I3).
Fields 4–21	Fields 2 and 3 are repeated if the activity given in field 1 affects multiple nodes. A zero in an even-numbered field indicates the end of the data on the card.

The last card of this type *must* have a zero in field 1.

DATA CARD 7

Used only if the project number is negative (field 2, data card 1).

Field 1	The run number for which tracing of the end-of-activity events should begin (I3).
Field 2	The run number for which tracing of the end-of-activity events should terminate (I3).

Multiple networks can be analyzed by stacking the data cards as described above, one after another. No blank cards should separate the data cards for each network. A blank card is required to indicate the end of all networks to be simulated.

The dimensions of the GERTS III program have been set to allow for a maximum of 999 nodes, 999 activities, 4 counter types, collections of statistics on 100/(number of counter types $+$ 1) nodes, and 300 parameter sets.

9.4 Examples of the Use of GERTS III

In this section we will illustrate the use of the GERTS III computer package. The examples chosen are taken from Pritsker and Burgess [1]. They are presented to illustrate the features of the GERTS III system. Other examples of the application GERTS III will be given in Chapters 10 and 11.

EXAMPLE 1.   ILLUSTRATION OF THE FEATURES OF GERTS III

Figure 9.4 shows the network to be analyzed in Example 1. The source node for the network is node 2, and the sink node is node 12. From node 2, three activities emanate which are performed simultaneously. These activities cause nodes 3, 4, and 5 to be realized. The activities emanating from nodes 3, 4, and 5 are all incident to node 6. The number of releases required to realize node 6 is three; therefore node 6 will be realized only when all three activities incident to node 6 are realized. For this example, we desire to obtain statistics on the time delay between the completion of the first activity incident to node 6 and the time node 6 is realized. To obtain these statistics, we define node 6 as a statistics node with delay statistics (code D) desired.

We also desire to collect statistics on the time required to go from node 7 to node 11. To accomplish this, we define node 7 as a mark node (node type 4) and node 11 as a statistics node with the statistics calculated on an interval basis (code I). The statistical quantities collected at node 11 will be the interval of time from the realization of node 7 to the realization of node 11. If alternative paths existed between node 7 and 11, the interval

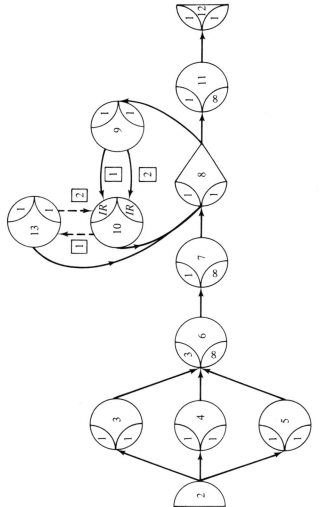

Figure 9.4. GERT network for Example 1.

315

node would collect statistics on the time required to traverse the separate paths.

Node 8 has a PROBABILISTIC output side so that either activity emanating from node 8 can be taken. For those situations where the feedback path is taken, it is desired to determine the time required to traverse the feedback path. To accomplish this we make node 8 a statistics node with statistics collected on time between realization of node 8 (code B). Nodes 9 and 10 are also defined as statistics nodes. For node 9 statistics are collected on its first realizations (code F) and for node 10 statistics are collected on all realizations (code A). Thus, this example includes all the types of statistical calculations that can be included within the GERTS III program. Since node 12 is a sink node, statistics will automatically be collected on it. For this example, statistics on the first realization of node 12 were specified. If the type of statistics desired is not specified by the input information, the GERTS III program assumes that statistics on first realization are desired, i.e., the default condition is the first realization.

Also included in this example is the network modification feature and the stopping of an activity in progress. If the top activity (activity 1) from node 9 to node 10 is realized first, then node 10 is replaced by node 13. If the bottom activity (activity number 2) is realized first, then node 10 remains in the network. If node 10 had been replaced by node 13, then node 10 is reinstated into the network. The network modification is implemented by assigning activity numbers to the branches between node 9 and node 10. When either of these activities is completed, the network modification is implemented and the other activity is stopped since node 10 has an *R* assigned to it. The removal of scheduled activities incident to a node applies to all realizations of the node.

A listing of the input cards for this example is shown in Table 9.1. The description of the network that is printed by the GERTS III program is shown in Tables 9.2 and 9.3. Table 9.4 presents a trace of a simulation run for this network. We will use this trace to describe the operating procedure of the GERTS III program in simulating the network shown in Fig. 9.4.

The simulation begins by scheduling end-of-activity completion events from each source node. For Example 1, the source node is node 2 and end-of-activity events are scheduled for the activities from node 2 to node 3, node 2 to node 4, and node 2 to node 5. To obtain the time for each of these events, samples are drawn from (1) a normal distribution using parameter set 1, (2) the Erlang distribution using parameter set 2, and (3) the uniform distribution using parameter set 3.

The trace of the simulation starts with the first end-of-activity event. This is the activity that is incident to node 5 and the event occurs at time 1.88. Since node 5 had its number of releases equal to 1, node 5 is realized and the activity from node 5 to node 6 can be initiated. An end-of-activity

Table 9.1. INPUT DATA FOR EXAMPLE 1.

```
ALL FEATURES  -1   5201973 500   11   40    1267          EX 1   10

·13  1   1   1    6   1   1                               EX 1   20

 2   1       D                                            EX 1   30
 3       1  1D                                            EX 1   40
 4       1  1D                                            EX 1   50
 5       1  1D                                            EX 1   60
 6   3   3   D    1        1   D                          EX 1   70
 7   4   1  1D                                            EX 1   80
 8   3   1  1P    1        1   B                          EX 1   90
 9   3   1  1D    27       5   F                          EX 1  100
10   3   1  1DR   30      2    A                          EX 1  110
11   3   1   D     8      2    I                          EX 1  120
12   2   1   D    35      2    A                          EX 1  130
13       1  1D                                            EX 1  140
 0                                                        EX 1  150

   10          0      100        1                        EX 1  160
    2          0      100        2                        EX 1  170
    3          0        5        5                        EX 1  180
    4          0      100        1                        EX 1  190
    5          0      100                                 EX 1  200
    6          0      100                                 EX 1  210
  .6425      -5.0     5.0      1.492                      EX 1  220
    8          0      100        6                        EX 1  230
    3          2        5                                 EX 1  240
    0                                                     EX 1  250
  1.358     0.0      100.0     0.218                      EX 1  260

    1     2   3   1   2                                   EX 1  270
    1     2   4   2   4                                   EX 1  280
    1     2   5   3   3                                   EX 1  290
    1     3   6   4   4                                   EX 1  300
    1     4   6   5   6                                   EX 1  310
    1     5   6   6   1                                   EX 1  320
    1     6   7  11   5                                   EX 1  330
    1     7   8   8   1                                   EX 1  340
    6     8   9   9   9                                   EX 1  350
    4     8  11   7   8                                   EX 1  360
    1     9  10   3   7       1                           EX 1  370
    1     9  10   9   3       2                           EX 1  380
    1    10   8  10   1                                   EX 1  390
    1    11  12   1   2                                   EX 1  400
    1    13   8   1   1   1                               EX 1  410
          0                                              EX 1  420

  1 10 13   0                                            EX 1  430
  2 13 10   0                                            EX 1  440
  0                                                      EX 1  450

  1   5                                                  EX 1  460
```

Table 9.2. ECHO CHECK FOR EXAMPLE 1.

GERT SIMULATION PROJECT −1 BY ALL FEATURES
DATE 5/20/ 1973
NETWORK DESCRIPTION
NODE CHARACTERISTICS

HIGHEST NODE NUMBER IS   13
NUMBER OF SOURCE NODES IS   1
NUMBER OF SINK NODES IS   1
NUMBER OF NODES TO REALIZE THE NETWORKS IS   1
STATISTICS COLLECTED ON   6 NODES
NUMBER OF PARAMETER SETS IS   11
INITIAL RANDOM NUMBER IS   1267   0.0

NODE	NUMBER RELEASES	NUMBER OF RELEASES FOR REPEAT	OUTPUT TYPE	REMOVAL DESIRED AT REALIZATION	STATISTICS BASED ON REALIZATIONS
2	0	9999	D		
3	1	1	D		
4	1	1	D		
5	1	1	D		
6	3	9999	D		D
7	−1	−1	D		
8	1	1	P		B
9	1	1	D		F
10	1	1	D	R	A
11	1	9999	D		I
12	1	9999	D		A
13	1	1	D		

SOURCE NODE NUMBERS
    2
SINK NODE NUMBERS
    12
STATISTICS COLLECTED ALSO ON NODES
    6   11   10   9   8

event for this activity is then scheduled by the program. At time 4.31, the activity on node 4 is completed as shown by the second line in the trace of Fig. 9.4. At time 7.88 the activity from node 5 to node 6 was completed, and we have the first activity incident to node 6 being completed. (To determine from the trace that this was the activity from node 5 to node 6, the attributes of the activity are examined.) Since node 6 is a delay node, the program records 7.88 as the time of first completion of an activity that is incident to node 6. At time 8.31, the activity from node 4 to node 6 is completed. At time 10.16 the activity incident to node 3 is completed and node 3 is realized. The activity from node 3 to node 6 is then scheduled and is completed at time 12.09. This is the third activity that is realized incident to node 6; hence, node 6 is realized. The time from the first activity completion on node 6 to the time that node 6 is realized is the delay time. This value is 4.21

Table 9.3. FURTHER ECHO CHECK FOR EXAMPLE 1.

ACTIVITY PARAMETERS

PARAMETER	PARAMETERS			
NUMBER	1	2	3	4
1	10.0000	0.0	100.0000	0.1000
2	2.0000	0.0	100.0000	2.0000
3	3.0000	0.0	5.0000	0.5000
4	4.0000	0.0	100.0000	1.0000
5	5.0000	0.0	100.0000	0.0
6	6.0000	0.0	100.0000	0.0
7	0.6425	−5.0000	5.0000	1.4920
8	8.0000	0.0	100.0000	0.6000
9	3.0000	2.0000	5.0000	0.0
10	0.0	0.0	0.0	0.0
11	1.3580	0.0	100.0000	0.2180

ACTIVITY DESCRIPTION

START NODE	END NODE	PARAMETER NUMBER	DISTRIBUTION TYPE	COUNT TYPE	ACTIVITY NUMBER	PROBABILITY
2	3	1	2	0	0	1.0000
2	4	2	4	0	0	1.0000
2	5	3	3	0	0	1.0000
3	6	4	4	−1000	0	1.0000
4	6	5	6	−1000	0	1.0000
5	6	6	1	−1000	0	1.0000
6	7	11	5	0	0	1.0000
7	8	8	1	0	0	1.0000
8	9	9	9	0	0	0.6000
8	11	7	8	0	0	0.4000
9	10	3	7	0	1	1.0000
9	10	9	3	0	2	1.0000
10	8	10	1	0	0	1.0000
11	12	1	2	0	0	1.0000
13	8	1	1	1	0	1.0000

NETWORK MODIFICATIONS

ACTIVITY	NODE	FILE NODE	FILE NODE	FILE NODE	FILE NODE	FILE NODE	FILE NODE
1	10	13					
2	13	10					

(12.09 − 7.88) and is one sample of the delay time associated with node 6. Next the activity from node 6 to node 7 is scheduled. This activity is completed at time 13.73.

Node 7 is a mark node and the time 13.73 is identified with the path of

Table 9.4. TRACING OF ACTIVITY COMPLETIONS FOR THE SIMULATION OF THE NETWORK IN EXAMPLE 1.

	TIME		NODE		ATTRIBUTES		RUN
AT TIME	1.88	ACTIVITY ON NODE	5	WITH ATTRIBUTES	3 3 0 0	WAS REALIZED ON RUN	1
AT TIME	4.31	ACTIVITY ON NODE	4	WITH ATTRIBUTES	2 4 0 0	WAS REALIZED ON RUN	1
AT TIME	7.88	ACTIVITY ON NODE	6	WITH ATTRIBUTES	6 1–1000 0	WAS REALIZED ON RUN	1
AT TIME	8.31	ACTIVITY ON NODE	6	WITH ATTRIBUTES	5 6–1000 0	WAS REALIZED ON RUN	1
AT TIME	10.16	ACTIVITY ON NODE	3	WITH ATTRIBUTES	1 2 0 0	WAS REALIZED ON RUN	1
AT TIME	12.09	ACTIVITY ON NODE	6	WITH ATTRIBUTES	4 4–1000 0	WAS REALIZED ON RUN	1
AT TIME	13.73	ACTIVITY ON NODE	7	WITH ATTRIBUTES	11 5 0 0	WAS REALIZED ON RUN	1
AT TIME	21.73	ACTIVITY ON NODE	8	WITH ATTRIBUTES	8 1 0 0	WAS REALIZED ON RUN	1
AT TIME	25.56	ACTIVITY ON NODE	9	WITH ATTRIBUTES	9 9 0 0	WAS REALIZED ON RUN	1
AT TIME	28.79	ACTIVITY ON NODE	10	WITH ATTRIBUTES	3 7 0 1	WAS REALIZED ON RUN	1
AT TIME	38.79	ACTIVITY ON NODE	8	WITH ATTRIBUTES	1 1 1 0	WAS REALIZED ON RUN	1
AT TIME	41.72	ACTIVITY ON NODE	9	WITH ATTRIBUTES	9 9 0 0	WAS REALIZED ON RUN	1
AT TIME	43.98	ACTIVITY ON NODE	10	WITH ATTRIBUTES	3 7 0 1	WAS REALIZED ON RUN	1
AT TIME	53.98	ACTIVITY ON NODE	8	WITH ATTRIBUTES	1 1 1 0	WAS REALIZED ON RUN	1
AT TIME	57.10	ACTIVITY ON NODE	9	WITH ATTRIBUTES	9 9 0 0	WAS REALIZED ON RUN	1
AT TIME	59.54	ACTIVITY ON NODE	10	WITH ATTRIBUTES	9 3 0 2	WAS REALIZED ON RUN	1
AT TIME	59.54	ACTIVITY ON NODE	8	WITH ATTRIBUTES	10 1 0 0	WAS REALIZED ON RUN	1
AT TIME	62.19	ACTIVITY ON NODE	9	WITH ATTRIBUTES	9 9 0 0	WAS REALIZED ON RUN	1
AT TIME	65.39	ACTIVITY ON NODE	10	WITH ATTRIBUTES	3 7 0 1	WAS REALIZED ON RUN	1
AT TIME	75.39	ACTIVITY ON NODE	8	WITH ATTRIBUTES	1 1 1 0	WAS REALIZED ON RUN	1
AT TIME	78.23	ACTIVITY ON NODE	9	WITH ATTRIBUTES	9 9 0 0	WAS REALIZED ON RUN	1
AT TIME	81.89	ACTIVITY ON NODE	10	WITH ATTRIBUTES	3 7 0 1	WAS REALIZED ON RUN	1
AT TIME	91.89	ACTIVITY ON NODE	8	WITH ATTRIBUTES	1 1 1 0	WAS REALIZED ON RUN	1
AT TIME	95.05	ACTIVITY ON NODE	9	WITH ATTRIBUTES	9 9 0 0	WAS REALIZED ON RUN	1
AT TIME	97.56	ACTIVITY ON NODE	10	WITH ATTRIBUTES	3 7 0 1	WAS REALIZED ON RUN	1
AT TIME	107.56	ACTIVITY ON NODE	8	WITH ATTRIBUTES	1 1 1 0	WAS REALIZED ON RUN	1
AT TIME	112.56	ACTIVITY ON NODE	11	WITH ATTRIBUTES	7 8 0 0	WAS REALIZED ON RUN	1
AT TIME	122.65	ACTIVITY ON NODE	12	WITH ATTRIBUTES	1 2 0 0	WAS REALIZED ON RUN	1
AT TIME	1.29	ACTIVITY ON NODE	4	WITH ATTRIBUTES	2 4 0 0	WAS REALIZED ON RUN	2
AT TIME	2.98	ACTIVITY ON NODE	5	WITH ATTRIBUTES	3 3 0 0	WAS REALIZED ON RUN	2

AT TIME	ACTIVITY ON NODE	WITH ATTRIBUTES				WAS REALIZED ON RUN
4.29	6	5	6-1000	0	0	2
8.98	6	6	1-1000	0	0	2
10.11	3	1	2	0	0	2
20.39	6	4	4-1000	0	0	2
21.97	7	11	5	0	0	2
29.97	8	8	1	0	0	2
32.77	9	9	9	0	0	2
35.39	10	9	3	0	2	2
35.39	8	10	1	0	0	2
38.47	11	7	8	0	0	2
48.63	12	1	2	0	0	2
3.57	5	3	3	0	0	3
9.57	6	6	1-1000	0	0	3
9.88	3	1	2	0	0	3
14.02	4	2	4	0	0	3
15.08	6	4	4-1000	0	0	3
20.02	7	5	6-1000	0	0	3
21.35	8	11	5	0	0	3
29.35	9	8	1	0	0	3
31.92	10	9	9	0	2	3
35.10	8	10	3	0	0	3
35.10	9	9	9	0	0	3
39.08	10	3	7	0	1	3
42.70	8	1	1	0	0	3
52.70	11	7	8	0	0	3
53.36	12	1	2	0	0	3
63.29	5	3	3	0	0	3
4.99	4	2	4	0	0	4
5.52	6	5	6-1000	0	0	4
7.52	3	1	2	0	0	4
10.10	6	6	1-1000	0	0	4
10.99	6	6	1-1000	0	0	4

Table 9.4. Continued

AT TIME 13.71 ACTIVITY ON NODE 6 WITH ATTRIBUTES 4 4-1000 0 WAS REALIZED ON RUN 4
AT TIME 15.03 ACTIVITY ON NODE 7 WITH ATTRIBUTES 11 5 0 WAS REALIZED ON RUN 4
AT TIME 23.03 ACTIVITY ON NODE 8 WITH ATTRIBUTES 8 1 0 WAS REALIZED ON RUN 4
AT TIME 24.43 ACTIVITY ON NODE 11 WITH ATTRIBUTES 7 8 0 WAS REALIZED ON RUN 4
AT TIME 34.35 ACTIVITY ON NODE 12 WITH ATTRIBUTES 1 2 0 WAS REALIZED ON RUN 4
AT TIME 1.04 ACTIVITY ON NODE 5 WITH ATTRIBUTES 3 3 0 WAS REALIZED ON RUN 5
AT TIME 2.91 ACTIVITY ON NODE 4 WITH ATTRIBUTES 2 4 0 WAS REALIZED ON RUN 5
AT TIME 3.91 ACTIVITY ON NODE 6 WITH ATTRIBUTES 5 6-1000 0 WAS REALIZED ON RUN 5
AT TIME 7.04 ACTIVITY ON NODE 6 WITH ATTRIBUTES 6 1-1000 0 WAS REALIZED ON RUN 5
AT TIME 10.13 ACTIVITY ON NODE 3 WITH ATTRIBUTES 1 2 0 WAS REALIZED ON RUN 5
AT TIME 18.45 ACTIVITY ON NODE 6 WITH ATTRIBUTES 4 4-1000 0 WAS REALIZED ON RUN 5
AT TIME 19.76 ACTIVITY ON NODE 7 WITH ATTRIBUTES 11 5 0 WAS REALIZED ON RUN 5
AT TIME 27.76 ACTIVITY ON NODE 8 WITH ATTRIBUTES 8 1 0 WAS REALIZED ON RUN 5
AT TIME 28.38 ACTIVITY ON NODE 11 WITH ATTRIBUTES 7 8 0 WAS REALIZED ON RUN 5
AT TIME 38.26 ACTIVITY ON NODE 12 WITH ATTRIBUTES 1 2 0 WAS REALIZED ON RUN 5

activities following node 7. The activity emanating from node 7 is then completed at time 21.73, and node 8 is realized. This value is recorded as the first time node 8 is realized since the time between realizations of node 8 is desired. From the trace, we see that node 9 is realized next at time 25.56. This indicates that the branching operation took the branch from node 8 to node 9 for this simulation of the network. Statistics are collected on node 9, which was realized at time 25.56. The two activities emanating from node 9 are then scheduled. Activity 1, the upper branch, is completed first at time 28.79. This causes node 10 to be realized, and the value of 28.79 is recorded as a time of realization of node 10. Since node 10 removes all activities scheduled to be completed that are incident to node 10, activity 2 is halted. Since activity 1 has been completed, node 10 is replaced by node 13 according to the prescribed network modification.

Branching from node 13 is now done. The trace indicates this by the attributes associated with the end-of-activity event on node 8, which are the attributes for the branch between node 13 and node 8. The time between realizations of node 8 is collected, and the current time is used as the last time node 8 was realized. Again the branching process selects the activity from node 8 to node 9, and the loop around node 8 is traversed again. On this second traversal of the loop, activity 1 again was completed before activity 2, and the branch from node 13 to node 8 is included in the network. On the third traversal of the loop, activity 2 was completed before activity 1, and the branch from node 10 to node 8 which involved no time delay is included in the network. Finally, at time 107.56, node 8 is realized, and the branching process directs that the activity from node 8 to node 11 be completed. Node 11 is realized at time 112.56. Since node 11 is an interval node, a value is calculated which represents the time to go from node 7 to node 11. In this case it is 98.78 (112.56 − 13.78). The activity from node 11 to node 12 is scheduled and completed at 122.65. At this time node 12 is realized. Since node 12 is the sink node of the network and since it only takes one realization of the sink node to realize the network, the network is realized. The value of 122.65 is then recorded as one sample of the time to realize node 12 or, equivalently, the time to realize the network. This completes one simulation run of the network.

Several comments on the statistics collected on nodes 8, 9, and 10 are in order. For run 1, node 8 was realized seven times; therefore, six values were calculated for the time between realizations of node 8. For node 9, statistics are collected on the time of first realization; therefore, only the value 25.56 is recorded as the appropriate sample on run 1 for node 9. For node 10, all realization times are collected since node 10 is an all-node. Thus, the values 28.79, 43.98, 59.54, 65.39, 81.89, and 97.56 are sample values regarding the realization of node 10.

In this example, all nine distribution types were utilized to obtain samples for the time required to perform an activity. In Table 9.4, a trace of four additional simulation runs is presented to indicate both the variability of the time required to perform an activity and the variability involved in the network structure due to the branching process and the network modification procedures.

The final GERTS summary report for Example 1 is presented in Table 9.5 for 500 simulations of the network. The statistics presented for node 12 represent the values associated with the completion time of the network. From Table 9.5, it is seen that node 12 has a probability of 1 of being realized as expected. The average time to realize node 12 was approximately 53.55 time units with a standard deviation of approximately 23.43 time units. In one simulation the network was realized in less than 30 time units, and in another simulation it required over 148 time units to realize the network. Since node 12 is realized only once in each simulation, there is no difference between statistics based on first realization and all realizations. In this simulation the branch from node 13 to node 8 was designated with a counter type 1. Statistics are automatically collected on the number of times that branch was taken prior to the realization of the node on which the statistics are collected. For node 12, the average number of times the branch from node 13 to node 8 was taken prior to the realization of the network was 0.894. In some cases, the branch was never taken, and in at least one simulation the branch was taken seven times before the network was realized.

Statistics on node 6 show that the time between the first completion of an activity on node 6 and the time node 6 is realized required almost 7 time units. For the count statistics listed under node 6, it is seen that the branch from node 13 to node 8 was never taken prior to the realization of node 6. This is as expected since that branch follows node 6. Other items of interest from the final GERTS summary report will now be described. The probability associated with nodes 9 and 10 represents the probability that either of these nodes were realized in any simulation run. We see that branching around the loop from node 8 occurred in 56.6% of the runs. Even though statistics for node 10 are collected for all realizations, the probability of realizing node 10 on a simulation run is the probability of ever realizing node 10 in that simulation run. If it is desired to obtain the average number of times node 10 was realized, this can be calculated from the number of observations divided by the number of simulation runs (697 divided by 500 for this example). The average time of realizing node 10 in a simulation run is the sum of all realization times of node 10 divided by the number of times node 10 is realized. This statistic is not an ordinary one for network models since it combines the time of first realization, second realization, and so on. Care must be taken when using these values.

Histograms for each of the statistics nodes are also presented in Table 9.5. Consider the histogram for node 12, where the lower limit of the second

Table 9.5. GERTS III SUMMARY REPORT FOR EXAMPLE 1.

GERT SIMULATION PROJECT −1 BY ALL FEATURES
DATE 5/ 20/ 1973
FINAL RESULTS FOR 500 SIMULATIONS

NODE	PROB./COUNT		MEAN	STD. DEV.	# OF OBS.	MIN.	MAX.	NODE TYPE
12	1.0000		53.5455	23.4290	500.	29.5463	148.3445	A
12		1	0.8940	1.2923	500.	0.0	7.0000	
6	1.0000		14.2856	3.8987	500.	4.2100	35.3449	D
6		1	0.0	0.0	500.	0.0	0.0	
11	1.0000		27.8881	23.1954	500.	8.0286	125.6182	I
11		1	0.8940	1.2923	500.	0.0	7.0000	
10	0.5660		46.5107	21.7802	697.	23.3519	132.5558	A
10		1	0.8838	1.2373	697.	0.0	6.0000	
9	0.5660		26.7693	3.7985	283.	21.7731	46.9250	F
9		1	0.0	0.0	283.	0.0	0.0	
8	1.0000		17.0576	7.1713	1197.	4.1900	44.8769	B
8		1	0.8881	1.2601	1197.	0.0	7.0000	

*HISTOGRAMS**

NODE	LOWER LIMIT	CELL WIDTH	FREQUENCIES										
12	35.00	2.00	90	46	41	29	14	14	17	22	14	36	12
			9	14	7	8	12	4	12	11	9	3	5
			5	5	6	4	4	5	2	4	4	32	
6	1.00	1.00	0	0	0	0	1	0	0	0	0	5	84
			70	74	64	35	27	35	30	27	8	12	5
			6	2	5	2	2	0	3	0	0	3	
11	8.00	2.00	0	97	52	73	7	12	13	2	15	22	25
			32	8	11	6	5	3	13	9	13	6	5
			1	2	7	4	2	5	10	3	3	34	
10	30.00	2.00	183	45	52	23	20	10	29	41	27	27	27
			10	15	12	14	23	10	7	11	7	4	8
			7	11	7	6	4	2	3	4	5	43	
9	27.00	0.50	181	12	6	8	10	6	9	8	16	4	3
			2	5	1	0	2	2	1	0	0	0	0
			1	2	0	0	0	0	1	1	0	2	
8	1.00	1.00	0	0	0	0	24	124	82	20	0	0	0
			0	0	4	63	174	159	45	2	56	80	73
			71	41	31	33	29	33	12	12	6	23	

cell is 35 and the cell width of each cell is 2. From the data presented, we see that in 90 of the 500 simulation runs, the realization time for node 12 and, hence, the network, was less than 35 time units. In 46 other simulation runs, the time to complete the network was between 35 and 37 time units. Other values can be read directly from Table 9.5. This example demonstrates that a great deal of data can be obtained from GERTS III.

EXAMPLE 2. ANALYSIS AND SEQUENCING OF SPACE EXPERIMENTS

We next consider an application of GERTS III to analyze the sequencing of space experiments. The example was suggested to Pritsker and Burgess by Mr. J. Ignizio at a conference on GERT.

The performance of experiments in space by a spacecraft crew are almost always severely constrained by time. Many experiments are usually proposed by the scientific community, and of those proposed a subset must be chosen for a given space mission. The sequencing of these experiments that can be completed is then accomplished.

A GERT network of the sequence of experiments will be developed which permits the assessment of the time required to perform the experiments. In addition, we will determine information regarding the number of experiments that can be completed in a specified period of time. By modifying the sequence of experiments (which involves modifying the GERT network), an analysis can be performed on proposals for different sequencing procedures.

We will assume that there are three possible outcomes from the performance of an experiment: (1) successful completion, (2) failure, and (3) inconclusive results. If an experiment is successfully completed, the next experiment in the sequence is performed. If a failure occurs, the experiment is scrubbed and the next experiment is then performed. If the results of an experiment are inconclusive, the experiment is repeated n times or until a success or failure occurs. The experiment is scrubbed if it is tried n times and the results are still inconclusive.

The GERT network for a three-experiment program is shown in Fig. 9.5. Node 2 is the start node and initiates a transfer to node 3. Node 3 represents the decision point for the first experiment. If the first experiment is successful, the activity from node 3 to node 4 is traversed. If the first experiment fails, then the activity from node 3 to node 19 is taken. The second experiment is started by transferring from node 19 to node 4. If the results of the first experiment are inconclusive, the activity from node 3 to node 10 is taken. The output of node 10 is DETERMINISTIC; hence, the first experiment is performed again and a signal to node 13 is sent to indicate that the first experiment has been performed once. Thus, for each experiment we will either transfer to node 4 or reach node 13. When node 13 is realized three times, the activity from node 13 to node 14 is traversed. This activity is labeled as activity 1, and it causes the network to be modified by replacing node 3 with node 7. After this occurs when node 3 is realized, node 7 is in the network and a transfer to node 4 is caused. A similar discussion holds

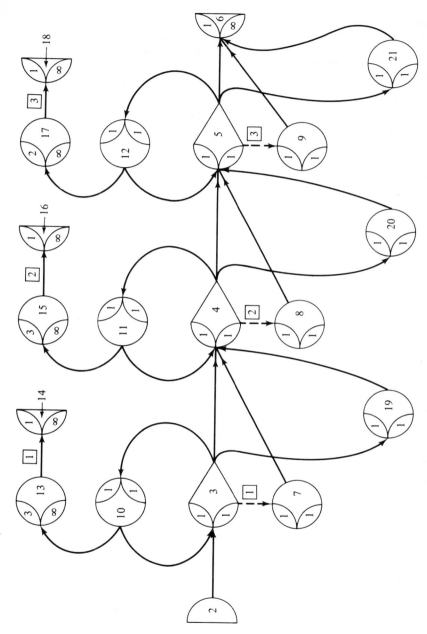

Figure 9.5. GERT network for the analysis and sequencing of space experiments.

327

for experiments 2 and 3. We see from the network that nodes 19, 20, and 21 represent the failure of experiments 1, 2, and 3, respectively. Nodes 14, 16, and 18 represent the outcomes that inconclusive results were obtained after the maximum number of experiments could be performed for experiments 1, 2, and 3, respectively.

Table 9.6 presents the data for the experimental characteristics to be analyzed. In Table 9.7, the input for Example 2 is presented. Table 9.8 presents a summary report describing the results from the GERTS III simulation of the sequence proposed for the space experiment program. From the output we see that experiment 1 failed 15.75% of the time and had inconclusive results 3.5% of the time. Therefore, it was successful 81% of the time. The time to complete experiment 1 is the time to reach node 4 of the network. Table 9.8 shows that, on the average, it took over 46 time units to reach node 4 with a standard deviation of over 18 time units. In some instances it took as little as 24.5 time units, and in others it took over 114 time units to complete the experiment. The number of times that experiment 1 was completed within given time intervals is presented in the histogram for node 4. Similar statistical quantities are available for the other nodes of the network.

Another interesting feature that we could have incorporated into the network model for the sequencing of experiments is the changing of the sequence depending on the results of some of the experiments. This would be accomplished through the network modification procedures of the GERTS III program.

Table 9.6. EXPERIMENT CHARACTERISTICS FOR EXAMPLE 2.

Experiment	Probability of Success	Probability of Failure	Probability of Inconclusive Results	Allowable Numbers of Repeats
1	0.6	0.1	0.3	3.
2	0.5	0.1	0.4	3
3	0.7	0.1	0.2	2

Experiment	Mean Time	Minimum Time	Maximum Time	Standard Deviation
1	10.0	5.0	20.0	2.0
2	20.0	15.0	25.0	1.0
3	15.0	10.0	30.0	3.0

Table 9.7. INPUT DATA FOR EXAMPLE 2.

SPACE EXPS 2 5201973 400 4 40 1267	1	EX 2 10
21 1 1 1 10 1	2	EX 2 20
2 1 0 D		EX 2 30
3 3 1 1P 5 1 A		EX 2 40
4 3 1 1P 25 3 A		EX 2 50
5 3 1 1P 37 3 A		EX 2 60
6 2 1 D 37 3 A		EX 2 70
7 1 1D		EX 2 80
8 1 1D		EX 2 90
9 1 1D		EX 2 100
10 1 1D	3	EX 2 110
11 1 1D		EX 2 120
12 1 1D		EX 2 130
13 3 D		EX 2 140
14 3 1 D 1 1 F		EX 2 150
15 3 D		EX 2 160
16 3 1 D 65 1 F		EX 2 170
17 2 D		EX 2 180
18 3 1 D 50 2 F		EX 2 190
19 3 1 1D 1 1 A		EX 2 200
20 3 1 1D 27 2 A		EX 2 210
21 3 1 1D 41 3 A		EX 2 220
0		EX 2 230
10 5 20 2		EX 2 240
20 15 25 1	4	EX 2 250
15 10 30 3		EX 2 260
0		EX 2 270
1 2 3 1 2		EX 2 280
6 3 4 2 2		EX 2 290
3 3 10 4 1		EX 2 300
1 3 19 4 1		EX 2 310
5 4 5 3 2		EX 2 320
4 4 11 4 1		EX 2 330
1 4 20 4 1		EX 2 340
7 5 6 4 1		EX 2 350
2 5 12 4 1		EX 2 360
1 5 21 4 1		EX 2 370
1 7 4 2 2		EX 2 380
1 8 5 3 2	5	EX 2 390
1 9 6 4 1		EX 2 400
1 10 3 1 2		EX 2 410
1 10 13 4 1		EX 2 420
1 11 4 2 2.		EX 2 430
1 11 15 4 1		EX 2 440

Table 9.7. Continued

1	12 5 3 2					EX 2 450
1	12 17 4 1					EX 2 460
1	13 14 4 1	1				EX 2 470
1	15 16 4 1	2				EX 2 480
1	17 18 4 1	3				EX 2 490
1	19 4 2 2					EX 2 500
1	20 5 3 2					EX 2 510
1	21 6 4 1					EX 2 520
	0					EX 2 530

1 3 7			EX 2 540
2 4 8		6	EX 2 550
3 5 9			EX 2 560
0			EX 2 570

Table 9.8. GERTS III SUMMARY REPORT FOR EXAMPLE 2.

```
             GERT SIMULATION PROJECT    2 BY   SPACE EXPS
                        DATE   5/20/1973
               **FINAL RESULTS FOR   400 SIMULATIONS**
```

NODE	PROB./COUNT	MEAN	STD. DEV.	# OF OBS.	MIN.	MAX.	NODE TYPE
6	1.0000	66.1562	21.8904	400.	36.7050	137.6164	A
21	0.1300	68.2321	21.8631	52.	37.9771	118.8453	A
20	0.1525	44.6018	16.2177	61.	25.6203	96.1071	A
19	0.1575	13.3597	6.6827	63.	5.5631	32.2204	A
18	0.0350	77.9196	22.1003	14.	52.1680	119.6703	F
16	0.0675	73.4611	7.2378	27.	66.5126	93.7269	F
14	0.0350	29.9889	3.0599	14.	25.3156	36.5681	F
5	1.0000	65.8379	21.5760	493.	36.7050	137.6164	A
4	1.0000	46.1740	18.7907	670.	24.5228	114.1765	A
3	1.0000	14.1291	7.5545	576.	5.0000	43.2401	A

```
                        **HISTOGRAMS**
```

NODE	LOWER LIMIT	CELL WIDTH	FREQUENCIES									
6	37.00	3.00	2	10	32	44	30	18	17	11	28	25	21
			17	18	12	11	14	..19	13	7	4	5	2
			6	5	4	8	1	5	3	3	0	5	
21	41.00	3.00	4	5	1	5	2	2	0	5	2	0	3
			5	2	1	1	0	4	2	1	3	0	1
			0	1	0	0	2	0	0	0	0	0	
20	27.00	2.00	4	5	7	4	5	1	2	2	3	3	2
			2	2	4	2	1	0	0	0	3	3	0
			2	0	2	0	0	0	1	0	0	1	

Table 9.8. Continued

19	1.00	1.00	0	0	0	0	0	2	3	4	7	8	10
			9	2	0	1	2	0	0	0	5	0	0
			2	3	0	0	0	2	0	1	0	2	
18	50.00	2.00	0	0	1	0	1	1	2	0	2	0	1
			0	0	0	0	0	0	0	1	0	1	1
			0	0	0	0	1	0	0	0	0	2	
16	65.00	1.00	0	0	4	1	1	2	6	2	4	1	0
			0	0	0	1	0	1	1	0	0	1	0
			0	0	0	0	0	0	1	1	0	0	
14	1.00	1.00	0	0	0	0	0	0	0	0	0	0	0
			0	0	0	0	0	0	0	0	0	0	0
			0	0	0	1	0	1	7	0	1	4	
5	37.00	3.00	2	12	39	49	41	24	21	15	36	35	27
			22	22	13	13	17	20	13	9	5	5	3
			10	5	6	9	1	6	4	4	0	5	
4	25.00	3.00	2	59	135	73	17	31	28	30	44	55	35
			8	20	15	15	14	23	10	4	8	8	4
			10	10	2	1	6	1	0	0	2	0	
3	5.00	1.00	0	11	20	41	59	74	81	55	37	13	11
			8	10	8	11	20	13	19	15	10	10	2
			1	4	10	3	2	3	5	2	3	15	

9.5 GERTS IIIQ

In modeling some systems there is a need for a node which provides a storage capability. To meet this need, a Q node was developed and added to the basic GERTS III program. The simulation program which includes the capability of modeling systems with this Q node has been labeled GERTS IIIQ. The Q node is the only new feature included in GERTS IIIQ.

9.5.1 Characteristics of GERTS IIIQ.
The graphical representation of the Q node is shown in Fig. 9.6. When an activity is completed that is incident to a Q node, two things can occur: (1) the activity following the Q node can be initiated, or (2) the number in the queue can be increased by 1. The activity emanating from a Q node represents a service activity. It is

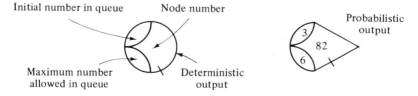

Figure 9.6. Queue nodes.

assumed in GERTS IIIQ that the service activity can serve only one item at a time. (The word item will be used to represent the concept of a customer, a transaction, etc.) If the service activity is currently serving an item, then the arriving item is put in the queue by increasing the number in the queue by one. The position of the arriving item in the queue can be done either on a first-in-first-out (FIFO) or last-in-first-out (LIFO) basis. The GERTS IIIQ program also permits the specification of a maximum number of items in a queue. When this maximum value is obtained and a new item arrives, the item can either balk from the system or balk to a node as specified by the analyst. If the item balks from the system, no new nodes are realized, and the path representing the item is completed. If the item balks to a node, the node is realized immediately, and the activities emanating from the node are scheduled. From the above we see that the concept of number of releases is not appropriate for a Q node.

For each Q node used in the GERTS IIIQ program, the following information is required:

1. Initial number of items in the queue.
2. The maximum number of items allowed in the queue.
3. The node an item balks to if it cannot join the queue.
4. Whether the items joining the queue are to be ranked according to the FIFO or LIFO priority procedure.

These characteristics are associated with a node and as such would be described on data card type 3 of the input cards.

The input format for the GERTS IIIQ program is identical to that described in Section 9.3, except that data card type 3 is modified as follows for each Q-node:

Field 2	Code for a Q node is 5.
Field 3	For a Q node, the initial number in the queue. If greater than zero, the service activity is assumed busy and an end-of-service-activity event is defined automatically.
Field 4	For a Q node: -1 to indicate maximum number in queue is 0; 0 to indicate no limit on number in the queue; otherwise the maximum number allowed in the queue.
Field 7, 8	Lower limit of cell 2 and width of each cell for statistics on the average number in the queue.
Field 10	Priority ranking procedure for the Q nodes.
	0—First-in-first-out (FIFO).
	1—Last-in-first-out (LIFO).
Field 11	Node that is transferred to when an activity is completed that is incident to the Q node and the maximum number allowed is in the queue (the node to which an item balks).

If the initial number in the queue is greater than zero, it is assumed that the server is busy. If it is desired to have 0 in the queue but the server busy, then an activity from a source node to the Q node will accomplish this result. No description of the activities incident to a Q node nor activities emanating from a Q node are required. The GERTS IIIQ program automatically identifies these activities when the activities for the network are inserted into the computer.

9.5.2 Restrictions associated with Q nodes. The major restriction with regard to Q nodes is that only one service activity can be associated with a Q node. However, a probabilistic output from a Q node is allowed where each of the activities emanating from the Q node can be considered to represent the same server. The other restrictions associated with a Q node are due to the desire to save core storage space and to alleviate the data input problem. These restrictions are:

1. A service activity cannot be an input to a Q node.
2. Activities which are either incident to a Q node or represent a service activity cannot have counter types associated with them.
3. If the maximum number allowed in the queue is not specified, a large value is assigned. If no queue is really desired, a -1 must be inserted for the maximum number allowed in the queue.

Statistics are collected on each Q node. The number of Q nodes plus the number of statistics nodes must be less than or equal to 100. To increase this number, the dimensions of selected arrays must be changed.

9.5.3 Statistics collected. The GERTS IIIQ program is written to simulate the same network a given number of times. For each simulation of the network, values are collected on the average number in each queue as opposed to the number in a queue. As an automatic output of the GERTS IIIQ program, statistical estimates are given for the average number in each queue, the average busy time or utilization of each server, and the average rate of balkers per unit of time for each Q node.

9.5.4 An example of GERTS IIIQ. The example presented in this section of the use of the GERTS IIIQ program is a single-channel queueing system.

EXAMPLE 3. A SIMPLE QUEUEING SITUATION

The GERT network representing a single-server, single-queue situation is presented in Fig. 9.7. Node number 2 is the source node and is used to

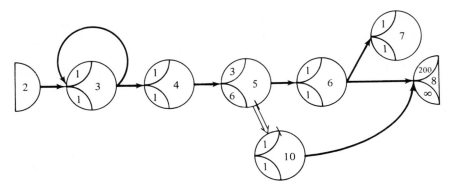

Figure 9.7. GERT network for a simple queueing situation.

activate node 3 which represents the generation of arrivals to the system. The output side of node 3 is DETERMINISTIC; hence, the branch from node 3 to node 4 and the branch from node 3 back to node 3 are realized every time node 3 is realized. The branch from node 3 to node 3 represents the time between the arrivals of items to the system. Since we desire to establish the time of arrival of items to the system, node 3 is made a mark node. Node 4 is inserted into the system in order to collect statistics on the times between arrivals of items. This node will provide a check on the interarrival time distribution represented by the branch from node 3 to node 3.

The output of node 4 is then the input to the Q node, node 5. It is assumed that there are three items in the queue initially and that the queue has a capacity of six items. The branch from node 5 to node 6 represents a service activity. If items are in the queue at time zero, then it is assumed that the service activity is on-going. The time required for the item to traverse this branch is the service time. Node 6 is used to collect statistics on the time an item spends in a system. Node 6 is identified as a statistics node with statistics collected on a time interval. Thus the time it takes for the item that was marked at node 3 until the time the item reached node 6 is a value that is collected and associated with node 6. The output side of node 6 is DETERMINISTIC, and the branches from node 6 to node 7 and node 6 to node 8 are both realized. Node 7 is a statistics node and collects samples on the time between realizations of node 6. This represents the time between departures of items from the system. Node 8 is a sink node and is used to specify the number of items required to realize the network. In this example, the number of releases for node 8 is set at 200. This corresponds to the number of items to be included in one simulation of the network.

Table 9.9 presents the input data for the network. Tables 9.10 and 9.11 illustrate the GERTS IIIQ echo check of the network. Note that the echo check indicates a count type of -1 for the branch from node 4 to node 5. A -1 is associated with the variable representing a count type for all activ-

Table 9.9. INPUT DATA FOR EXAMPLE 3.

SIMPLE QUEUE	3	5201973	50	3	25		1267		1		EX 3	10
10	1	1	1	4					2		EX 3	20

2	1								EX 3 30
3	4								EX 3 40
4	3	1	1		1		3	B	EX 3 50
5	5	3	6		1		2	10	EX 3 60
6	3	1	1		1		3	I	EX 3 70
7	3	1	1		1		1	B	EX 3 80
8	2200		1000		50	A			EX 3 90
10	1	1							EX 3 100
0									EX 3 110

(Parameter set 3)

0					EX 3 120
60	0		75	1	EX 3 130
5	0		75	1	EX 3 140

(Parameter set 4)

1	2	3	1	1	EX 3 150
1	3	3	2	4	EX 3 160
1	3	4	1	1	EX 3 170
1	4	5	1	1	EX 3 180
1	5	6	3	4	EX 3 190
1	6	7	1	1	EX 3 200
1	6	8	1	1	EX 3 210
1	10	8	1	1	EX 3 220
0					EX 3 230

(Parameter set 5)

Table 9.10. ECHO CHECK FOR EXAMPLE 3.

GERT SIMULATION PROJECT   3 BY SIMPLE QUEUE
DATE   5/20/1973
NETWORK DESCRIPTION
NODE CHARACTERISTICS

HIGHEST NODE NUMBER IS   10
NUMBER OF SOURCE NODES IS   1
NUMBER OF SINK NODES IS   1
NUMBER OF NODES TO REALIZE THE NETWORK IS   1
STATISTICS COLLECTED ON   4 NODES
NUMBER OF PARAMETER SETS IS   3
INITIAL RANDOM NUMBER IS   1267   0.0

NODE	NUMBER RELEASES	NUMBER OF RELEASES FOR REPEAT	OUTPUT TYPE	REMOVAL DESIRED AT REALIZATION	STATISTICS BASED ON REALIZATIONS
2	0	9999	D		
3	−1	−1	D		
4	1	1	D		B
6	1	1	D		I

Table 9.10. Continued

7	1	1	D	B
8	200	9999	D	A
10	1	1	D	

| | | QUEUE NODES | | | |
NODE	INITIAL # IN QUEUE	MAXIMUM # ALLOWED	OUTPUT TYPE	NODE FOR BALKERS	PRIORITY SCHEME
5	3	6	D	10	FIFO

SOURCE NODE NUMBERS
2
SINK NODE NUMBERS
8
STATISTICS COLLECTED ALSO ON NODES
7   6   4

Table 9.11. FURTHER ECHO CHECK FOR EXAMPLE 3.

"ACTIVITY PARAMETERS"

| PARAMETER NUMBER | PARAMETERS | | | |
	1	2	3	4
1	0.0	0.0	0.0	0.0
2	6.0000	0.0	75.0000	1.0000
3	5.0000	0.0	75.0000	1.0000

"ACTIVITY DESCRIPTION"

START NODE	END NODE	PARAMETER NUMBER	DISTRIBUTION TYPE	COUNT TYPE	ACTIVITY NUMBER	PROBABILITY
2	3	1	1	0	0	1.0000
3	3	2	4	0	0	1.0000
3	4	1	1	0	0	1.0000
4	5	1	1	−1	0	1.0000
5	6	3	4	−5	0	1.0000
6	7	1	1	0	0	1.0000
6	8	1	1	0	0	1.0000
10	8	1	1	0	0	1.0000

ities incident to Q nodes. For branch 5–6, the count-type variable is given the negative of the source node number. Note that these values are not inserted as input information but are automatically set in the GERTS IIIQ program. The storage savings resulting from using the same variables for two different purposes is significant. In Table 9.10, the code −1 is given for the number of releases for node 3. This identifies node 3 as a mark node. In the section on queue nodes in Table 9.10, node 5 is identified as a Q node with the queue priority being first-in-first-out (FIFO). The 10 in the node for balkers column indicates that items balk to node 10 if the queue is filled

Table 9.12. GERTS IIIQ SUMMARY REPORT FOR EXAMPLE 3.

GERT SIMULATION PROJECT   3 BY SIMPLE QUEUE
DATE   5/20/1973
FINAL RESULTS FOR 50 SIMULATIONS

NODE	PROB./COUNT	MEAN	STD. DEV.	# OF OBS.	MIN.	MAX.	NODE TYPE
8	1.0000	1195.4866	77.5555	50.	1043.5808	1412.6975	A
7	1.0000	6.3787	6.3492	9355.	0.0	62.3682	B
6	1.0000	15.9314	24.0739	9355.	0.0012	284.9275	I
4	1.0000	5.9883	5.9670	9943.	0.0	56.4771	B

QUEUE NODES

NODE	MEAN	STD. DEV.	# OF OBS.	MIN.	MAX:	
5	1.8260	0.4464	50.	0.8555	2.7119	AVERAGE NUMBER IN THE QUEUE
5	0.7870	0.0588	50.	0.5975	0.9061	AVERAGE BUSY TIME OF PROCESSOR
5	0.0111	0.0061	50.	0.0	0.0263	AVERAGE BALKERS PER UNIT TIME

HISTOGRAMS

NODE	LOWER LIMIT	CELL WIDTH	FREQUENCIES										
8	1000.00	50.00	0	1	3	11	14	9	8	2	1	1	0
			0	0	0	0	0	0	0	0	0	0	0
			0	0	0	0	0	0	0	0	0	0	
7	1.00	1.00	1368	1134	1040	812	685	605	558	452	401	342	289
			259	226	160	137	122	110	93	91	72	76	43
			38	40	32	18	16	13	20	20	15	68	
6	1.00	3.00	553	1838	1593	1237	944	659	472	364	225	184	141
			123	118	83	77	62	63	55	36	41	46	33
			26	39	32	24	14	11	20	13	20	209	
4	1.00	3.00	1544	3317	1932	1252	790	429	279	156	96	55	28
			25	16	15	5	1	0	2	0	1	0	0
			0	0	0	0	0	0	0	0	0	0	
5	1.00	0.20	2	1	4	9	7	13	5	1	5	3	0
			0	0	0	0	0	0	0	0	0	0	0
			0	0	0	0	0	0	0	0	0	0	

when they arrive. The capacity of the queue is printed under the column heading "maximum number allowed." The balking is graphically represented by a double line from node 5 to node 10.

The final results for fifty simulations of the simple queueing network are given in Table 9.12. These results are given for the input information presented in Table 9.9. The interarrival time distribution was exponential, and the service times were also exponentially distributed (Erlang with $k = 1$). From the values obtained for node 4, it is seen that the interarrival time distribution is close to the theoretical distribution. The time spent in the system by a customer is collected at node 6. The average value is 15.93. The standard deviation, minimum, maximum, and a histogram for the time spent in the system is also shown in Table 9.12. For exponential interarrival times with an unlimited queue, it is well known that the departure process is also exponential with the same parameter as for the arrival process. With a limited queue this is not the case, and one would expect slightly higher average times between departures. This is confirmed by the statistics collected on node 7. The statistics collected on node 8 represent the time required to process 200 items, including those items that balked.

Statistics collected on the Q node indicate that the average of the average number in the queue was 1.826. A histogram for this variable is shown under node 5 in the histogram section of the final report. The standard deviation of the average number in the queue is 0.4464. The high and low values observed for the average number in the queue are also recorded in Table 9.12. The average of the fraction of the time the server or processor was busy was 0.787.

Figure 9.8 shows the network for a queueing situation described above in which the queue is empty initially and the server is idle. Figure 9.9 illustrates the network in which the queue is empty and the server is busy.

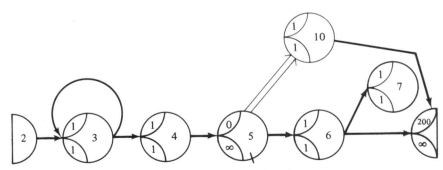

Figure 9.8. GERT network for simple queueing situation with the server initially idle.

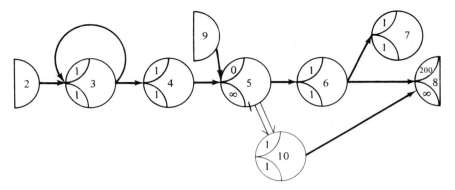

Figure 9.9. GERT network for simple queueing situation with the server initially busy and no one in the queue.

9.6 GERTS IIIC and GERTS IIIR

As discussed earlier, GERTS III was also modified to accommodate cost and resource information. These systems are referred to as **GERTS IIIC** and **GERTS IIIR**. We will not discuss these systems in detail in this text. The interested reader is referred to Pritsker [1].

9.6.1 GERTS IIIC program. For GERTS IIIC the GERTS III program was modified to permit the calculation of total cost expended until a node of the network is realized. The cost information for each activity is a start-up cost and a cost per unit time. These two cost parameters are appended to the parameters associated with a branch of the network. Total cost expended until a node is realized is calculated for each node of the network designated as a statistics node. The total cost expended includes the start-up cost for all activities started prior to the realization of the node and the variable cost expended to the time the node is realized.

9.6.2 The GERTS IIIR program. GERT networks considered previously do not allow for the situation in which there are limited resources available to perform a project. When this is the case, the activities of the project compete for the resources. To introduce resources into the GERTS program, the resource requirements for each activity are added to the description of each activity. The maximum number of each type of resource that is available must also be specified. The introduction of resource requirements and resource limitations introduces a scheduling problem into the GERT framework.

With resource limitations, it is no longer possible to schedule an activity whose start node has been realized since resource requirements may not be available. Also, when an end-of-activity event occurs, resources that were used on that activity are now available for use on other activities, and these activities may be considered for scheduling. Thus, when end-of-activity events occur, a selection is to be made from among those activities that can be started (the activities which have all their predecessor activities completed). This selection should be made on the basis of the performance measures associated with the completion of the project.

There has been a large amount of research performed on resource allocation problems for projects described in terms of PERT networks. No general conclusions about a method for scheduling for such projects has been developed, although several good rules have been proposed [6]. It has been found that the performance associated with a rule is dependent on the structure of the network. Since the GERT network structure is much more complex than the PERT network structure, it is expected that the scheduling task will be more difficult. The purpose of the GERTS IIIR program is to provide a vehicle with which research can be performed in the development of scheduling rules for GERT networks.

EXERCISES

1. For Example 3, determine the effect on the LIFO scheduling rule.

2. For Example 2, discuss the network modification necessary to change the sequence of experiments depending upon the results of some of the experiments.

3. Modify Example 3 to allow queueing at a second node following node 5. You are thus modeling a series of queues. (Remember that two Q nodes cannot be incident.)

4. Determine the relationship of queue length and arrival rate in Example 3.

In Exercises 5–10 model the systems using GERTS III, determine the appropriate statistics to collect, and run your system on the computer.

5. Exercise 6 in Chapter 8

6. Exercise 9 in Chapter 8

7. Exercise 10 in Chapter 8

8. Exercise 12 in Chapter 8

9. Exercise 16 in Chapter 8

10. Exercise 15 in Chapter 8

11. Consider the system described in Exercise 7 in Chapter 8. Using GERTS IIIQ, model the system to allow queueing at the traffic lights.

REFERENCES

1. Pritsker, A. A. B. and R. R. Burgess, *The GERTS Simulation Programs: GERTS III, GERTS IIIQ, GERTS IIIC and GERTS IIIR.* NASA/ERC Contract NAS-12-2113, Virginia Polytechnic Institute (June, 1970).

2. Pritsker, A. A. B., *User's Manual for GERT Simulation Program.* NASA/ERC Grant NGR-03-001-034, Arizona State University (July, 1968).

3. Pritsker, A. A. B., *Definitions and Procedures for GERT Simulation Program.* NASA/ERC Grant NGR-03-001-034, Arizona State University (July, 1968).

4. Pritsker, A. A. B. and P. C. Ishmael, *GERT Simulation Program II.* NASA/ERC Contract NAS-12-2035 (June, 1969).

5. Pritsker, A. A. B. and P. J. Kiviat, *Simulation with GASP II.* Englewood Cliffs, New Jersey, Prentice-Hall, Inc., 1969.

6. Wiest, J. D., "A Heuristic Model for Scheduling Large Projects with Limited Resources." *Management Science*, Vol. 13, No. 6 (February, 1967), pp. B-359-377.

DISCIPLINE-ORIENTED
APPLICATIONS OF GERT

In this and the next chapter, we will view a number of areas where GERT analysis has proven that it is useful. This chapter will concern itself with a number of areas of application such as queueing systems, inventory systems, reliability systems, sampling and quality control, project management, and management decision making.

The next chapter will deal with specific projects to which GERT has been applied. It is hoped that by presenting the material in this manner the reader will see the potential for this technique in this chapter and then see how others have used the material presented to analyze specific problem areas in the next chapter.

10

10.1 Queueing Analysis

10.1.1 Description of the queueing system to be analyzed. Although GERTS makes it possible to analyze systems involving infinite queues, the discussion in this section will be limited to the investigation of problems with limited queues. One type of problem which falls into this classification is the "repairman" problem, which can be described as follows. It is assumed that there are m independent identical machines, supported by n identical spare machines. There is a repair facility that is assumed to be capable of repairing P machines simultaneously. This facility could be imagined to be P repairmen. The time between failures and the repair times for all machines are independent identically distributed random variables. The queue discipline is as follows. If all repairmen are busy, each new failure joins a waiting line and waits until a repairman is free.

The system can be described by a Markov-type process. The states of this process will be called the number of machines undergoing repair and waiting for repair. If the failure and repair times are distributed exponentially, an $M/M/P^*$ system is generated. Due to the forgetfulness property of exponential distribution, the $M/M/P$ is fairly easy to model.

If each machine in service has a mean life of $1/\lambda$ and each serviceman takes an average time of $1/\mu$ to fix a machine, then the state-dependent μ_i's and λ_i's can be defined for the repairman system with m operating machines, n spares, and P repairmen as follows, assuming $P < m$:

i	λ_i	μ_i
0	$m\lambda$	0
1	$m\lambda$	μ
2	$m\lambda$	2μ
.	.	.
.	.	.
.	.	.
$P-1$	$m\lambda$	$(P-1)\mu$
P	$m\lambda$	$P\mu$
$P+1$	$m\lambda$	$P\mu$
.	.	.
.	.	.
.	.	.
$n-1$	$m\lambda$	$P\mu$
n	$m\lambda$	$P\mu$
$n+1$	$(m-1)\lambda$	$P\mu$
.	.	.
.	.	.
.	.	.
$m+n-1$	λ	$P\mu$
$m+n$	0	$P\mu$

Realizing that if there are i machines down, the mean time before a repair occurs is $1/\mu_i$ and the mean time before another machine breaks down is $1/\lambda_i$, we can interpret the state-dependent μ_i's and λ_i's as follows:

1. If $i = 0$, then there are no machines down; thus $\mu_1 = 0$. Also, m machines are operating; thus $\lambda_1 = m\lambda$.

---

*This is Kendall's shorthand notation for queueing system. The notation has the following meaning:

ARRIVAL DISTRIBUTION/SERVICE DISTRIBUTION/NUMBER OF SERVERS

where

M = Poisson Arrivals or Exponential Service
D = Constant Service Time
G = General Service Distribution

for example, an $M/M/P$ system is one with Poisson Arrivals, exponential service, and P servers, and an $M/D/I$ system is one with Poisson Arrivals, Constant Service Time, and 1 server.

2. If $i = P + 1$, then there is a $P + 1$ machine down. But there are only P repairmen to work on the machines; thus $\mu_{P+1} = P\mu$. Since $P < n$, there are still m machines operating (and $n - (P + 1)$ reserve machines); thus $\lambda_{P+1} = m\lambda$.

3. If $i = n + 1$, then there are more machines down than there are repairmen to work on them; so $\mu_{n+1} = P\mu$. We now have more machines down than we have spares, so there are only $(m + n) - (n + 1) = m - 1$ machines operating; thus $\lambda_{n+1} = (m - 1)\lambda$.

4. Finally, if $i = m + n$, then there are more machines down than there are repairmen; so $\mu_{m+n} = P\mu$. Since there are no machines operative, $\lambda_{m+n} = 0$.

Using these state dependents μ_i and λ_i, we observe that the time to move from any node i to either node $i + 1$ or node $i - 1$ is an exponential distribution with mean $1/(\mu_i + \lambda_i)$. The probability to move to node $i + 1$ will be $\lambda_i/(\lambda_i + \mu_i)$, while the probability to move to node $i - 1$ is $\mu_i/(\lambda_i + \mu_i)$. The GERT representation of the $M/M/P$ system is shown in Fig. 10.1. An alternative approach to modeling the $M/M/P$ repairman problem by approximating it as a birth-and-death process is given by Whitehouse [43].

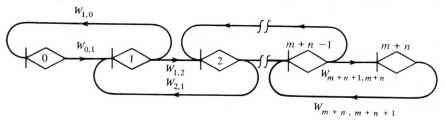

where $W_{i,j}(s) = \lambda_i/(\lambda_i + \mu_i)(1 - s/\lambda_i - \mu_i)^{-1}$ if $j > i$

$= \mu_i/(\lambda_i + \mu_i)(1 - s/\lambda_i - \mu_i)^{-1}$ if $j < i$

Figure 10.1. Model of a $M/M/P$ repairman system.

10.1.2 Information available from the GERT representation of a queueing situation. The investigation of the information available from the GERT representation of an $M/M/1$ repairman problem with $m = 2$ and $n = 1$ will now be undertaken. Machines will fail at the rate of once each time period, and it will take the repairman one-half of a time period to repair a machine on the average. The state-dependent λ's and μ's will thus be:

i	0	1	2	3
λ_i	2	2	1	0
μ_i	0	2	2	2

The GERT representation of this system is shown in Fig. 10.2. Using this as an example, consider the information available from the GERT network. Unless otherwise stated, it will be assumed that the process starts in state 0 (all machines operative).

TIME TO PLANT FAILURE

This is defined as the time until the plant or operation is not completely operative. It corresponds to state $n + 1$, that is, all the spare machines plus one are being serviced or waiting for service. A representation of this will be $M_{0,n+1}(s)$, which is the moment generating function of the time to go from 0 to state $n + 1$. To obtain this information we solve the graph in Fig. 10.2 for the first passage time from node 0 to node $n + 1$.

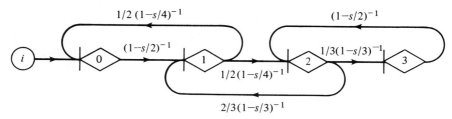

Figure 10.2. Model of the $M/M/1$ repairman system.

In the problem being considered, $M_{0,2}(s)$ is the quantity of interest:

$$M_{0,2}(s) = \frac{0.5(1 - s/2)^{-1}(1 - s/4)^{-1}}{(1 - 0.5(1 - s/2)^{-1}(1 - s/4)^{-1})}$$

In operations where there are more than one machine in operation at one time, another quantity of interest is the *time to total failure*. This is represented by $M_{0,m+n}(s)$ and it is defined as the time until the system is completely inoperative. In the example being represented, this corresponds to $M_{0,3}(s)$:

$$M_{0,3}(s) = \frac{0.167(1 - s/2)^{-1}(1 - s/4)^{-1}(1 - s/3)^{-1}}{1 - 0.5(1 - s/2)^{-1}(1 - s/4)^{-1} - 0.33(1 - s/3)^{-1}(1 - s/4)^{-1}}$$

DURATION OF FAILURE

This represents the consecutive time that a plant will not be operating at full capacity. For the system under discussion, this is equivalent to $M_{n+1,n}(s)$, and it represents the time from which the process first becomes partially inoperative $(n + 1)$ until the service facility first makes the plant completely operative again (n).

In the example, $M_{2,1}(s)$ represents the duration of failure:

$$M_{2,1}(s) = \frac{0.67(1 - s/3)^{-1}}{(1 - 0.33(1 - s/3)^{-1}(1 - s/2)^{-1})}$$

BUSY PERIOD AND IDLE TIME FOR THE SERVICE FACILITY

When the process is in state 0, all machines and spares are in workable condition and the service facility is idle. In cases where there are multiple service facilities, the service facility will not be completely operative until state P is reached.

Defining idle time as the consecutive time that the service facility has no work, we can observe that this is represented by $M_{0,1}(s)$. In the example under consideration:

$$M_{0,1}(s) = \left(1 - \frac{s}{2}\right)^{-1}$$

The complement of this quantity is the consecutive time the service facility is busy. It is called *busy time* and is equivalent to $M_{1,0}(s)$. In the system under discussion:

$$M_{1,0}(s) = \frac{0.5(1 - s/4)^{-1}(1 - 0.33(1 - s/3)^{-1}(1 - s/2)^{-1})}{1 - 0.33(1 - s/4)^{-1}(1 - s/3)^{-1} - 0.33(1 - s/3)^{-1}(1 - s/2)^{-1}}$$

STEADY-STATE PROBABILITIES

The steady-state probabilities are the probabilities that, at any time after the system has reached steady state, the process will be at a given state. The system has a regenerative nature in that all returns to a given point in the system have a common distribution. The steady-state probability for a given state is equal to the expected time in the given state during a regeneration of the system, divided by the expected total time of the regeneration.

The expected time of regeneration is equal to the first moment of MGF representing the time to return to any node in the system. This calculation is facilitated if we represent the system as shown in Fig. 10.3. The mean recurrence time is:

$$\left.\frac{dM_{0,0'}(s)}{ds}\right|_{s=0}$$

$$M_{0,0'}(s) = \frac{0.5(1 - s/2)^{-1}(1 - s/4)^{-1}(1 - 0.33(1 - s/3)^{-1}(1 - s/2)^{-1})}{1 - 0.33(1 - s/3)^{-1}(1 - s/4)^{-1} - 0.33(1 - s/3)^{-1}(1 - s/2)^{-1}}$$

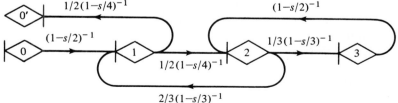

Figure 10.3. Modification of the $M/M/1$ repairman system to facilitate the finding of steady-state probabilities.

and

$$\left.\frac{dM_{0,0\prime}(s)}{ds}\right|_{s=0} = 1.75$$

To find the portion of time spent in each state, set the MGF equal to one for all paths which do not emanate from the node of interest and proceed as before.

Consider the time spent in node 3; the graph is shown in Fig. 10.4.

$$M_{0,0\prime}(s) = \frac{0.5(1 - 0.33(1 - s/2)^{-1})}{(1 - 1/3 - 1/3(1 - s/2)^{-1})}$$

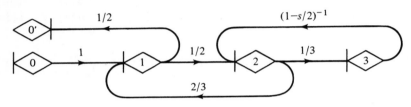

Figure 10.4. Modification of the *M/M/*1 repairman problem to find the time spent in state 3 during the regeneration of the system.

and

$$\left.\frac{dM_{0,0\prime}(s)}{ds}\right|_{s=0} = 0.25$$

Thus, the steady-state probability that the process is in state 3 is 0.25/1.75 = 1/7. The steady-state probabilities of states 0, 1, and 2 are all 2/7.

These results can be interpreted as follows:

1. The probability is 1/7 that the system is inoperative,
2. The probability that the service facility is operating is 5/7, etc.

Of course, the probabilities would be the same regardless of where the GERT network was broken, but the mean recurrence time would probably be different depending upon where the graph was broken. For example, if the graph was broken at node 3, the mean recurrence time would be 3.50 and time spent in node 3 would be 1/2. The steady-state probability of being in state 3 is (0.5)/(3.5) = 1/7, which is the same result we obtained by breaking the graph at node 0.

TIME TO REGENERATION GIVEN THAT THE SYSTEM DOES NOT FAIL

In the previous section, the MGF of the recurrence time for the system was developed in connection with the development of the steady-state probabilities. A quantity of similar interest is the distribution of time until regeneration, given that the system didn't fail to some degree. For total failure, all paths that lead into state $n + m$ can be tagged with a z tag, and the MGF of the system can be found for the recurrence with z equated to zero.

For the example being studied, consider Fig. 10.5 and:

$$W_{0,0'}(s \mid \text{no failure}) = W_{0,0'}(s, z)|_{z=0} = \frac{0.5(1 - s/2)^{-1}(1 - s/4)^{-1}}{1 - 0.33(1 - s/3)^{-1}(1 - s/4)^{-1}}$$

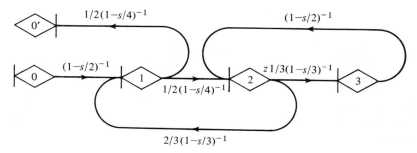

Figure 10.5. $M/M/1$ repairman problem adapted to calculate the regeneration time, given the system does not totally fail.

and

$$M_{0,0'}(s \mid \text{no failure}) = \frac{W_{0,0'}(s, z)|_{z=0}}{W_{0,0'}(s, z)|_{\substack{s=0 \\ z=0}}} = \frac{0.67(1 - s/2)^{-1}(1 - s/4)^{-1}}{(1 - 0.33(1 - s/3)^{-1}(1 - s/4)^{-1})}$$

DISTRIBUTION OF THE NUMBER OF MACHINES FAILING BEFORE COMPLETE FAILURE OF THE PLANT

All elements in the graph going from state $i \rightarrow k$ where $k > i$ are tagged with the tag, $e^{(k-i)c}$. Note that in the $M/M/1$ process $k - i$ will always be one. The distribution of the number of machine failures until the plant fails is then equal to $M_{0, m+n}(s, c)|_{s=0}$. Consider the graph shown in Fig. 10.6.

$$M_{0,3}(s, c)|_{s=0} = \frac{0.167e^{3c}}{(1 - 0.5e^c - 0.33e^c)}$$

A similar result can be obtained for the number of repairs until failure if we label elements in the graph going from state $i \rightarrow k$ for $i > k$, with the tag $e^{(i-k)c}$, and then solve as before.

EXPECTED NUMBER OF RECOVERIES BEFORE TOTAL FAILURE

Since the system essentially starts over again every time it reaches the 0 state, a quantity of interest might be the expected number of recoveries before time of failure. We can investigate this by placing an e^c tag on all

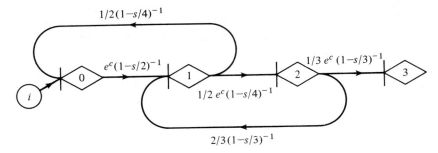

Figure 10.6. $M/M/1$ repairman problem adapted to find the distribution of the number of machines failing before complete failure of the plant.

paths entering state 0 from another state. The moment generating function of the expected number of recoveries is $M_{0,m+n}(s, c)|_{s=0}$. For the graph as shown in Fig. 10.7:

$$M_{0,3}(s, c)|_{s=0} = \frac{0.167}{(1 - 0.5e^c - 0.33)}$$

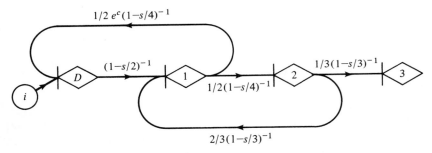

Figure 10.7. $M/M/1$ repairman problem adapted to obtain the distribution of the number of recoveries before total failure.

Another quantity of interest might be the distribution of the number of changes in state before failure. This can be investigated if we place the e^c tags on all transitions from one state to another. The MGF of the number of changes in state will equal $M_{0,m+n}(s, c)|_{s=0}$. The graph to solve this problem is given in Fig. 10.8.

$$M_{0,3}(s, c)|_{s=0} = \frac{0.167e^{3c}}{(1 - 0.5e^{2c} - 0.33e^{2c})}$$

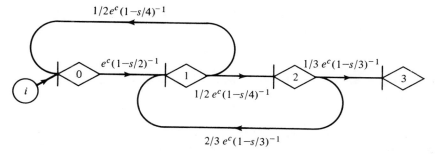

Figure 10.8. $M/M/1$ repairman problem adapted to find the distribution of the number of changes in state before total failure.

PROBABILITY OF A FAILURE BEFORE THE SYSTEM RETURNS TO STATE 0, GIVEN THAT YOU ARE IN A PARTICULAR STATE I

This problem reduces to a system in which nodes 0 and $n + m$ are considered to be trapping nodes. Thus the probability to get from state i to state $n + m$ is $W_{i,n+m}(s)|_{s=0}$. This will meet the given conditions of the problem because the system could not have returned to state 0 before reaching the state $n + m$ or the process would have been trapped in state 0.

This is a rather trivial problem for the example under consideration, but the information is quite important. Assume that the system being considered is in state 1. Then the probability of failure before returning to state zero is $W_{1,3}(s)|_{s=0}$.

$$W_{1,3}(s)|_{s=0} = \tfrac{1}{4}$$

OTHER INFORMATION

Many other things could be found from this graph. No mention has been made of the conditional distribution given a particular number of passages in a particular path in this system. These are easily found by the methods presented in Chapter 8.

The expected number of repairmen busy in a time period can be found by placing a counter on all paths representing the number of servicemen working if the process was in that state. The tag equals e^{ic} for $0 \le i \le k$, and it equals e^{kc} for $k \le i \le m + n$. The expected number of servicemen busy is:

$$\frac{\dfrac{\partial M_{0,0\prime}(s, c)}{\partial c}\Big|_{\substack{s=0 \\ c=0}}}{\dfrac{\partial M_{0,0\prime}(s, c)}{\partial s}\Big|_{\substack{s=0 \\ c=0}}}$$

By observation, this calculation reduces to:

$$\sum_{i=1}^{k} ip_i + \sum_{i=k+1}^{n+m} kp_i$$

where the p_i are the steady-state probabilities derived in an earlier portion of this chapter.

10.1.3 Systems to which the GERT approach is applicable.

In the preceding portion of this section, the information available from a GERT network representation of a queueing problem has been established. The system considered was a simple $M/M/1$ process. It is not too surprising that this system is easily handled by GERT because it is a Markov process, and there is a strong relationship between GERT and Markov. In the remainder of this section, we will discuss other types of queueing systems that can be represented by GERT. Once a queueing system can be represented as a GERT network, all the information that was established for the $M/M/1$ process can be found for that system by essentially the same methods. There may have to be minor changes in the procedures suggested, but the logic is the same as for the $M/M/1$ situation.

Two approaches will be used to develop the graph of a queueing system; one gives the exact representation of the system, while the other, which depends upon the concept of imbedded Markov chains, gives a reasonable approximation of the system. We will concentrate on $M/G/1$ repairman problems and special cases of this type of system. The modeling of other queueing systems will basically follow the approaches shown here for the $M/G/1$ system. Section 10.1.4 will demonstrate how we can use the GERTS system to analyze queueing systems.

THE EXACT REPRESENTATION OF THE FINITE QUEUEING SYSTEM

Figure 10.9 shows an exact representation of an $M/G/1$ repairman problem with $m = 1$ and $n = 2$. The nodes represent the number of machines down at a given time. Two nodes are required to represent two machines down because the time to go from 2 to 3 is different from the time to go from 2' to 3. Node 2 indicates two machines down following a failure, and node 2' indicates two machines down following a repair. The key to the graph shown in Fig. 10.9 is the realization that the interarrival distribution of our breakdowns is exponential. We can therefore take advantage of the forgetfulness property of this distribution.

The W_{ij} terms as shown in Fig. 10.9 are equal to p_{ij}, and the MGF of the t_{ij} are derived below. Let $t_r =$ repair time; $t_f =$ time between failures.

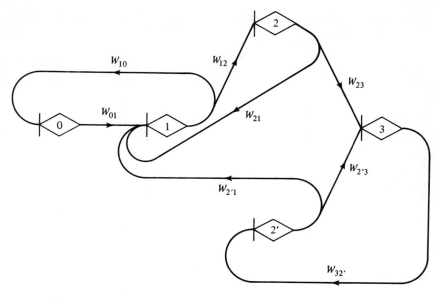

Figure 10.9. Exact representation of the $M/G/1$ repairman problem.

$$p_{01} = 1; \qquad t_{01} = t_f$$

$$p_{10} = P[t_f \geq t_r]; \qquad t_{10} = t_r \,|\, t_f \geq t_r$$

$$p_{12} = P[t_f < t_r]; \qquad t_{12} = t_f \,|\, t_f < t_r$$

$$p_{21} = p[2t_f \geq t_r; \qquad t_f < t_r]; \qquad t_{21} = t_r - t_f \,|\, 2t_f \geq t_r, \quad t_f < t_r$$

$$p_{23} = P[2t_f < t_r; \qquad t_f < t_r]; \qquad t_{23} = t_f \,|\, 2t_f \leq t_r, \quad t_f < t_r$$

$$p_{32'} = 1; \qquad t_{32'} = t_r - 2t_f \,|\, 2t_f \leq t_r, \quad t_f > t_r$$

$$p_{2'1} = P[t_f \geq t_r]; \qquad t_{2'1} = t_r \,|\, t_f \geq t_r$$

$$p_{2'3} = P[t_f < t_r]; \qquad t_{2'3} = t_f \,|\, t_f < t_r$$

This notation can be generalized if we subscript f and r, viz., $t_{f_0}, t_{f_1}, t_{r_0}$ etc. Increasing the number of nodes will follow the same pattern as presented here.

To demonstrate the calculations involved in finding W_{ij}'s needed, consider a special case of the $M/G/1$ system; namely, the $M/D/1$ system.

$$f_{t_r}(t) = \begin{cases} 1; & t = \dfrac{1}{\mu} \\ 0; & \text{otherwise} \end{cases} \qquad\qquad f_{t_r}(t) = \begin{cases} \lambda e^{-\lambda t}; & t > 0 \\ 0; & \text{otherwise} \end{cases}$$

$$p_{12} = P\left[t_f < \frac{1}{\mu}\right] = 1 - e^{-\lambda/\mu} \qquad f_{t_{12}}(t) = f_{t_f}|_{t_f \leq 1/\mu}(t)$$

$$= \begin{cases} \dfrac{\lambda e^{-\lambda t}}{p_{12}}; & t < \dfrac{1}{\mu} \\ 0; & \text{otherwise} \end{cases}$$

$$p_{10} = e^{-\lambda/\mu}$$

$$f_{t_{10}}(t) = \begin{cases} 1; & t = \dfrac{1}{\mu} \\ 0; & \text{otherwise} \end{cases}$$

$$p_{2'3} = p_{12}$$

$$f_{t_{2'3}}(t) = f_{t_{12}}(t)$$

$$p_{2'1} = p_{10}$$

$$f_{t_{2'1}}(t) = f_{t_{10}}(t)$$

$$p_{23} = 1 - \lambda \frac{e^{-\lambda/\mu}}{\mu e^{-\lambda/\mu}}$$

$$f_{t_{23}}(t) = \begin{cases} \dfrac{\lambda(\lambda t)e^{-\lambda t}}{p_{23}}; & t < \dfrac{1}{\mu} \\ 0; & \text{otherwise} \end{cases}$$

$$p_{21} = 1 - p_{23}$$

$$f_{t_{21}}(t) = \begin{cases} \dfrac{\lambda e^{-\lambda(1/\mu - t)}}{p_{21}}; & t \leq \dfrac{1}{\mu} \\ 0; & \text{otherwise} \end{cases}$$

$$p_{32'} = 1$$

$$f_{t_{32'}}(t) = \begin{cases} \dfrac{\lambda[\lambda(1/\mu - t)[e^{-\lambda/1/\mu - t)}]}{1 - e^{-\lambda/\mu} - \lambda/\mu(e^{-\lambda/\mu})}; & \\ & t \leq \dfrac{1}{\mu} \\ 0; & \text{otherwise} \end{cases}$$

To apply GERT in this manner, we need the MGF of the truncated exponential and gamma distributions.

$$M_{L_1}(s) = \int_0^{1/\mu} e^{st} f(t)\, dt$$

$$M_{L_2}(s) = \int_{1/\mu}^{\infty} e^{st} f(t)\, dt$$

Let $\rho = \lambda/\mu$, E designate exponential, and G designate gamma:

$$M_{E_1}(s) = \left[\frac{1 - e^{s/\mu - \rho}}{1 - e^{-\rho}}\right]\left[1 - \frac{s}{\lambda}\right]^{-1}$$

$$M_{E_2}(s) = e^{s/\mu}\left[1 - \frac{s}{\lambda}\right]^{-1}$$

$$M_{G_1}(s) = \left[\frac{1}{1 - e^{-\rho}\rho e^{-\rho}}\right]\left[\left(1 - \frac{s}{\lambda}\right)^{-2} - \rho\left(1 - \frac{s}{\lambda}\right)^{-1}e^{s/\mu - \rho} \right.$$
$$\left. - \left(1 - \frac{s}{\lambda}\right)^{-2}e^{s/\mu - \rho}\right]$$

$$M_{G_2}(s) = \left[\frac{1}{e^{-\rho} + \rho e^{-\rho}}\right]\left[\rho\left(1 - \frac{s}{\lambda}\right)^{-1}e^{s - \mu - \rho} + \left(1 - \frac{s}{\lambda}\right)^{-2}e^{s/\mu - \rho}\right]$$

APPROXIMATE SOLUTION OF THE FINITE QUEUEING SYSTEM

Kendall [20] developed a method of reducing non-Markovian queueing processes to Markov chains. This approach, called the method of imbedded Markov chains, will allow many queueing problems to be approximated by means of GERT.

Parzen [29] included the following example which describes the concept behind Kendall's idea:

Consider a box office with a single cashier at which the arrivals of customers are events of Poisson type with intensity λ. Suppose that the service times of successive customers are independent identically distributed random variables. For $n > 1$, let x_n denote the number of persons waiting in line for service at the moment the nth person to be served (on a given day) has finished being served. The sequence (x_n) is a Markov chain. We show this by showing that the conditional distribution of x_{n+1}, given the values of $x_1, x_2, \ldots, x_n$, depends only on the value of x_n. Let U_n denote the number of customers arriving at the box office during the time that the nth customer is being served. We may then write:

$$x_{n+1} = x_n - \delta(x_n) + U_{n+1}$$

where we define:

$$\delta(x_n) = 1 \quad \text{if} \quad x_n > 0$$
$$= 0 \quad \text{if} \quad x_n = 0$$

In other words, the number of persons waiting for service when the $(n + 1)$th customer leaves depends on whether the $(n + 1)$th customer was in the queue when the nth customer departed service. If $(x_n) = 0$, then $x_{n+1} = U_{n+1} + x_n$, while if $\delta(x_n) = 1$, $x_{n+1} = U_{n+1} + x_n - 1$. Since U_{n+1} is independent of $x_1, \ldots, x_{n-1}$, it follows that, given the value of x_n, one need not know the values of $x_1, \ldots, x_{n-1}$ to determine the conditional probability distribution of x_{n+1}.

The queueing situation described here is recognized as the $M/G/1$ case. The imbedded Markov process described can be represented by a transition matrix as follows.

For $j = 0$:

$$P_{0,k}(n, n + 1) = \Pr(U_{n+1} = k) = a_k$$

For $j > 0$:

$$P_{j,k}(n, n + 1) = \Pr(U_{n+1} = k - j + 1) = a_{k-j+1}$$

where a_k is the probability that k customers arrive during the service of a customer.

The transition matrix will, therefore, appear as follows:

$$
P = \begin{bmatrix}
a_0 & a_1 & a_2 & \cdots \\
a_0 & a_1 & a_2 & \cdots \\
0 & a_0 & a_1 & \cdots \\
0 & 0 & a_0 & \cdots \\
\cdot & \cdot & \cdot & \\
\cdot & \cdot & \cdot & \\
\cdot & \cdot & \cdot &
\end{bmatrix}
$$

where:

$$
a_k = \int_0^\infty \frac{((e^{-\lambda t})(\lambda t)^k)}{(k!)}\, dF_s(t)
$$

the distribution function of service times.

For the repairman problem, a slightly different transition matrix for its imbedded Markov chain will be used as an approximation because its queue length is of limited size. Since we are only observing the network at the instant a machine leaves service, the highest state we will have is $m + n - 1$ for the repairman problem. The P matrix becomes:

$$
P = \begin{bmatrix}
a_{0,0} & a_{0,1} & a_{0,2} & \cdots & A_0 \\
a_{1,0} & a_{1,1} & a_{1,2} & \cdots & A_1 \\
0 & a_{2,0} & a_{2,1} & \cdots & A_2 \\
0 & 0 & a_{3,0} & \cdots & A_3 \\
\cdot & \cdot & \cdot & & \cdot \\
\cdot & \cdot & \cdot & & \cdot \\
\cdot & \cdot & \cdot & & \cdot \\
0 & \cdots & a_{m+n-1,0} & \cdots & A_{m+n-1}
\end{bmatrix}
$$

where:

$$
a_{i,k} = \int_0^\infty \frac{((e^{-\lambda_i t})(\lambda_i t)^k)}{(k!)}\, dF_s(t)
$$

λ_i is the failure rate which is a function of the number of machines operating at the time of the last service.

$$
A_0 = \sum_{k=m+n}^{\infty} a_{i,k} = 1 - \sum_{k=0}^{m+n-1} a_{i,k}
$$

and

$$
A_k = \sum_{j=m+n-k}^{\infty} a_{i,j} = 1 - \sum_{j=0}^{m+n-1-k} a_{i,j} \quad \text{for} \quad 1 \le j \le m+n-1
$$

The GERT network is defined not only by the Markov transition matrix, but also by the MGF of the time to take the path. For this problem, the MGF on all branches except those emanating from state 0 will be the service time distribution. This is because the P matrix gives the probability of moving from one state to another during a service period. If one is in state 0, we have to wait for an arrival before we can start servicing it. Therefore, the appropriate distribution is the sum of the service distribution and the arrival distribution.

As an example, consider a $M/D/1$, which is a special case of the $M/G/1$, with a service period of 6-time units. Assume that $m = 1$ and $n = 3$. The interarrival time is exponential with mean equal to:

$$a_{i,k} = \frac{e^{-\lambda_i t}(\lambda_i t)^k}{k!} = \frac{e^{-0.6}(0.6)^k}{k!} \quad \text{for} \quad i = 0, 1, 2, \text{ and } 3.$$

The P matrix of the imbedded Markov chain for the system is:

$$P = \begin{bmatrix} a_{0,0} & a_{0,1} & a_{0,2} & A_0 \\ a_{1,0} & a_{1,1} & a_{1,2} & A_1 \\ 0 & a_{2,0} & a_{2,1} & A_2 \\ 0 & 0 & a_{3,0} & A_3 \end{bmatrix}$$

which reduces to;

$$P = \begin{bmatrix} 0.55 & 0.33 & 0.10 & 0.02 \\ 0.55 & 0.33 & 0.10 & 0.02 \\ 0 & 0.55 & 0.33 & 0.12 \\ 0 & 0 & 0.55 & 0.45 \end{bmatrix}$$

The moment generating function of the time to move from state 1, 2, or 3 is e^{6s}. The time to move from state 0 is $e^{6s}(1 - s/10)^{-1}$ which is the moment generating function of the service time plus the time waiting for an arrival.

The GERT network can now be drawn as shown in Fig. 10.10. The major weakness in this method is that there will not be a state $n + m$ which seems to imply that all the machines would never be down at once. The weakness could be partially alleviated if we realize that one would move to node $m + n$ from node $m + n - 1$ if there is an arrival before a service is completed. In fact, node $m + n$ would be reached from any node i if there were $m + n - i$ arrivals before a service. If we want to incorporate this feature into the graph, the appropriate probabilities and distributions must be developed in much the same manner that was used to develop the exact solution graph.

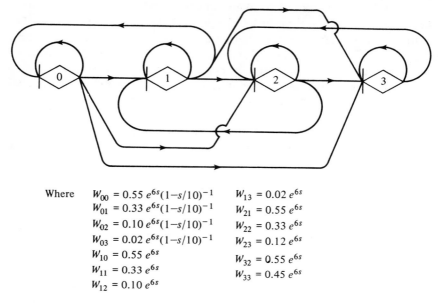

Where
$$W_{00} = 0.55\, e^{6s}(1-s/10)^{-1} \qquad W_{13} = 0.02\, e^{6s}$$
$$W_{01} = 0.33\, e^{6s}(1-s/10)^{-1} \qquad W_{21} = 0.55\, e^{6s}$$
$$W_{02} = 0.10\, e^{6s}(1-s/10)^{-1} \qquad W_{22} = 0.33\, e^{6s}$$
$$W_{03} = 0.02\, e^{6s}(1-s/10)^{-1} \qquad W_{23} = 0.12\, e^{6s}$$
$$W_{10} = 0.55\, e^{6s}$$
$$W_{11} = 0.33\, e^{6s} \qquad\qquad\qquad W_{32} = 0.55\, e^{6s}$$
$$W_{12} = 0.10\, e^{6s} \qquad\qquad\qquad W_{33} = 0.45\, e^{6s}$$

Figure 10.10. GERT approximation of a $M/D/1$ repairman system using the imbedded Markov chain concept.

References that should help the reader model systems other than the $M/G/1$ by the approximate methods are 9, 20, and 29.

10.1.4 Modeling queueing systems with GERTS. The advent of the GERTS III system discussed in Chapter 9 has made it possible to conveniently model most queueing systems. In this section, we will consider one such system—a conveyor system of the class studied by Pritsker [32], Burbridge [1], and Disney [3].

The conveyor system has units arriving at a service facility from some outside source. The service facility may have provisions for the storage of a limited number of units prior to service. If the units cannot enter the service area due to all the storage space being occupied, they recirculate and reenter the arrival stream at some later time. If the units can enter the service area, they must then wait their turn for service.

We will assume that the system has three servers. We also assume that the service time for each server is normal with a mean of 1 time unit and a standard deviation of 0.5. There is room for storage of 2 units in front of each server. Units arrive with an exponential interarrival time with a mean equal to 0.5. The travel time from station to station is 1 time unit, and recycle time is 25 time units. This example is modeled in Fig. 10.11. Queue nodes 11,

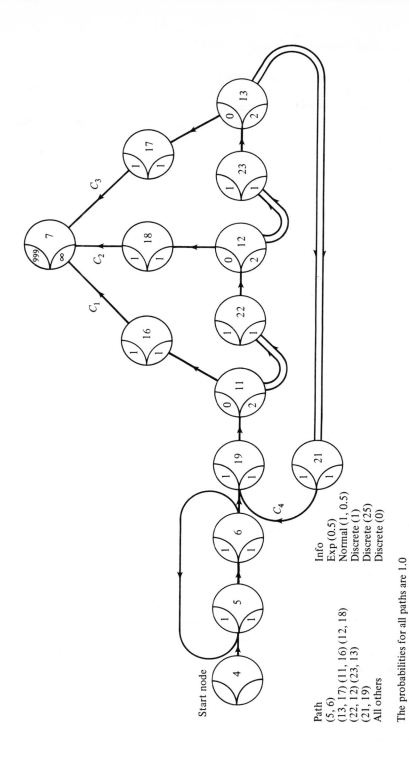

Start node

Path | Info
(5, 6) | Exp (0.5)
(13, 17) (11, 16) (12, 18) | Normal (1, 0.5)
(22, 12) (23, 13) | Discrete (1)
(21, 19) | Discrete (25)
All others | Discrete (0)

The probabilities for all paths are 1.0

Figure 10.11. GERTS model of a conveyor system.

12, and 13 represent the queues preceding the service facilities. The service times are represented by paths (11, 16), (12, 18), and (13, 17). Path (5, 6) represents the interarrival time distribution of items. When units find a facility full, they take path (22, 12), (23, 13), or (21, 19) which represents movement on the conveyor. The model will simulate until node 7 is released, which represents the servicing of 1000 units. Counters C_1, C_2, C_3, and C_4 have been placed on paths (16, 7), (18, 7), (17, 7), and (21, 19), respectively. The output statistics of interest are those collected on:

C_1, C_2, C_3, which are the number of units serviced by service facilities 1, 2, and 3;

C_4, which is the number of units recycled; and

Queue nodes 11, 12, and 13, which are the queue characteristics preceding each service facility.

A GERTS run gives the following results:

$$C_1 = 479.0$$
$$C_2 = 368.8$$
$$C_3 = 151.2$$
$$C_4 = 33.2$$

	Q_{11}	Q_{12}	Q_{13}
Avg. number in the queue:	1.405	0.740	0.219
Avg. busy time of the processor:	0.975	0.748	0.301

10.2 Analysis of Inventory Systems*

10.2.1 Introduction. In this section we will consider the application of GERT to the analysis of inventory systems. As in the queueing section, we will first discuss a very simple model with which the information available from a GERT model will be illustrated. Some of the information available is (1) the distribution of the times between receipt of orders, (2) the average inventory size, and (3) the distribution of costs. The application of GERT to more complicated systems such as the continuous review (S, s) and periodic review (S, s) is presented next. The application of GERT seems appropriate for a limited group of inventory systems. It is only in those cases where the number of units stocked is small and they are of relatively high value, such

---

*The information in this section has been adapted from material in Reference 45 and is used with the permission of the American Production and Inventory Control Society.

as aircraft parts or major appliances. A brief discussion of the use of GERTS is also presented. We will see that GERTS allows us to analyze a wider class of inventory models. Two GERTS models are given in Section 10.2.3. Much of the information in this section is drawn from papers by Scarf [37] and Whitehouse [45].

10.2.2 Information available from the GERT model of the inventory system.

Consider a model in which the number of items in storage is observed every M weeks. If the inventory is found to be equal to or less than a given quantity, s, enough items are ordered to bring the inventory up to full capacity, S. The delivery time is instantaneous. The distribution of the number of items received during a review period is assumed to be constant for all periods. If an item is requested and there is none in inventory, the buyer will wait until the stock is replenished.

The states of the GERT network will be equivalent to the stock level of the inventory when it is reviewed. A negative state indicator is equivalent to a stock-out with customers awaiting items. The MGF of the time on the paths of the GERT network will represent the time between reviews. If $S = 3$, $s = 0$, $M = 1$, and the probability of the number of arrivals during a review period is $\Pr(0) = 0.5$, $\Pr(1) = 0.4$, and $\Pr(2) = 0.1$. The GERT network for this system is shown in Fig. 10.12.

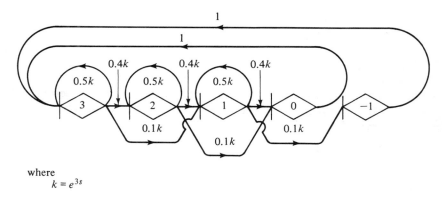

where
$$k = e^{3s}$$

Figure 10.12. GERT representation of an inventory system.

DISTRIBUTION OF THE TIME BETWEEN RECEIPT OF ORDERS

All paths going from a node of lower numerical value to one of higher numerical value represent the receipt of an order. If we redraw the graph so that these nodes flow into a dummy node 3′, the MGF of the time between orders will be $M_{S,S'}(s)$. The new graph for the problem under consideration is shown in Fig. 10.13.

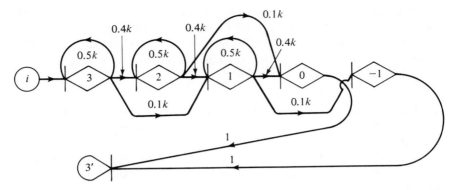

where
$$k = e^{3s}$$

Figure 10.13. GERT representation of an inventory system adapted to find the distribution of the time between receipt of orders.

$$M_{3,3'}(s) = \frac{0.16e^{6s}(0.1e^{3s} + 0.4e^{3s})}{(1 - 0.5e^{3s})^3} + \frac{0.1e^{3s}(0.4e^{3s} + 0.1e^{3s}) + 0.04e^{6s}}{(1 - 0.5e^{3s})^2}$$

$$= \frac{0.09e^{6s} + 0.035e^{9s}}{(1 - 0.5e^{3s})^3}$$

The expected time between receipt of orders is equal to:

$$\left. \frac{dM_{3,3'}(s)}{ds} \right|_{s=0} = 9.83 \text{ weeks}$$

The distribution of the number of inspections between orders can also be found easily by tagging all elements representing an inspection period with an e^c tag. $M_{S,S'}(s, c)|_{s=0}$ will be the MGF of the number of inspections. For our problem:

$$M_{3,3'}(s, c)|_{s=0} = \frac{0.09e^{2c} + 0.035e^{3c}}{(1 - 0.5e^c)^3}$$

AVERAGE INVENTORY SIZE

Consider the graph as it appears in the previous section. If, in the moment generating functions of this GERT network, the s is replaced by a qs terms, where q represents average inventory during the period described by the path, the first derivative of the $M_{S,S'}(s)$ of this graph with respect to s, evaluated at $s = 0$, will be equal to the expected value of the sum of the inventory levels for the period between receiving orders. Dividing this by the expected duration between orders, we obtain the average inventory. For our problem, the network will appear as in Fig. 10.14.

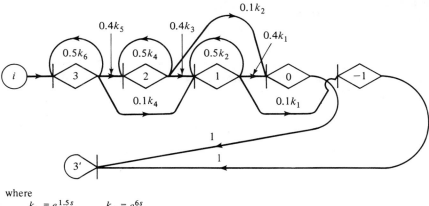

where
$$k_1 = e^{1.5s}, \qquad k_4 = e^{6s},$$
$$k_2 = e^{3s}, \qquad k_5 = e^{7.5s},$$
$$k_3 = e^{4.5s}, \text{ and } k_6 = e^{9s}.$$

Figure 10.14. GERT representation of an inventory system adapted to find the average inventory in the system.

$$M_{3,3'}(s) = \frac{0.08e^{13.5s} + (1 - 0.5e^{6s})(0.05e^{7.5s}) + (1 - 0.5e^{3s})(0.04e^{10.5s})}{(1 - 0.5e^{9s})(1 - 0.5e^{6s})(1 - 0.5e^{3s})}$$

$$\left. \frac{dM_{3,3'}(s)}{ds} \right|_{s=0} = 17.74$$

Distribution of Costs and Average Annual Cost

Costs in the inventory model fall into three categories: those dependent on the inventory size, those dependent upon orders, and those dependent upon stock-outs. The cost of inventory space, investment charges, and taxes are a function of the inventory size. The s in all MGF on the paths will be replaced by qks, where q is the average inventory during the realization of the path and k is the combined cost of inventory space, taxes, and investment per unit per time period. The cost of ordering and the purchase price are both functions of the ordering in the model. All paths in the model moving from a node of lower numerical value to one of higher value represent the receipt of an order. Each of these elements will be tagged with an $e^{(DP+N)s}$ tag, where D is the number of items received in an order, P is the purchase price per unit, and N is the cost of placing and processing an order. There is often a penalty cost connected with stock-outs. This may be a fixed penalty or a penalty cost which is a function of the demand during the stock-out period. If the cost is a constant, it can be represented by an e^{Rs} tag, where R is the penalty cost, on the paths which represent receipt of an order that emanates from a node of negative numerical value. If it is a function of the size of the stock-out, an e^{BTs} tag is used on the paths representing a receipt of an order that emanates from nodes which have a negative value. B is the absolute value of the numerical value of the negative node, and T is the unit penalty cost.

For our example, assume $P = \$10$, $k = \$0.10$, $N = \$5$, and $R = \$25$. The adaptation of the model to consider cost is shown in Fig. 10.15. The MGF of the cost per reorder period is:

$$M_{3,3}(s) = \frac{0.011e^{70.45s} + 0.01e^{70.25s} + 0.04e^{35.35s} + 0.04e^{35.25s} + 0.024e^{35.45s}}{(1 - 0.5e^{0.3s})(1 - 0.5e^{0.2s})(1 - 0.5e^{0.1s})}$$

The expected cost during the reorder period is:

$$\left. \frac{dM_{3,3}(s)}{ds} \right|_{s=0} = \$47.01$$

Since the average length of a reorder period is 9.83 weeks, the average annual cost is:

$$\$47.01\left(\frac{52}{9.83}\right) = \$249$$

THE PROBABILITY OF STOCK-OUT BEFORE A REPLENISHMENT

The probability of a stock-out before the stock is replenished is also an interesting quantity which can be found by considering nodes -1 and $3'$ as terminal nodes (Fig. 10.16). This is equivalent to finding the probability of going from node 3 to node -1. For example:

$$W_{3,-1}(s) = \frac{0.016e^{9s} + 0.01e^{6s}(1 - 0.5e^{3s})}{(1 - 0.5e^{3s})^3}$$

$$P_{3,-1} = W_{3,-1}(s)|_{s=0} = 0.168$$

Therefore, for this particular inventory system, approximately 17% of the inventory cycles will lead to stock-out.

OTHER INFORMATION

The four items discussed in this section appear to be the ones most often investigated, but there is much more information available from the GERT network. The stationary probabilities of being in a given state can be found by the methods discussed in Chapter 8.

If a lead time exists, another interesting quantity is the distribution of the successive inspection times that a stock-out condition exists. This is equal to the MGF of the time to move from node 0 to node 1.

Having shown the possible information available from the GERT model of the inventory system, the next thing we will consider is the modeling

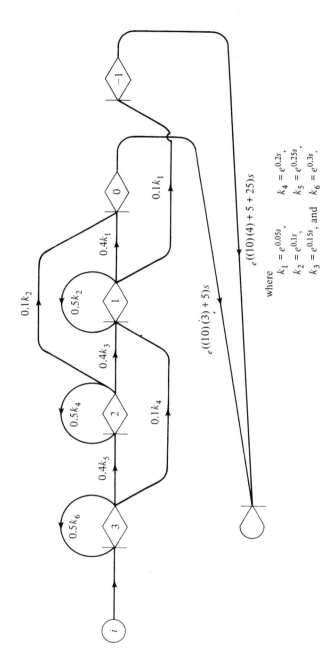

Figure 10.15. GERT representation of an inventory system adapted to find the average annual cost.

where

$$k_1 = e^{0.05s}, \qquad k_4 = e^{0.2s},$$
$$k_2 = e^{0.1s}, \qquad k_5 = e^{0.25s},$$
$$k_3 = e^{0.15s}, \text{ and} \quad k_6 = e^{0.3s}.$$

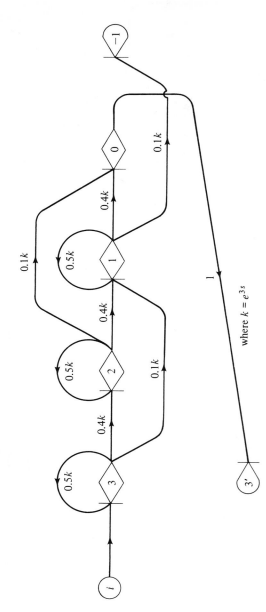

Figure 10.16. GERT representation of an inventory system adapted to find the probability of stockout before a replenishment.

of some popular inventory systems which are discussed in the literature. The information found by the methods of this section will not be found for each of these models. In most cases the methods discussed in this section will be directly applicable, but occasionally, due to peculiarities in the systems, slight adaptations might be necessary.

10.2.3 GERT models of various inventory systems.

Some of the models considered here are studied using the concept of imbedded Markov chains discussed in Section 10.1.3. In all cases, back orders will be assumed to be lost. It is easy to extend the system to one with limited back orders, but extending the case to include unlimited back orders will create an infinite model. The problem of considering infinite graphs has not yet been studied for GERT networks.

CONTINUOUS REVIEW (S, s) POLICY WITH POISSON DEMANDS AND CONSTANT LEAD TIME

In this model the inventory is reviewed after each sale, assuming there are unit sales. If the items in stock plus those on order are equal to s, then $S - s$ items are ordered. After a constant time period, the order arrives and replenishes the stock. Since the model under consideration ignores back orders, if s is less than $S - s$, there can be only one outstanding order at a time. This assumption will be made. Since the orders arrive in a Poisson manner, the interarrival time is exponential. Also of interest will be the number of arrivals during the time to receive an order. This defines the state of the inventory just before the reorder quantity is received. If the constant lead time is T time units and the mean of the Poisson arrival distribution is λ, then the number of orders during the lead is defined by a Poisson distribution with mean λT. After the order is received, the state of the system will be increased by $S - s$ units.

Consider the example in which $S = 5, s = 2, \lambda = 0.1$, and $T = 10$. The probability of the number of arrivals during the lead time is a Poisson distribution with mean of 1.

$$\Pr(0) = 0.37$$
$$\Pr(1) = 0.37$$
$$\Pr(n > 1) = 0.26$$

The GERT representation of this system is shown in Fig. 10.17.

CONTINUOUS REVIEW (S, s) POLICY WITH POISSON DEMANDS AND ARBITRARY LEAD TIME

This is basically the same problem that was just discussed with the exception that the lead time is no longer a constant. Therefore, the probability of

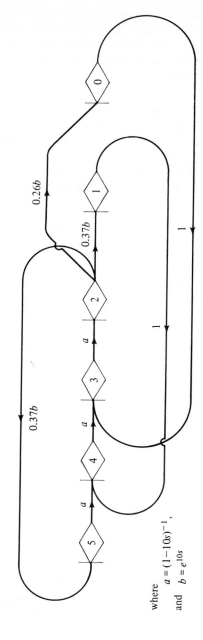

Figure 10.17. GERT representation of a continuous review (S, s) inventory policy with Poisson demand and constant lead time.

where
$a = (1 - 10s)^{-1}$,
and $b = e^{10s}$

368

the number of orders during the lead time can no longer be found by means of the Poisson distribution. But this is basically the same problem that we studied in Section 10.2.3, using imbedded Markov chains for queueing systems in which the number of exponentially spaced arrivals during a service time of arbitrary duration was investigated. The probability of k arrivals during the serivice period was found to be:

$$a_k = \int_0^\infty \frac{(e^{-\lambda t})(\lambda t)^k}{(k!)} \, dF_s(t)$$

where $F_s(t)$ is the distribution function of the service times and λ is the arrival rate.

If $F_s(t)$ is defined as the lead time distribution and λ is defined as the rate at which items are sold, it is observed that a_k is equal to the probability that is being sought. Considering the problem discussed in the previous section with lead time distribution represented by $\phi(t)$, the GERT representation is shown in Fig. 10.18 where:

$$a_k = \int_0^\infty \frac{(e^{-0.1t})(0.1t)^k}{(k!)} \phi(t) \, dt \quad \text{for} \quad k = 0, 1$$

$$A_2 = 1 - \sum_{i=0}^{1} a_i$$

and $M_\phi(s)$ is the MGF of $\phi(t)$.

PERIODIC REVIEW (S, s) POLICY WITH POISSON DEMANDS AND CONSTANT LEAD TIME

In this model the inventory level is checked at periodic intervals. If the level is equal to or less than s, a quantity necessary to bring the stock level up to S is ordered. The model must consider the number of sales that arrive during the review period. This will be described by a Poisson distribution with mean equal to the product of the review time, R, and the rate that items are sold, λ. The number of orders that arrive during the lead time will be described by a Poisson distribution with mean equal to the product of the lead time T, and the rate that the items are sold, λ. The MGFs of the review time and lead time will be used as the time portion of activities in this model.

Consider modeling a system where $S = 4$, $s = 1$, $T = 10$, $R = 3$, and $\lambda = 0.1$. The number of sales during a review time will be described by a Poisson distribution with mean 0.3. $\Pr(0) = 0.74$, $\Pr(1) = 0.22$, $\Pr(2) = 0.03$, and $\Pr(3) = 0.01$. The sales during the lead time will be described by a Poisson distribution with mean equal to 1. The model will appear in GERT format as shown in Fig. 10.19. It would also be easy to consider the

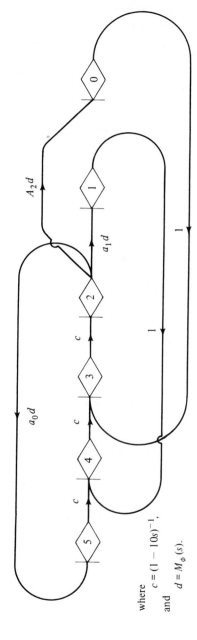

Figure 10.18. GERT representation of a continuous review (S, s) inventory policy with Poisson demands and arbitrary lead time.

where $c = (1 - 10s)^{-1}$,
and $d = M_\phi (s)$.

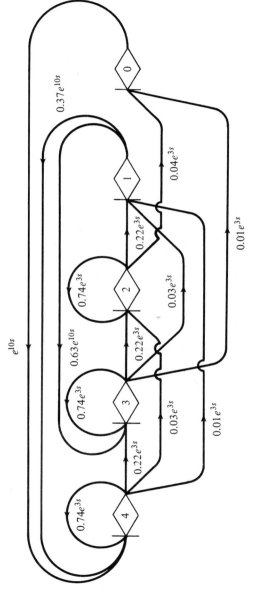

Figure 10.19. GERT representation of a periodic review (S, s) policy with Poisson demands and constant lead time.

371

lead time as a random variable by using the approach discussed in the previous section.

OTHER MODELS

The models presented in this section are just a sample of the possible systems that can be studied by GERT. In this section some recent articles will be mentioned which should aid the reader in investigating inventory systems with GERT.

Morse [34] considers the following version of an S, s policy, the study of which is possible by means of GERT. When inventory falls to s, an order of size $S - s$ is requested. If inventory on hand is depleted to zero while the order is outstanding, an additional order of size s is placed.

Karlin and Fabens [18] consider another variation which can also be investigated by means of GERT. Orders are placed whenever stock falls below s, but the order size is a multiple of $S - s$—the largest multiple that keeps the stock size less than or equal to S after ordering. In another article, Fabens [7] investigates the S, s policy by using a semi-Markov model which he originally developed for investigating batch queues with Poisson inputs. The idea of imbedded Markov chains is used. In this article and another which he co-authored with Karlin [8], he discusses the inclusion of seasonal trends in his model. These ideas should be helpful in the modeling of inventory systems by GERT.

Scarf [38] found the imbedded Markov chain of a model with an arbitrary interarrival time of orders and negative exponential time lag. This can be used advantageously in some GERT models.

Morse [24] does much work on the use of Markov processes in the stationary analysis of inventory systems. It is suggested that the user of GERT interested in modeling inventory systems read his second chapter; it contains many interesting ideas which should prove helpful.

Zehna [47] discusses the use of imbedded Markov processes to investigate different methods of depleting inventory. LIFO and FIFO models were considered. These imbedded chains could be usefully included in GERT models if the concept of depletion policy is deemed important in the particular model.

Masse [22] included the concept of discrete strategies in investigating some concepts in inventory theory. He considers the decision process for inventory systems that can be described by Markov processes. Thus, an investigation of his work should prove interesting in the consideration of optimizing inventory systems represented by GERT.

Naddor [28] describes an inventory system in which the forecast of future demand is a part of the inventory model. He describes the system by means of a Markov chain. This is an extremely interesting model which could be investigated by GERT. The concept of forecasting could also be included in the other models.

Johnson [17] suggests some points that might be useful to someone considering the use of graphical models for infinite horizon problems.

Other graphical approaches to the analysis of inventory systems have been suggested by Howard [15] and Elmaghraby [5].

Other articles that might be of help to the user have been written by Gaver [11, 12], Morse [25], and Derman [2].

Before closing this discussion, there is another large class of problems that can be considered by means of GERT. These problems fall under the classification of the *theory of dams* or *reservoir storage*. The problem is similar to inventory problems except that there is a limited available storage area for the commodity. The input into the system is not, in most cases, controllable, as in the case of inventory theory. There are applications for this theory aside from reservoir storage. Problems of this nature should be very susceptible to investigation by GERT. Those interested in this subject are referred to the book by Moran [23].

10.2.4 GERTS models of inventory systems.

In this section the usefulness of GERT in the analysis of inventory systems has been investigated. Many systems can be modeled by GERT and much information can be obtained by the analysis of the GERT model. The models, however, seem to become very complicated, and a computer-type solution would be a necessity for any reasonably sized model.

Two examples of modeling inventory systems using GERTS will be presented to illustrate its application in the inventory area.

AN INVENTORY SYSTEM WITH VARIABLE LEAD TIME

Consider the case of a retail company which stocks a product that has a demand per week which is normal, a mean of 100 units, and a standard deviation of 10 units. The reorder quantity for this particular product has previously been established at 404 units; however, the company's inventory policy is such that the inventory level is determined only at the end of the week. (Only if less than 404 units are on hand at the end of the week does the company reorder more of the product.) If the lead time required to receive shipment from the supplier is a random variable such that:

Lead Time (in weeks)	Probability
2	0.15
3	0.50
4	0.30
5	0.05

what is the probability that the company has a stock-out of this particular product?

There are three separate variables interacting in a random manner which must be represented in any GERTS model that could realistically simulate this problem. The first is the inventory level at the time the reorder is placed. The number of units on hand can be 404 or less, but it is extremely difficult to describe the actual distribution. The second is the number of units which customers demand during each week of the replenishment period. The third is the number of weeks until the new shipment of units arrives. A model which incorporates the interaction of all three of these random variables is shown in Fig. 10.20.

The parameter along each path is the number of units. The combination of paths (5, 6) and (2, 3) is used to simulate the random effect that the policy of checking only at the end of each week has on the number of parts in inventory when the reorder takes place. Path (5, 6) generates weekly demands for the product. When node 3 is realized, there are exactly 404 units left in inventory, and the realization of node 3 triggers a modification, A_1, which replaces node 6 with node 7 the next time node 6 is realized. Path (3, 4) keeps track of the number of units left in inventory. Node 8 accounts for the probabilistic nature of the lead time required before the new order arrives by varying the number of releases supplied to node 11. Path (9, 10) generates customer demand during the lead time, and each time node 10 is realized, another release is acquired by node 11. Thus, if node 11 is realized, it must mean that the new shipment has arrived. If node 4 is realized, it must be that the customer demand generated by path (9, 10) was greater than the number of units in inventory when the reorder was placed. And since node 11 has not been realized, no new units have arrived. Thus,

Prob. (node 11 is realized) = Prob. (new shipment arrives before
inventory level equals zero)

and

Prob. (node 4 is realized) = Prob. (stock-out occurs)

The GERTS run for this system yields a probability of stock-out of 0.42.

ANALYSIS OF IN-PROCESS INVENTORY IN AN ASSEMBLY LINE

The analysis of many assembly line situations involves the analysis of in-process inventory. Consider the following example similar to one suggested

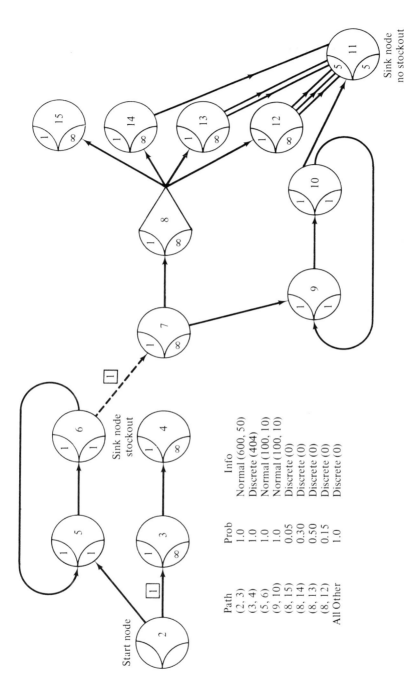

Figure 10.20. GERTS model of an inventory system.

Path	Prob	Info
(2, 3)	1.0	Normal (600, 50)
(3, 4)	1.0	Discrete (404)
(5, 6)	1.0	Normal (100, 10)
(9, 10)	1.0	Normal (100, 10)
(8, 15)	0.05	Discrete (0)
(8, 14)	0.30	Discrete (0)
(8, 13)	0.50	Discrete (0)
(8, 12)	0.15	Discrete (0)
All Other	1.0	Discrete (0)

by Pritsker and Kiviat [31]:

> The maintenance facility of a large manufacturer performs two operations. These operations must be performed in series; operation 2 always follows operation 1. The units that are maintained are fairly bulky, and there is limited space available for the maintenance facility. Currently, only eight units, including units being worked on, can be handled at one time. A proposed design leaves space for two units between work stations for operations 1 and 2, and space for four units before work station 1. Current company policy subcontracts maintenance of a unit if it cannot gain access to the queue preceding work stations 1 or 2.

> Historical data indicate that the time interval between requests for maintenance is exponentially distributed with a mean of 0.4 time units. Service times are approximately exponentially distributed, with the first station requiring on the average 0.50 time units, and the second service station, 0.25 time units. It takes 0.2 time units to move a part from work station 1 to 2.

This problem can be modeled using GERTS as shown in Fig. 10.21. The queue nodes 11 and 12 are used to model the queueing behavior of the in-process inventory in front of the two stations in the maintenance facility. If queue node 11 has four items in its queue, an arriving part will be shunted along path (11, 8) which is representative of subcontracting of maintenance. Similarly, if queue node 12 is full, the unit is shunted to path (12, 9). Paths (11, 6) and (12, 7) represent the service time at service facilities 1 and 2. Path (6, 12) is the transit time from service facility 1 to service facility 2. The interarrival time of units is represented by path (4, 5). The simulation is terminated by path (3, 10), exactly 1000 time units after the start of the run. Counters C_1, C_2, C_3, and C_4 are placed on paths (8, 13), (9, 13), (6, 12), and (7, 13), respectively. Interval statistics are collected between node 5 and nodes 8, 9, and 7 to measure the time in the system for parts subcontracted at the first step, for parts subcontracted at step 2, and for parts completely serivced.

The statistics of interest for this problem are those related to:

C_1–the number of units shunted to subcontract for total maintenance.

C_2–the number of units shunted to subcontract for the second portion of maintenance.

C_3, C_4–the number of units serviced by facilities 1 and 2.

Queue nodes 11, 12–the in-process inventory preceding service facilities 1 and 2.

Interval statistics–Time in the system for various parts.
on Nodes 8, 9, and 7

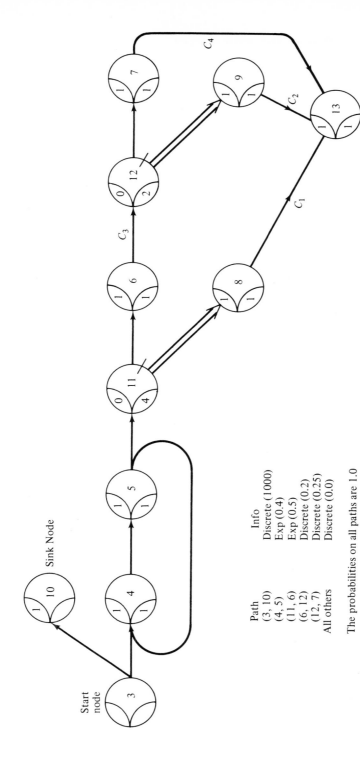

The probabilities on all paths are 1.0

Path	Info
(3, 10)	Discrete (1000)
(4, 5)	Exp (0.4)
(11, 6)	Exp (0.5)
(6, 12)	Discrete (0.2)
(12, 7)	Discrete (0.25)
All others	Discrete (0.0)

Figure 10.21. GERTS model of a production line with queueing.

377

The GERTS run gives the following results:

$$C_1 = 486$$
$$C_2 = 28$$
$$C_3 = 2091$$
$$C_4 = 2063$$

	$QN11$	$QN12$
Average in-process inventory at	2.202	0.1394
Busy time at	0.907	0.447
Expected time in the system:		
for parts subcontracted at step 1 = 0		
for parts subcontracted at step 2 = 3.23		
for parts not subcontracted = 2.22		

10.3 Analysis of Reliability Systems*

10.3.1 Introduction. In this section, we will show the flexibility of the GERT's logical nodes in the modeling of reliability systems. Then we will demonstrate that with the W function, W generating function, and counters much more information is available from the GERT model of a system than would be available from a comparable flowgraph model. Next, a complicated model will be studied in detail. Finally, some applications of GERTS will be considered. Much of the material in this section was derived from an article by Whitehouse [44].

10.3.2 Flexibility of GERT's node structure. It is very important to be able to determine the reliability of an entire system of parts if the reliability of each is known. The example given will be for switching systems in which Boolean algebraic techniques are now used.

There are two basic combinations of parts, series and parallel. For the series parts pictured in Fig. 10.22, both X and Y must operate for the circuit to operate. The Boolean transmission function between terminals a and b

Figure 10.22. Series switch circuit.

---

*The material in this section was adapted from material appearing in Reference 44 and has been used with the permission of *Technometrics*.

is:

$$T_{a,b} = X \cap Y$$

For the parallel combination of parts pictured in Fig. 10.23, if either X,

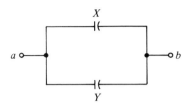

Figure 10.23. Parallel switch circuit.

Y, or both óperate, the transition between a and b will be realized. The Boolean representation of this is:

$$T_{a,b} = X \cup Y$$

It is possible to represent any combination of parts in the Boolean representation just discussed. For a further discussion of this subject, the reader is referred to Roberts [36]. ·

The reliability engineer is interested in the probability that continuity exists between a and b, and this is defined as the *reliability* of the network. The Boolean algebraic representation of the system is converted into probabilistic terms by the following two well-known expressions in probability calculus:

$$\Pr(X \cup Y) = \Pr(X) + \Pr(Y) - \Pr(X \cap Y)$$

and

$$\Pr(X \cap Y) = \Pr(X)\Pr(Y)$$

assuming X and Y independent. These probabilities are identical to those obtained for traversing series and parallel combinations of INCLUSIVE-OR nodes in GERT theory. It would seem reasonable that the reliability of a combination of switches could be investigated by representing the system as a GERT network with an INCLUSIVE-OR node placed at the intersection of all components. The probability on each transmittance is the probability that the particular component represented by the transmittance works. The MGF of the time is equated to 1 since the major interest at this point is reliability of the system.

Consider the bridge circuit in Fig. 10.24 presented by Roberts [36]. This may be represented by an INCLUSIVE-OR network as shown in Fig.

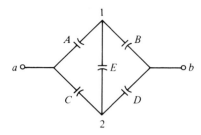

Figure 10.24. Complex switch circuit.

10.25, where p_i is the probability that component i functions. The probability of going from a to b is the reliability of the system. This system can be represented as an EXCLUSIVE-OR system as shown in Fig. 10.26, where $\bar{p}_k = 1 - p_k$ and node $\bar{n}, m$ is the realization of node m but not n in the INCLUSIVE-OR network. To understand Fig. 10.26, consider node 1, $\bar{2}$. This node can only be reached from node a if component A works and component C doesn't; thus the probability $p_A \bar{p}_C$. There are three possible destinations from node 1, $\bar{2}$. If component E works, then nodes 1 and 2 will be realized in the INCLUSIVE-OR network, and we move from node 1, $\bar{2}$ to node 1, 2 with probability p_E. If component E doesn't work, the realization of the total network is dependent upon whether component B works or not. Thus, we move to node b with probability $p_B \bar{p}_E$ and to node $\bar{b}$ with probability $\bar{p}_B \bar{p}_E$. The remainder of Fig. 10.26 is derived in a similar manner.

If the component engineer gives his assurance that all the components have the same reliability, then:

$$W_{a,b}(s)|_{s=0} = R = 2p^2 + 2p^3 - 5p^4 + 2p^5$$

and, if

$$p = 0.95, \qquad R = 0.995$$

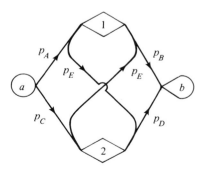

Figure 10.25. GERT representation of a complex switch circuit with INCLUSIVE-OR nodes.

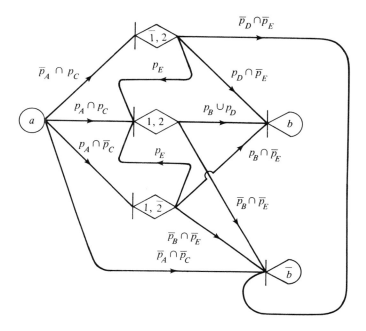

Figure 10.26. GERT representation of a complex switch circuit with EXCLUSIVE-OR nodes.

If the component is used over and over again, the distribution of the number of firings before a failure can be found by investigating the simple graph shown in Fig. 10.27:

$$M_{i,2}(s) = \frac{(1-R)}{(1-Re^s)}$$

$M_{i,2}(s)$ is the new MGF of the distribution of the number of successful firings of the device.

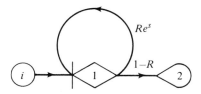

Figure 10.27. GERT representation of a reliability system adapted to find the distribution of successful uses before failure.

Roberts [36] introduces some other ideas in the application of Boolean algebra in logical design. He suggests that Boolean variables can be assumed to be inputs and outputs to black boxes. He also suggests three basic types of black boxes:

1. The logical product box which provides an output only if all inputs into the box are realized.
2. The logical sum box which provides an output if any of the inputs entering the box are realized.
3. The negation box which provides a signal only if the input to the box is not realized.

The logical product node is the same as the GERT AND node, and the logical sum box is the same as the INCLUSIVE-OR node. There is no node in GERT equivalent to the negation, but it is possible to model the negation box using GERTS.

10.3.3 Information available from GERT models.

To demonstrate the advantage of GERT models over standard flowgraph models of reliability systems, we will consider a system we have already analyzed in Chapter 6.

For systems with expensive parts, attempts are often made to salvage or recover the part for future use. For such systems, we want to determine the expected life of one of these parts in terms of both time and number of uses. Also the probability of a successful operation using such recovered parts is an important performance measure. Happ [14] proposed an application of flowgraph theory to the recovery of a rocket after a test, assuming that there is no deterioration in the system. Happ was able to calculate the probability of various states of the system. This model will be considered, with the MGF of the time of test, time of recovery, and repair of the rocket included in the model. It is shown in Fig. 10.28 with elements defined as follows.

Events:

i = initial flight.
f = any flight.
g = a successful flight.
b = an unsuccessful flight.
d = nonrecovery after g.
e = nonrecovery after b.

Probabilities:

S = probability of a successful flight.
R = probability of a recovery after g.
T = probability of a recovery after b.

MGF:

$M_S(s)$ = MGF of a flight time.
$M_R(s)$ = MGF of the time of recovery after g.
$M_T(s)$ = MGF of the time of recovery after b.

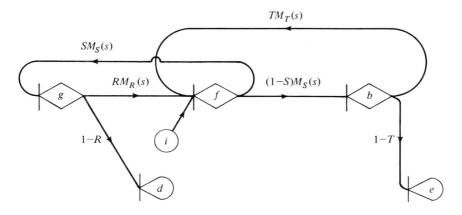

Figure 10.28. GERT representation of a rocket recovery problem.

Using the W function, W generating function, and counters, the following are examples of some of the additional information available for this system.

THE LIFE OF THE ROCKET

Applying Mason's rule to Fig. 10.28, we find that the MGF of the life of the rocket is equal to $W_{i,d}(s) + W_{i,e}(s)$.

$$\frac{S(1 - R)M_S(s) + (1 - S)(1 - T)M_S(s)}{1 - (RS)M_R(s)M_S(s) - T(1 - S)M_S(s)M_T(s)}$$

THE NUMBER OF FLIGHTS UNTIL NONRECOVERY

·The MGF of the number of flights can be found if we place an e^c tag on all elements entering the f node and then solve:

$$M(s, c) = [W_{i,e}(s, c) + W_{i,d}(s, c)]|_{s=0}$$

$$M(s, c) = \frac{e^c(1 - R)S + e^c(1 - S)(1 - T)}{1 - SRe^c - (1 - S)Te^c}$$

The MGF of the number of successes and failures can be found in a like manner.

THE PROBABILITY OF AT LEAST ONE SUCCESS

The probability of at least one success with a given rocket can be found if we tag all transmittances entering the g node with a z tag. Solving the graph as before and equating both z and s to zero, we obtain the probability of no successes. The complement of this quantity is the desired probability:

$$1 - W(s, z)|_{s=0}^{z=0} = 1 - \left(\frac{((1 - S)(1 - T))}{(1 - (1 - S)T)}\right)$$

THE FIRST PASSAGE TIME TO A SUCCESS

The MGF of the time until a successful flight can be found if we tag all transmittances entering the g node with a z tag. The MGF of interest is

$$M_{i,g}(s \mid 1) = \frac{W_{i,g}(s \mid 1)}{P(1)} = \frac{W_{i,g}(s \mid 1)}{W_{i,g}(s \mid 1)_{s=0}}$$

where

$$W_{i,g}(s \mid 1) = \frac{\partial W(s, z)}{\partial z}\bigg|_{z=0}$$

and

$$W_{i,g}(s, z) = \frac{SzM_S(s)}{1 - (RS)zM_R(s)M_S(s) - T(1 - S)M_S(s)M_T(s)}$$

If one is interested in the MGF of the nth passage time of a success, the MGF of interest is $M(s \mid n)$.

OTHER INFORMATION

Using the techniques presented in Chapter 8, we can find the conditional network probabilities and times, counts on elements in a network, renewal times to elements of a network, and correlations among elements of the network.

10.3.4 Analysis of a complicated reliability model. To further demonstrate the advantages of GERT analysis consider the following model.

Lloyd and Lipow [21] have suggested a very interesting model to which GERT is very applicable. Suppose a device is being developed for a given application. The application is such that the device, when put into operation, either succeeds or fails to accomplish what it is designed to do. Suppose further that there is only one thing that can go wrong with the device, and the device will eventually fail due to this fault. The whole purpose of the development effort on the device is to discover what the cause of failure is and then attempt to redesign or fix the device so that it won't fail at all. Assume repair either fixes the device or not, i.e., the probability $1 - p$ of a defective operation is constant until the device is completely fixed and always works. The development effort then consists of repeated trials of the device. If the device operates successfully on any given trial, the designer or development engineer decides to make no redesign action. He proceeds to the next trial on the chance that he has already fixed the device and that its probability of failure is zero. If it fails on any given trial, the engineer goes to work on it and has a probability, a, of fixing the device permanently prior to the next trial.

Lloyd and Lipow do not consider the time of repair or time of trial, but this can easily be incorporated into the GERT model of this system. Define:

$M_r(s)$ = MGF of the repair time including the trial immediately following the repair.

$M_t(s)$ = MGF of trial time.

There are three outcomes that are possible from a given trial: (1) the trial is successful, given that the device is faulty, (2) the trial is a failure, given that the device is faulty, and (3) the trial is successful, given that the device is fixed. These outcomes will represent the events of the GERT network. The model is now easily represented by GERT as shown in Fig. 10.29. The MGF of the time until the device is completely dependable, i.e., will never fail again, is equal to $M_{S,3}(s)$:

$$M_{S,3}(s) = \frac{a(1 - p)M_t(s)M_r(s)}{1 - pM_t(s) - (1 - a)(1 - p)M_r(s)}$$

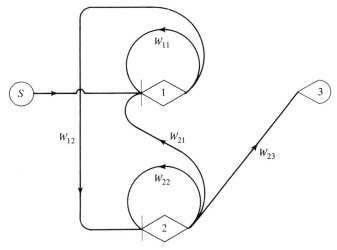

where
$$W_{11} = pM_t(s),$$
$$W_{12} = (1-p)M_t(s),$$
$$W_{21} = (1-a)pM_r(s),$$
$$W_{22} = (1-a)(1-p)M_r(s)$$
and $$W_{23} = aM_r(s).$$

Figure 10.29. GERT representation of a reliability repair model.

Some examples of the information available from this model using counters are:

1. The MGF of the number of trials necessary until the device is operative is found by tagging all elements with an e^c tag, solving the graph for $M_{S,3}(s, c)$, and then equating $s = 0$. The MGF of the number of trials until the device is dependable is:

$$M_{S,3}(s, c)_{s=0} = \frac{a(1 - p)e^{2c}}{1 - pe^c - (1 - a)(1 - p)e^c}$$

2. The MGF of the number of failures can be investigated if we tag all the elements entering the nodes representing a failure (in this case node 2) with an e^c tag and then solve for $M_{S,3}(s, c)$:

$$M_{S,3}(s, c)|_{s=0} = \frac{a(1 - p)e^c}{1 - p - (1 - a)(1 - p)e^c}$$

Lloyd and Lipow state that it would apparently be difficult to consider devices that had more than one type of failure. However, as they point out, Weiss [41] considers models of this nature.

Aside from mathematical manipulation which, in theory could be handled by the computer, it is relatively easy to consider problems with multiple failure modes by means of GERT. Consider the following in which there are two modes of failure. Define:

Events:

1 = success, given that both modes broken.
2 = failure, given that both modes broken.
3 = success, given that mode 1 fixed.
4 = failure, given that mode 1 fixed.
5 = success, given that mode 2 fixed.
6 = failure, given that mode 2 fixed.
7 = success, given that both modes fixed.

Probabilities:

p_1 = probability of mode 1 causing failure.
p_2 = probability of mode 2 causing failure.
a_1 = probability of fixing mode 1.
a_2 = probability of fixing mode 2.

The GERT model is shown in Fig. 10.30. This model can now be solved by the methods discussed for the simpler example in which there was only one mode of failure.

10.3.5 GERTS analysis of reliability systems. One type of reliability problem which GERTS handles quite well is the determination of the mean

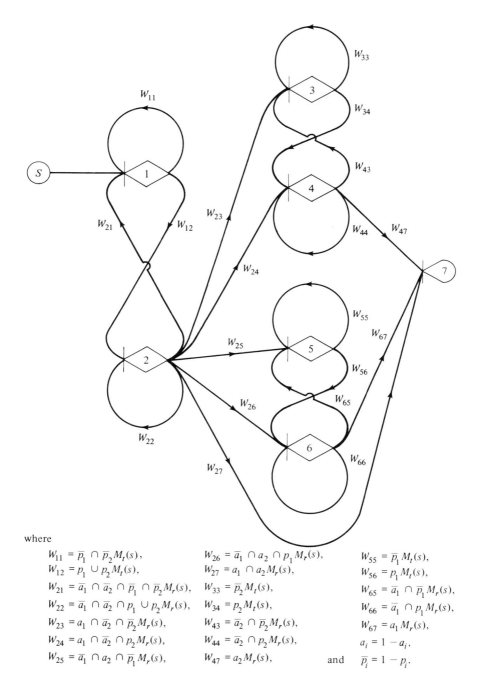

where

$W_{11} = \bar{p}_1 \cap \bar{p}_2 M_t(s),$

$W_{12} = p_1 \cup p_2 M_t(s),$

$W_{21} = \bar{a}_1 \cap \bar{a}_2 \cap \bar{p}_1 \cap \bar{p}_2 M_r(s),$

$W_{22} = \bar{a}_1 \cap \bar{a}_2 \cap p_1 \cup p_2 M_r(s),$

$W_{23} = a_1 \cap \bar{a}_2 \cap \bar{p}_2 M_r(s),$

$W_{24} = a_1 \cap \bar{a}_2 \cap p_2 M_r(s),$

$W_{25} = \bar{a}_1 \cap a_2 \cap \bar{p}_1 M_r(s),$

$W_{26} = \bar{a}_1 \cap a_2 \cap p_1 M_r(s),$

$W_{27} = a_1 \cap a_2 M_r(s),$

$W_{33} = \bar{p}_2 M_t(s),$

$W_{34} = p_2 M_t(s),$

$W_{43} = \bar{a}_2 \cap \bar{p}_2 M_r(s),$

$W_{44} = \bar{a}_2 \cap p_2 M_r(s),$

$W_{47} = a_2 M_r(s),$

$W_{55} = \bar{p}_1 M_t(s),$

$W_{56} = p_1 M_t(s),$

$W_{65} = \bar{a}_1 \cap \bar{p}_1 M_r(s),$

$W_{66} = \bar{a}_1 \cap p_1 M_r(s),$

$W_{67} = a_1 M_r(s),$

$a_i = 1 - a_i,$

and $\quad \bar{p}_i = 1 - p_i.$

Figure 10.30. GERT representation of a two-level reliability repair model.

time before failure of a system. Most problems of this type can be modeled if we let certain paths represent the time before failure of individual components, and then adjust the number of releases required for realization of the sink nodes, and the paths leading into sink nodes, to give the system the proper operating characteristics. For example, consider the power source for a complex transmitter which consists of a primary generator, *A*, and three secondary generators *B*, *C*, and *D*. The time before failure of the four generators are random variables; *A* is exponentially distributed with a mean of 1000 hours, and *B*, *C*, and *D* are normally distributed with a mean of 800 hours and a standard deviation of 100 hours. If *A* has not failed, and either *B*, *C*, or *D* are still operating, then the power level will be sufficiently high for the transmitter to operate; However, if *A* has ceased to function, then all three of the secondary generators must work if the transmitter is to operate. What is the expected time before the transmitter fails to function?

Figure 10.31 shows the GERTS model of the power source just de-

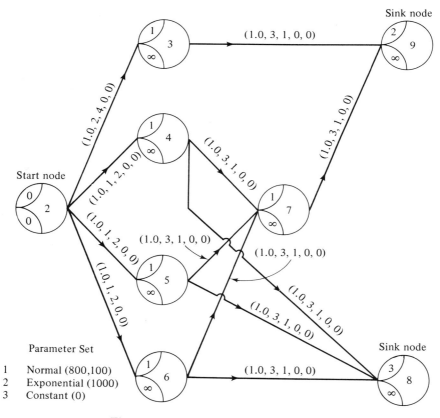

Figure 10.31. GERTS reliability model.

scribed. Paths (2, 3), (2, 4), (2, 5), and (2, 6) generate the time before failure of the individual components where nodes 3, 4, 5, and 6 represent the failure of *A*, *B*, *C*, and *D*, respectively. Node 8 represents one of the possible ways the system can fail, namely, components *B*, *C*, and *D* fail. The other possible mode of failure—component *A* and one of the secondary generators fail—is represented by node 9. To ensure that one of the releases on node 9 comes along path (3, 9), node 7 (which can only be realized once) is inserted in the network.

FINAL RESULTS FOR 100 SIMULATIONS

Node	Prob./Count	Mean	Std. Dev.	Min.	Max.
8	0.26	858.9324	481.1085	135.28	1764.59
9	0.74	815.4226	506.4171	52.69	2000.00

Mean time before failure
$$\text{of transmitter} = (0.26)(858.9324) + (0.74)(815.4226)$$
$$= 827.5 \text{ hours}$$

GERTS can also be useful in evaluating different maintenance policies on a cost basis. Consider the case of a company which has maintenance problems with a certain complex piece of equipment. This equipment contains four identical vacuum tubes that have been the cause of trouble. The problem is that the tubes fail fairly frequently, thereby forcing the equipment to be shut down while replacements are made. The current practice is to replace tubes only when they fail. However, a proposal has been made to replace all four tubes whenever any one of them fails in order to reduce the frequency with which the equipment must be shut down. The objective is to compare these two alternatives on an expected cost basis.

The pertinent data are the following. For each tube, the operating time until failure has approximately a normal distribution, with a mean of 600 hours and a standard deviation of 100 hours. The equipment must be shut down for one day in order to replace a tube, or for two days in order to replace all four tubes. The cost associated with shutting down the equipment and replacing tubes is $300 per day plus $50 for each new tube. Based on 3000 hours of simulated operation, which is the preferred alternative on a cost basis?

For the purpose of discussion of this problem, the following definitions will be made:

A_1 = policy which replaces all tubes whenever one fails.

A_2 = policy which replaces each tube individually.

Figure 10.32 shows the GERTS network which examines both policies at

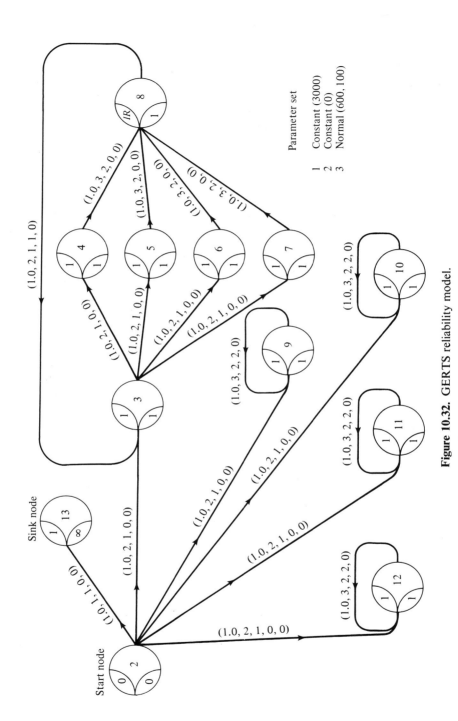

Figure 10.32. GERTS reliability model.

390

the same time. The section of the network consisting of nodes 3, 4, 5, 6, 7, and 8 is used to evaluate policy A_1. The paths leading into node 8 generate the time before failure of the four vacuum tubes, and node 8 is realized when the first tube fails. When node 8 is realized, path (8, 3) releases node 3 and also triggers counter C_1. It is of great importance to note that node 8 is designated a removal node and thus, once it is realized, all other paths leading into it are removed. Once node 3 is released, the entire procedure repeats itself; hence, the number of times counter C_1 is realized is the number of times the set of four vacuum tubes is replaced using policy A_1 for 3000 hours of simulated operation. Nodes 9, 10, 11, and 12 are used to evaluate policy A_2. Each of these nodes represents the failure of one of the vacuum tubes. Each time a tube fails, counter C_2 is increased by one; the number of times counter C_2 is realized is the number of times the equipment has to be shut down to replace a tube during the 3000 hours of simulated operation under policy A_2. The results of the simulation are:

FINAL RESULTS FOR 100 SIMULATIONS

Node	Prob./Count	Mean
13	1.0	3000.00
13	1	5.53
13	2	18.14

Cost per shutdown under policy $A_1 = (2)(300) + (4)(50) = \$800$

Cost per shutdown under policy $A_2 = 300 + 50 = \$350$

Expected cost of A_1 for 3000 hours $= (800)(5.53) = \$4424$

Expected cost of A_2 for 3000 hours $= (350)(18.14) = \$6349$

It seems reasonable to adopt policy A_1.

10.4 Modeling Quality Control Sampling Plans Using GERT

10.4.1 Introduction. Quality control has become the first point of attack in methods improvement. A properly designed quality control system can reduce the losses from rejections, scrap, and reworked production to a very low percentage of total output and hold that level. Cost is reduced and output is increased. The backbone of quality control is the design of effective sampling plans.

In this section various approaches to modeling sampling systems will be discussed; the modeling of various sampling plans using both GERT and GERTS will be considered.

Fry [10], Powell [30], and Mullen [27] applied GERT directly in the area of quality control but to a limited extent. They studied Dodge's continuous sampling plan CSP-1 and proposed some modifications to this plan. In addition, Whitehouse [42] has used GERT to model such quality control systems as lot-acceptance sampling plans, Bayesian models, and control charts.

10.4.2 A single sampling plan. One of the major fields of statistical quality control is *acceptance sampling*. The purpose of acceptance sampling is to determine a course of action, not to estimate lot quality. Acceptance sampling prescribes a procedure that will give a specified risk of accepting lots of given quality. In other words, acceptance sampling yields quality assurance.

A single sampling plan [4] is designated by two numbers, n and c. A sample of size n is taken from a given lot. If it contains c or less defective units, it is accepted. Otherwise, it is rejected. An example is presented to show how GERT can be used in a single sampling plan.

EXAMPLE 1

Let $n = 100$, $c = 2$. Two kinds of information can be obtained from GERT. Let p be the lot fraction defective.

(a) The system is shown in Fig. 10.33. Nodes 0, 1, 2, and 3 represent the

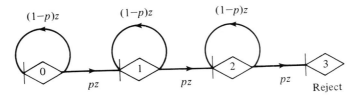

Figure 10.33. Single sampling plan (EXCLUSIVE-OR type).

number of defective units. Node 3 also represents an absorbing state (i.e., the state in which the lot is rejected). Each branch of the network is multiplied by z. From Mason's rule we get:

$$W_{C,R}(z) = \frac{p^3 z^3}{1 - 3(1 - p)z + 3(1 - p)^2 z^2 - (1 - p)^3 z^3}$$
$$= a_3 z^3 + a_4 z^4 + \cdots + a_j z^j + \cdots$$

The generating function can be obtained either by division or from the W

function as follows:

$$a_j = \frac{1}{j!} \cdot \frac{\partial^j W_{0,R}}{\partial z^j}\bigg|_{z=0}$$

Thus, we can get:

$$\text{Probability of rejecting a lot} = \text{Pr} = \sum_{i=3}^{100} a_i$$

$$\text{Probability of accepting a lot} = 1 - \text{Pr}$$

This approach illustrates only that GERT is applicable to this type of problem; however, it is not a practical way to get these probabilities from the above computation. Actually we obtain these probabilities from the GERTS model shown in Fig. 10.34 for $p = 0.03$. Node 2 is the starting node. Nodes 3 and 6 are sink

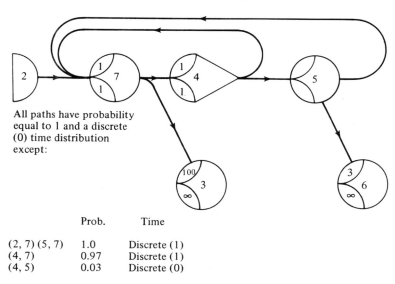

All paths have probability
equal to 1 and a discrete
(0) time distribution
except:

	Prob.	Time
(2, 7) (5, 7)	1.0	Discrete (1)
(4, 7)	0.97	Discrete (1)
(4, 5)	0.03	Discrete (0)

Figure 10.34. GERTS model of a single sampling plan.

nodes. Node 3 represents the acceptance of a lot, and node 6 represents the rejection of a lot. If path (7, 3) is realized 100 times, then node 3 will be realized and the lot will be accepted. Nodes 4, 5, and 6 are used to determine if individual parts are acceptable. Path (4, 5) will be realized if a bad part is sampled. This will cause path (5, 6) to be realized. If this path is realized three times, node 6 will be released designating the rejection of the lot. Paths (4, 7) and (5, 7) trigger a new part through the inspection system. Paths (2, 7), (4, 7), and (5, 7) have a discrete distribution of 1 unit to give a measure of the number of units sampled before the lot is accepted or rejected.

GERTS gives us the following results:

Node	Prob./Count	Mean	Std. Dev.	Min.	Max.
3	0.4250	100.0000	0.0000	100.0000	100.0000
6	0.5750	59.9217	23.0884	14.0000	99.0000

From the above results, we can see:

Probability of accepting a lot = 0.425

Probability of rejecting a lot = 0.575

Expected number of units passed until lot is accepted = 100.0

Expected number of units passed until lot is rejected = 59.9217

Standard deviation of units passed until lot is accepted = 0.0

Standard deviation of units passed until lot is rejected = 23.0884

Minimum number of units passed until lot is accepted = 100.0

Minimum number of units passed until lot is rejected = 14.0

Maximum number of units passed until lot is accepted = 100.0

Maximum number of units passed until lot is rejected = 99.0

The results observed from the operating characteristic (OC) curve for the sampling inspection plan $n = 100$, $c = 2$, $p = 0.03$ [4] are as follows:

Probability of acceptance = 0.43

Probability of rejection = 0.57

Thus, we can see that the probabilities obtained from the above two methods are very close. However, GERTS II can give us the expected values and standard deviation of the distribution while the OC curve does not provide us such information. The GERTS model in Fig. 10.34 can be used for any single sampling plan by setting the number of releases on node 3 equal to n and on node 6 equal to $c + 1$.

(b) Suppose only c is known. We would like to know the expected number of samples before rejecting the lot. For the loop system shown in Fig. 10.35(a), standard flowgraph operation would reduce to the graph shown in Fig. 10.35(b) as discussed in Chapter 8.

Therefore for $c = N$, the system shown in Fig. 10.36(a) can be reduced to the graph shown in Fig. 10.36(b). Thus:

$$W_{0,N+1}(c) = \left(\frac{pe^c}{1 - (1 - p)e^c}\right)^{N+1}$$

$$P_{0,N+1} = W_{0,N+1}(c)|_{c=0} = 1$$

$$M_{0,N+1}(c) = \frac{W_{0,N+1}(c)}{P_{0,N+1}} = W_{0,N+1}$$

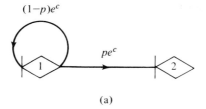

(a)

(b)

Figure 10.35. Basic loop GERT system.

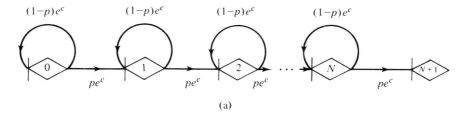

(a)

(b)

Figure 10.36. Single sampling plan for $c = N$.

$$\left.\frac{dM_{0,N+1}(c)}{dc}\right|_{c=0} = \frac{N+1}{p}$$

= expected number of samples before rejection
for $c = N$

The second moment can be computed by:

$$\frac{d^2M_{0,N+1}(c)}{dc^2} = \left.\frac{d}{dc}\left(\frac{(N+1)(pe^c)^{N+1}}{(1-(1-p)e^c)^{N+2}}\right)\right|_{c=0}$$

$$= \frac{(N+1)(N-p+2)}{p^2}$$

The variance for the distribution may then be computed:

$$\sigma_{0,N+1}^2 = \frac{(N+1)(N-p+2)}{p^2} - \left(\frac{N+1}{p}\right)^2$$

$$= \frac{(N+1)(1-p)}{p^2}$$

10.4.3 Double sampling plan.

Duncan [4] stated the plan as follows: A double sampling plan is designated by five numbers n_1, n_2, c_1, c_2, and c_3, with c_1 being less than c_2 and c_2 being less than or equal to c_3. A sample of size n_1 is taken from a given lot. If it contains c_1 or less defective units, it is immediately accepted. If it contains more than c_2 defective units, it is immediately rejected. If the number of defective units is greater than c_1 but not more than c_2, a second sample of size n_2 is taken. If there are c_3 or less defective units in the combined samples, the lot is accepted. If there are more than c_3 defective units, the lot is rejected. Frequently c_2, is taken equal to c_3.

The probability of either acceptance or rejection on first sampling can be obtained from the method proposed in the previous section. The next example shows how to compute the probability of either acceptance or rejection on combined samples.

EXAMPLE 2

Let $n_1 = 50$, $n_2 = 100$, $c_1 = 1$, $c_2 = c_3 = 3$, and p be the lot fraction defective. The probabilities of accepting and rejecting on combined samples can be obtained from the GERT network as shown in Fig. 10.37. Nodes 0, 1, 2, 3, and 4

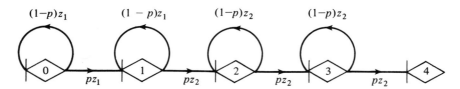

Figure 10.37. The GERT network of Example 2.

represent the number of defective units. Node 4 represents an absorbing state (i.e., the state in which combined samples are rejected). The first three branches of the network are multiplied by z_1, and the other branches are mutliplied by z_2 to distinguish the second sampling from the first sampling. From Mason's rule, we get:

$$W_{0,4}(z) = P^4 z_1 z_2^3 / [1 - 2(1-p)z_1 - 2(1-p)z_2 + (1-p)^2 z_1^2$$
$$+ 4(1-p)^2 z_1 z_2^2 + (1-p)^2 z_2^2 - 2(1-p)z_1^2 z_2$$
$$- 2(1-p)z_1 z_2^2 + (1-p)^4 z_1^2 z_2^2]$$

$$= K_1 z_1 z_2^3 + K_2 z_1^2 z_2^3 + K_3 z_1 z_2^4 + \cdots$$

$$= \sum_{j=1}^{\infty} \sum_{k=1}^{\infty} K_j z_1^j z_2^k$$

The terms that satisfy conditions $j \leq 50$ and $j + k = 150$ are as follows:

$$K_1 z_1 z_2^{149}, K_2 z_1^2 z_2^{148}, K_3 z_1^3 z_2^{147}, \ldots, K_{49} z_1^{49} z_2^{101}$$

Thus:

Probability of rejection on combined samples $= P_r = \sum_{L=1}^{49} K_L$

Probability of acceptance on combined samples $= 1 - P_r$

Since this is not a practical approach, we can easily obtain these probabilities with GERTS. To compare the results with Duncan's [4], another example is presented as follows.

EXAMPLE 3

Let $c_1 = 2$, $c_2 = c_3 = 6$, $n_1 = 50$, $n_2 = 100$, and $p = 0.08$. A GERTS model is formulated as shown in Fig. 10.38. Node 2 is the starting node. Nodes 4, 5, 6, and 7 are used to determine if individual parts are acceptable. Path (4, 5) will be taken if a bad part is sampled. This will cause path (5, 6) to be realized. If this path is realized, $(c_1 + 1)$ times node 6 is realized. This triggers path (6, 7) which includes activity 1 which effectively causes the second sampling to occur. The sampling will reject if node 7 is released $(c_2 - c_1 + 1)$ times or if node 12 is released $(c_3 + 1)$ times—these both feed into node 14. Due to activity 2 on path (8, 15), node 7 will be bypassed by node 16 when the second portion of the sampling begins. Path (3, 8) is released every time a sample is taken; node 8 will be released after n_1 samples. If activity 1 has not been activated, node 11 will be realized; this represents the acceptance of the lot after the first stage of sampling. Otherwise, path (9, 10) is taken. If this occurs $(n_2 + 1)$ times before node 14 is realized, then the lot is accepted after the second sampling. Paths (2, 3), (4, 3), and (5, 3) have discrete distribution of 1 unit to represent the sampling of units.

The computer output of GERTS is shown below:

Node	Prob./Count	Mean	Std. Dev.	Min.	Max.
11	0.2125	52.3529	15.2477	50.0000	150.0000
14	0.7875	74.9651	23.3705	32.0000	145.0000

From these results, we can see that:

Probability of acceptance on combined samples $= 0.2125$

Probability of rejection on combined samples $= 0.7875$

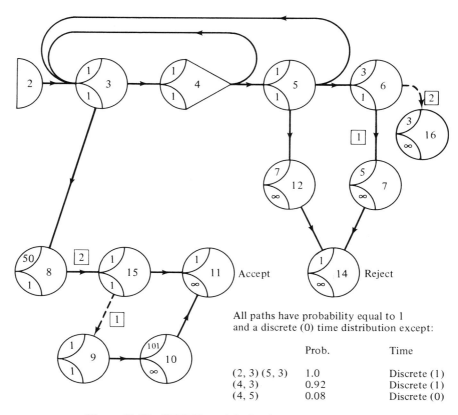

All paths have probability equal to 1
and a discrete (0) time distribution except:

	Prob.	Time
(2, 3) (5, 3)	1.0	Discrete (1)
(4, 3)	0.92	Discrete (1)
(4, 5)	0.08	Discrete (0)

Figure 10.38. GERTS model of a double sampling plan.

The results observed from OC curves [4] for the double sampling plan $n_1 = 50, n_2 = 100, c_1 = 2, c_2 = c_3 = 6$ are as follows:

Probability of acceptance on combined samples = 0.23

Probability of rejection on combined samples = 0.77

Thus, the results obtained from the above two methods are quite close. However, GERTS III can give us additional information such as the mean and standard deviation of the distribution.

The model shown in Fig. 10.38 is not dependent upon the fact that $c_2 = c_3$, and it could be used to analyze any double sampling plan.

10.4.4 Military Standard 105 D plan. Military Standard 105 D is a complex sampling plan. The purpose of this plan is to constrain the supplier so that he will produce at AQL (acceptable quality level) quality. Military

Standard 105 D is thus indexed with respect to a series of AQLs. It is also necessary in applying Military Standard 105 D to decide on the *inspection level*. This determines the relationship between the lot size and the sample size. Three general levels of inspection are offered. Level II is designated as normal. Level I may be specified when less discrimination is needed, and level III, when more discrimination is needed.

For a specified AQL and inspection level, and a given lot size, Military Standard 105 D gives a normal sampling plan that is to be used as long as the supplier is apparently producing a product of AQL quality. It also gives a tightened plan to which a shift is to be made if there is evidence that quality has deteriorated, and a reduced plan if the quality is running especially good. For a particular plan, switching procedures might be as follows:

1. *Normal to tightened.* When normal inspection is in effect, tightened inspection shall be instituted when two out of five consecutive lots or batches have been rejected on original inspection.
2. *Tightened to normal.* When tightened inspection is in effect, normal inspection shall be instituted when five consecutive lots or batches have been considered acceptable on original inspection.
3. *Normal to reduced.* When normal inspection is in effect, reduced inspection shall be instituted when the preceding ten lots or batches have been considered all acceptable on original inspection.
4. *Reduction to normal.* When reduced inspection is in effect, normal inspection shall be instituted when a lot or batch is rejected.

In the event that ten consecutive lots or batches remain on tightened inspection, inspection under provisions of this document should be discontinued pending action to improve the quality of submitted material.

A GERT network can be formulated to describe the behavior of this sampling plan. The development will be considered in subgraphs, e.g., the normal, tightened, and reduced phases. Define p_F as the probability of a faulty lot and p_G as the probability of a good lot.

NORMAL PHASE

Figure 10.39 shows the normal phase of the Military Standard 105 D plan. Node N represents entrance into the normal phase from either the tightened or reduced phase. Nodes 0, 1, 2, and 3 represent the possibility of getting 0 through 3 consecutive good lots at the beginning of the normal phase. Nodes 0, 1..., and 10 represent the number of successive good lots since the last faulty lot was discovered. If a faulty lot is discovered in state 0, 1, 2, or 3, it means that 2 out of 5 consecutive lots have been found faulty and we should enter tightened inspection. Node T represents the exit from the normal phase to the tightened phase. If node 10 is reached, this means that ten consecutive good lots have been found and we should move to the reduced phase. Node R represents this movement.

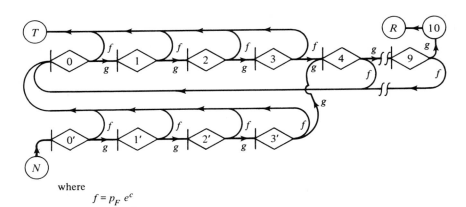

where

$$f = p_F \, e^c$$
$$g = p_G \, e^c$$

Figure 10.39. Normal phase of the Mil. Std. 105D plan.

REDUCED PHASE

The reduced phase is modeled in Fig. 10.40. Node *R* is the entrance to the reduced phase from the normal phase. Node 1 represents the reduced sampling procedure. The process remains in this state until a faulty lot is

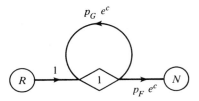

Figure 10.40. Reduced phase of the Mil. Std. 105D plan.

found, and then the normal testing phase is entered again. Node *N* represents the exit from the reduced testing phase.

TIGHTENED PHASE

A GERT model as shown in Fig. 10.41 can be used to model the tightened phase. Node *T* represents entrance into the tightened phase from the normal phase. Nodes 0, 1, 2, 3, 4, and 5 represent 0, 1, 2, 3, 4, and 5 consecutive good lots, respectively. If node 5 is reached, the process returns to the normal phase as designated by node *N*. If the process remains in this phase for more than ten lots, the process should be stopped for review. This is represented by node *S*.

These three graphs could be combined into one graph if we connect

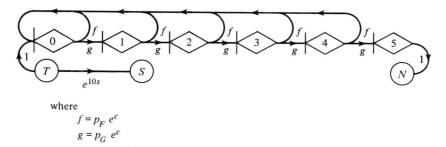

where

$$f = p_F \; e^c$$
$$g = p_G \; e^c$$

Figure 10.41. Tightened phase of the Mil. Std. 105D plan.

the common nodes (R, N, and T) of the graphs. The same information is available from this graph as from the graphs of previously described systems.

10.4.5 Dodge's continuous sampling plan CSP-1.
Dodge's continuous sampling plan CSP-1 is a plan of sampling inspection for a product consisting of individual units (parts, subassemblies, finished articles, etc.) manufactured in quantity by an essentially continuous process. The plan operates as follows:

1. An inspector selects a predetermined f percent (or fraction) of the product in such a manner as to assure an unbiased sample.
2. When a defect is found, a predetermined clearing sequence of i subsequent and consecutive units of product must be found free of defects.
3. Upon finding i units free of defects, the inspector resumes sampling the fraction.

If during a period of clearing i units, a defective unit is found, the count must start over. This is a rectifying plan, i.e., all defective units found are to be corrected or replaced by good units.

The expected number of units passed under the sampling procedure before a defect is found can be obtained from a GERT model as shown in Fig. 10.42. Suppose the process is in statistical control, so that the probability of any incoming unit being defective can be considered constant (p), and the probability of any unit being good is $1 - p = q$. Node U represents a unit not under inspection. Node I represents a unit under inspection. Node D represents the detailing state. From Mason's rule, we obtain:

$$W_{U,D}(c) = \frac{fpe^c}{1 - (1 - f)e^c - f(1 - p)e^c}$$

$$P_{U,D} = W_{U,D}(c)|_{c=0} = 1$$

$$M_{U,D}(c) = \frac{W_{U,D}(c)}{P_{U,D}} = W_{U,D}(c)$$

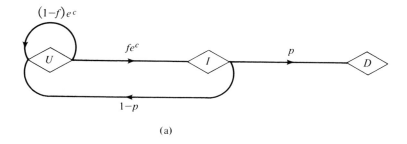

(a)

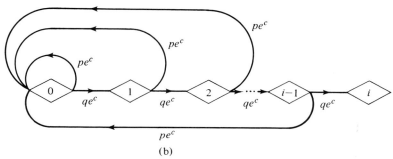

(b)

Figure 10.42. Dodge's continuous sampling plan CSP1.

$$\left.\frac{dM_{U,D}(c)}{dc}\right|_{c=0} = \frac{(fp)^2 + fp(1-fp)}{(fp)^2}$$

$$= \frac{1}{fp} = \text{expected number of units passed under the sampling procedure before a defect is found}$$

To determine the expected number of units that will be inspected by the detailer while attempting to clear i units, the GERT network shown in Fig. 10.42(b) will be used. From Mason's rule we get:

$$W_{0,i}(c) = \frac{(qe^c)^i}{1 - [pe^c + pe^c qe^c + pe^c(qe^c)^2 + \cdots + pe^c(qe^c)^{i-1}]}$$

$$= \frac{(qe^c)^i}{1 - pe^c[1 + qe^c + (qe^c)^2 + \cdots + (qe^c)^{i-1}]}$$

$$= \frac{(qe^c)^i}{1 - pe^c[1 - (qe^c)^i]/(1 - qe^c)}$$

$$= \frac{(1 - qe^c)(qe^c)^i}{1 - qe^c - pe^c[1 - (qe^c)^i]}$$

$$\because P_{0,i} = W_{0,i}(c)|_{c=0} = 1$$

$$\therefore M_{0,i}(c) = W_{0,i}(c)$$

$$\frac{dM_{0,i}(c)}{dc}\bigg|_{c=0} = \frac{[1 - q - p(1 - q^i)][iq^i - (i + 1)q^{i+1}] + (1 - q)q^i[1 - (i + 1)pq^i]}{(1 - q - p(1 - q^i))^2}$$

$$= \frac{q^i(1 - q)(1 - q^i)}{((1 - q)q^i)^2}$$

$$= \frac{1 - q^i}{pq^i} = \text{expected number of units inspected during the detailing state}$$

The result agrees with that of Dodge [4].

10.5 Project Management of Research and Development Projects*

10.5.1 Introduction. Pritsker and Enlow [6, 33] have shown that GERT and GERTS have great flexibility in the project management areas. The material in this section relies heavily on their work. First, a GERT model of a PERT network will be presented. Next a profitability analysis of research and development will be modeled. A model of the research and development process is next attempted using GERT. Finally, the idea generation or brainstorming in research and development type of system is modeled.

10.5.2 GERTS model of a PERT network. As discussed in Chapter 3, network methods have become popular as aids in the management of projects. One of these techniques is PERT. The PERT node is realized when all paths incident to it are realized. The PERT approach to analyzing networks is subject to a number of assumptions, one of which is that the longest (or critical) path will completely define the project duration. Van Slyke [40] used simulation to show that, if the times associated with the activities of a project are random variables, then the critical path will depend upon the particular observations of activity time. He showed that the PERT approach consistently underestimated the mean time to finish a project and overestimated the variance.

*The material in this section is adapted from References 6, 13, and 33 and is used with the permission of the American Institute of Industrial Engineers, Dr. Pritsker, and Dr. Enlow.

The PERT network in Fig. 10.43 can can be modeled in GERTS as shown in Fig. 10.44. The times in both figures are assumed to be normal. The number of releases for all GERTS nodes is set equal to number of paths incident to the node. Thus, node 6 has three releases. The statistics of interest will be those collected on node 6, and they represent the mean, variance, and histogram of the time to finish the project. Pritsker [33] has shown that GERTS gives the project manager great flexibility over standard project management techniques. The GERTS approach allows the inclusion of logical OR nodes, feedback, and network modification.

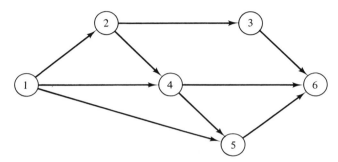

Figure 10.43. PERT network.

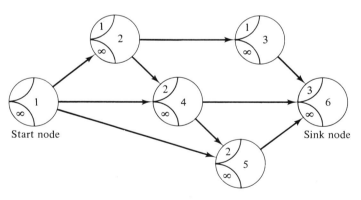

Figure 10.44. GERTS model of the PERT network in Fig. 10.43.

10.5.3 Analysis of research and development expenditures. In a recent article, Graham [13] discussed the problem of profit profitability analysis of research and development expenditures. Pritsker GERT-charted the problem and suggested that this might prove an interesting application of GERT.

Graham points out that investments related to production of goods are often controlled closely. There are many techniques of varying complexity and accuracy available for such studies. It appears, however, that little work in the area of research and development has been accomplished. It is therefore likely that expenditures are too great or too small. Funds may also have been misdirected. Evaluation in this area has been recognized, but it is often attacked intuitively.

The most straightforward methods of evaluating the likelihood of success of research and development is a network approach. Graham proposed a network approach very similar to Eisner's work. It is essentially a PERT system with decision blocks. Graham [13] considered the chart shown in Fig. 10.45, where the elements of the graph are defined by Graham as follows.

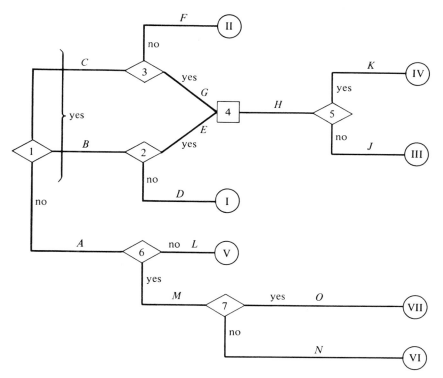

Figure 10.45. R & D expenditures.

Events:

1. Feasibility study indicates whether electrical control of high temperature system is feasible.
2. Determination of the suitability of ac control.
3. Determination of the suitability of dc control.
4. Optimum integration of ac/dc units achieved.

5. and 7. Unit economic feasibility of the design.
6. Determination of feasibility of pneumatic control.

Activities:

(*A*) Pneumatic feasibility study.
(*B*) ac control investigation.
(*C*) dc control investigation.
(*D, F, J, K, L, N,* and *O*) Report writing.
(*E* and *G*) Investigation of optimum ac/dc integration.
(*H* and *M*) Economic analysis.

Outcomes:

(I, II, III, V, VI) Project dropped.
(IV, VII) Project into production and marketed.

This network represents a simplified development program for an electrical control of a high-temperature system. Event 1 represents a feasibility study to determine if electrical control of the high temperature is feasible or not. If it isn't, pneumatic control is checked for feasibility. If pneumatic control is not feasible, the project is dropped. If the pneumatic control is found feasible, it is investigated to see if its unit price is within the potential market price. If the unit price is too expensive, the project is dropped; otherwise, the project is produced and marketed as a pneumatic control device. If electrical control is found feasible, ac and/or dc control would be considered. If both ac and dc are found unsuitable, the project is dropped. If ac and/or dc are found acceptable, the optimum integration of ac/dc circuits is achieved. After this integration, the unit is investigated to determine if the unit's cost is within the potential market price. If it isn't, the project is dropped; otherwise the project is produced and marketed as an electrically controlled device.

Thus, outcomes I, II, III, V, and VI represent the project being dropped, while outcome IV represents marketing electrical control, and outcome VII represents marketing a pneumatic control device.

In this graph each transition is defined by a probability of taking a given path, the time to perform the path, and the cost of the operation, as shown in Fig. 10.46.

Graham proceeds by brute force methods to solve this network for probability of success, expected cost, and expected time of the network. His results are somewhat confusing in that he suggests that combinations of outcomes I, II, III, and IV can happen simultaneously. Certain of these combinations imply that the project is both acceptable and unacceptable.

If we represent this system as a GERT network, outcomes I and II will be combined into one output, which implies that both ac and dc have been found unsuccessful, and therefore the project is aborted. The section of the

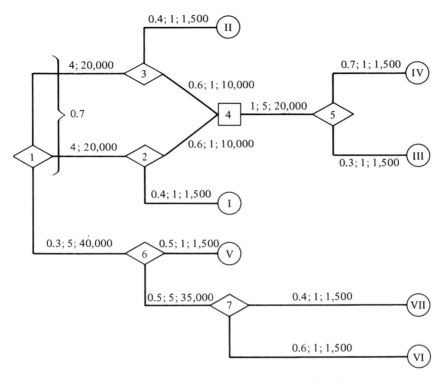

Figure 10.46. R & D project with costs and time shown.

graph between nodes 1 and 4 is represented by three transitions: (1) acceptance of both ac and dc, (2) acceptance of ac but not dc, and (3) acceptance of dc but not ac. The system can now be represented as in Fig. 10.47, where s represents time units and m represents thousands of dollars.

$$M_{1,E}(s, m) = 0.7e^{4s+40m}(0.16e^{s+3m}(0.36e^{s+20m} + 0.48e^{s+11.5m})(e^{5s+20m})$$

$$\times (0.7e^{s+1.5m} + 0.3e^{s+1.5m}))$$

$$+ 0.3e^{5s+40m}(0.5e^{s+1.5m} + 0.5e^{5s+35m})$$

$$\times (0.4e^{s+1.5m} + 0.6e^{s+1.5m}))$$

The expected cost of the project will equal $\partial M_{1,E}(s, m)/(\partial m)|_{m=0}^{s=0}$.

$$\left. \frac{\partial M_{1,E}(s, m)}{\partial m} \right|_{\substack{s=0 \\ m=0}} = 67.582$$

This is interpreted as 67.582 thousands of dollars, which checks with the published work of Graham [12].

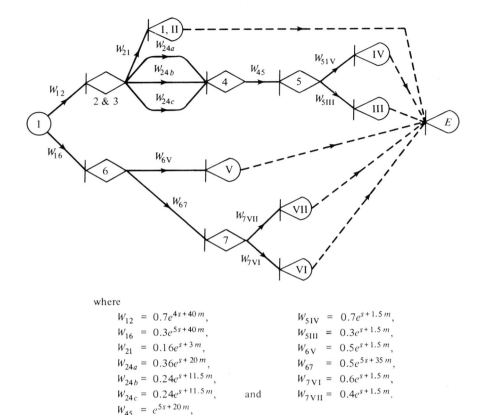

where

$$W_{12} = 0.7e^{4s+40\,m},$$
$$W_{16} = 0.3e^{5s+40\,m},$$
$$W_{21} = 0.16e^{s+3\,m},$$
$$W_{24a} = 0.36e^{s+20\,m},$$
$$W_{24b} = 0.24e^{s+11.5\,m},$$
$$W_{24c} = 0.24e^{s+11.5\,m}, \quad \text{and}$$
$$W_{45} = e^{5s+20\,m},$$

$$W_{5IV} = 0.7e^{s+1.5\,m},$$
$$W_{5III} = 0.3e^{s+1.5\,m},$$
$$W_{6V} = 0.5e^{s+1.5\,m},$$
$$W_{67} = 0.5e^{5s+35\,m},$$
$$W_{7VI} = 0.6e^{s+1.5\,m},$$
$$W_{7VII} = 0.4e^{s+1.5\,m}.$$

Figure 10.47. GERT representation of a R & D project.

The expected time of the project will equal $\partial M_{1,E}(s, m)/(\partial s)\,|_{\substack{s=0 \\ m=0}}$.

$$\left.\frac{\partial M_{1,E}(s, m)}{\partial s}\right|_{\substack{s=0 \\ m=0}} = 9.58 \text{ months}$$

This quantity is the expected duration of the project, which was not calculated by Graham.

The probability of marketing an electronic control device is $W_{1,IV}(s, m)\,|_{\substack{s=0 \\ m=0}}$.

$$W_{1,IV}(s, m)\,|_{\substack{s=0 \\ m=0}} = (0.7)(0.84)(1.0)(0.7) = 0.4116$$

The probability of marketing a pneumatic device is equal to $W_{1,VII}$ $(s, m)\,|_{\substack{s=0 \\ m=0}}$.

$$W_{1,VII}(s, m)\,|_{\substack{s=0 \\ m=0}} = (0.3)(0.5)(0.4) = 0.0600$$

The study tells the planner that 47.16% of the time a successful control device will be marketed. Graham proceeds in his article to perform classical engineering economy studies on the project, which will not be considered here.

GERT has been observed to be a powerful tool in the type of studies suggested by Graham. It allows for rigorous network construction procedures. Although not mentioned in this study, the distribution of time and costs could be considered in problems of this nature.

10.5.4 A GERT model of the research and development processes. Basically the research and development process consists of the following five milestones: (1) completion of problem definition, (2) completion of research activity, (3) acceptance of a proposed solution, (4) completion of a prototype, and (5) implementation of the solution.

Figure 10.48 presents a general network model of the activities involved in achieving the first three milestones. Since a hierarchical network development procedure will be used, the activities are defined in broad terms. The network of Fig 10.48 illustrates three attempts at obtaining a solution for a given definition of a problem. If all three solutions are unacceptable, then either a redefinition of the problem will be made or a new need will be explored and the researcher will essentially give up on the previous problem. Note that the sink node in Fig. 10.48 is an EXCLUSIVE-OR node since it is possible for only one of the three branches incident to the node to be realized.

Figure 10.49 presents a GERT representation of the activities involved in problem definition. On the chart is shown a creative thought process following the establishment of the need. As shown, there are four separate efforts involved in attempting to define the problem. On the output side of the node following creative thought, a probabilistic node is used to indicate that the problem is either defined or not defined, based on the creative thought efforts. If any one of the efforts results in a problem definition, then the node "problem definition proposed" will be reached. Thus an INCLUSIVE-OR node is required at that point. Only if all four of the efforts do not result in a problem definition will the "no definition formulated" be realized. Hence, an AND node is required.

The point *A* on Fig. 10.49 represents a possible regeneration point of the problem definition process. If the characteristics of the activities involved in problem definition do not change based on previous attempts at problem definition, then a return to the original start node can be made, and the network need not be repeated as shown on the bottom half of Fig. 10.49. Since learning occurs in the research and development process, it is more reasonable to indicate a repeat with new parameters of the activities involved in problem definition. This lack of regeneration points in the research and development process, we believe, has hindered many analysis attempts.

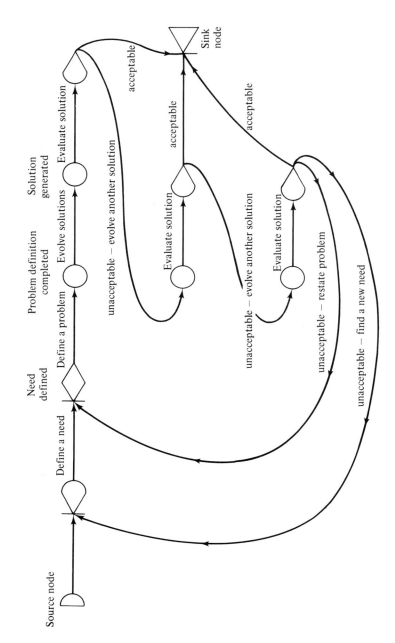

Figure 10.48. GERT research model.

Source node

Define a need

Need defined

Define a problem

Problem definition completed

Evolve solutions

Solution generated

Evaluate solution

acceptable

Sink node

unacceptable – evolve another solution

Evaluate solution

acceptable

unacceptable – evolve another solution

Evaluate solution

acceptable

unacceptable – restate problem

unacceptable – find a new need

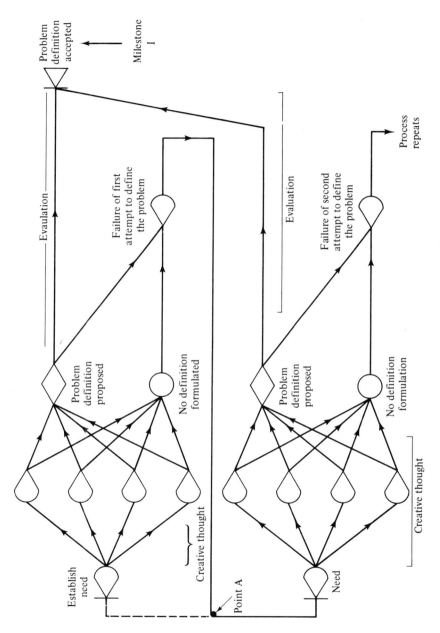

Figure 10.49. GERT problem definition.

411

In Fig. 10.50 is shown one possible representation of the research activity for one researcher involved in proposing solutions. Also shown in Fig. 10.50 is the evaluation procedure modeled in network form for considering both the time and cost considerations involved in the proposed solution.

Figure 10.51 shows the network for generating and evaluating solutions serially. Thus Fig. 10.51 includes both milestones 2 and 3. Tying the detailed elements together results in the network shown in Fig. 10.52 which illustrates one possible network for representing the scientific method approach to planning research and development projects.

10.5.5 GERT model of brainstorming in the research and development process. To analyze the research and development process from the *idea generation* point of view, the concepts of brainstorming have been modeled in network form. The intent is not to present a general model of brainstorming, but to clarify our concept of brainstorming and to provide a vehicle by which we can communicate to others what brainstorming means to us. The relationship of brainstorming to research and development projects and to idea generation in particular is assumed.

Again, we will go through the hierarchical method for developing networks. Figure 10.53 gives two levels of networks describing brainstorming. First, brainstorming is divided into three processes—idea generation, proposal of a concept, and evaluation of a concept. At the next level, each of these processes is broken down into slightly finer detail illustrating the concepts that an idea can be dropped or picked up, that at some point evaluation of an idea may be sought, and that acceptance and rejection of the idea is a possibility. (Note in our model of brainstorming that we are including the eventual output of the brainstorming session in our definition of brainstorming. Thus we have communicated this fact through the network model.) The next step is to detail each of the processes according to the activities involved.

At this point there are many alternatives that can be modeled. In the idea generation process, we can consider stages in which ideas are proposed, elaborated on, and either dropped or continued to be elaborated on. Also we can conceive of the idea generation process as being sequential, where one participant (researcher) builds on his own or other's ideas in a sequential fashion. Another approach is to consider a simultaneous idea generation process in which the participants are generating ideas and the participant who speaks first has his idea on the floor. An analogy to the processing of signals may clarify the difference between simultaneous and sequential idea generation. In the sequential procedure there there is only one signal being processed at a time, whereas in the simultaneous idea generation model, multiple signals are contained within the system. For the detailed model, the sequential

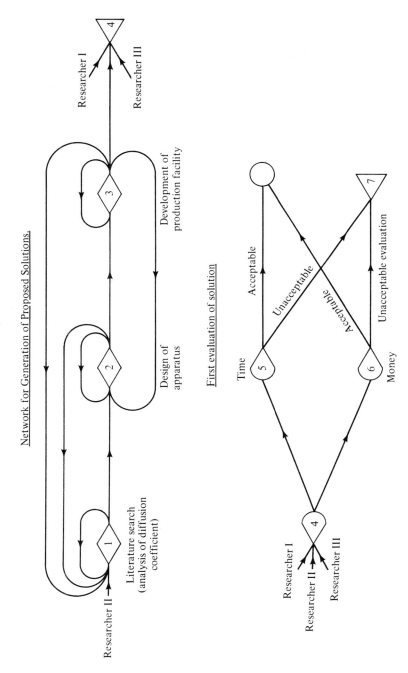

Network for Generation of Proposed Solutions.

Researcher II

Literature search
(analysis of diffusion
coefficient)

Design of
apparatus

Development of
production facility

Researcher I

Researcher III

First evaluation of solution

Time

Acceptable

Unacceptable

Acceptable

Unacceptable evaluation

Money

Researcher I

Researcher II

Researcher III

Figure 10.50. Network for generation of proposed solutions.

413

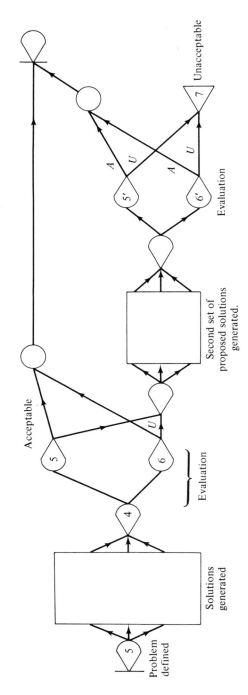

Figure 10.51. Series proposals.

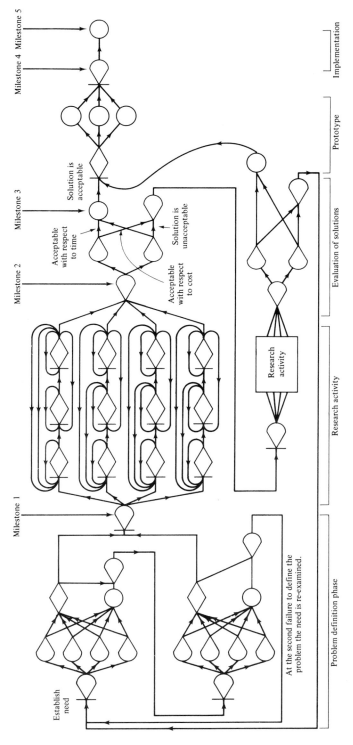

Figure 10.52. A GERT network representation of an example R & D project.

415

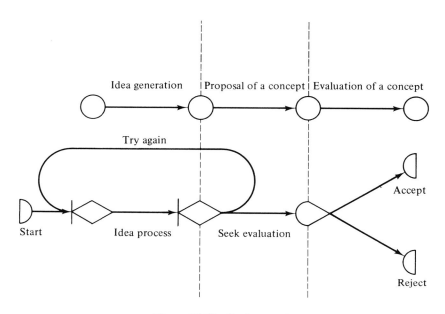

Figure 10.53. Brainstorming.

case will be considered with two stages and feedback within each stage and from the second stage to the first stage permitted.

The evaluation will be sought when one of the participants recommends a concept based on the idea process. The network model of the request for evaluation permits the probability of seeking evaluation to be different for each participant and conditioned by the participant who originated the idea. With regard to the evaluation of a concept by the participants, the network model should reflect the decision of each participant based on the proposer of the concept. In addition, the rules for accepting or rejecting a concept must be established. Two such rules are unanimous acceptance by each partici-pant, and a majority of the participants accepting. The network model developed will be based on the majority voting principle. Figure 10.54 shows the network model which has these conditions.

GERTS was used to analyze the network given in Fig. 10.54. Any branch incident to node 2, 3, or 4 is an idea branch and represents the activity "gen-eration of an idea at the first stage of idea generation." The start node is node 43, and the first idea is generated by researchers 1, 2, and 3 with equal probability. Activities representing second-stage idea generation are repre-sented by all branches incident to node 19, 20, or 21. The dropping of an idea or a return from the second stage to the first stage is done through node 25, 26, or 27. Remaining in stage one is accomplished by passing through node 16, 17, or 18; remaining in stage two is accomplished through node

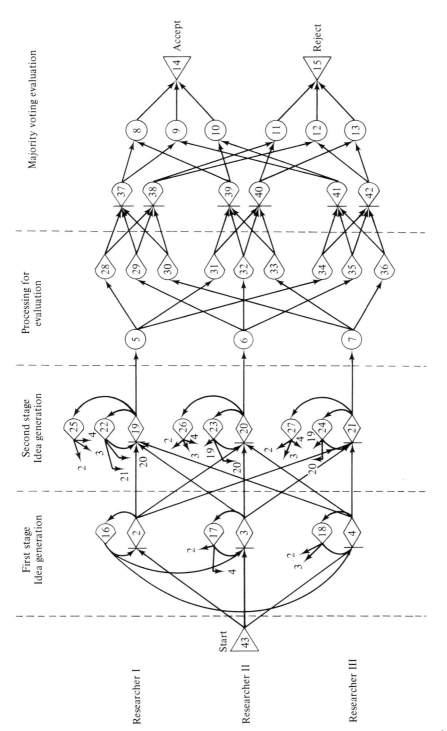

Figure 10.54. Voting considerations.

22, 23, or 24. The suggesting of a concept based on the idea generation stages is given by nodes 5, 6, and 7 for researchers 1, 2, and 3, respectively. Nodes 28, 29, and 30 represent the evaluation by researcher 1 of a concept suggested by researchers 1, 2, and 3, respectively. Nodes 28, 31, and 34 represent the evaluation of a proposed concept by researcher 1 by researchers 1, 2, and 3, respectively. Nodes 37, 39, and 41 represent the acceptance of a concept by researchers 1, 2, and 3, respectively, whereas nodes 38, 40, and 42 represent rejection of the concept by researchers 1, 2, and 3, respectively. Since majority voting is required, at least two acceptances or two rejections are required to accept or reject a concept. Thus two of the acceptance nodes 37, 39, and 41 must be realized in order for the concept to be accepted. This is shown by the three nodes 8, 9, and 10 and, eventually, the realization of node 14 if node 8, 9, or 10 is realized. A similar analysis holds for the rejection node 15.

To present some quantitative results from the network, several runs were made with GERTS. Figure 10.55 is a summary of preliminary analysis of

Project Number	Probability		Number of Ideas				Time Required			
	Accept	Reject	Accept		Reject		Accept		Reject	
			μ	σ	μ	σ	μ	σ	μ	σ
10	0.495	0.505	8.62	6.89	7.89	6.93				
20	0.385	0.615	12.83	9.77	11.10	9.92				
21	0.360	0.640					765.	398.	706.	383.
22	0.392	0.608					632.	266.	648.	325.
30	0.403	0.597	15.42	13.08	16.45	17.06				
31	0.368	0.632					880.	562.	824.	511.
32	0.370	0.630					839.	487.	828.	435.

Code: Project (*ij*) where if *i* = 1. Three equal researchers
2. One critical researcher
3. Two critical researchers
j = 0. Count number of ideas
1. Exponentially distributed times
2. Normally distributed times

Figure 10.55. Results obtained from the GERT simulation program.

Fig. 10.54. A critical researcher is one who does not follow up on other's ideas nor on his own ideas frequently, but causes the idea generation process to continually revert back to the first stage. He also does not suggest a concept as frequently as the other researchers. When he does suggest a concept there there is a high probability of it being accepted by him and the other researchers.

To count the number of ideas required before acceptance or rejection, a count is kept on the number of times branches incident to nodes 2, 3, and 4 and 19, 20, and 21 are realized. To obtain the time required to accept or reject a concept, random variables representing the times to generate an idea and the time to evaluate an idea are inserted on the proper branches. The exponentially distributed times employed had a mean time of 30 time units for idea generation and a mean time of 300 time units for evaluation. When normally distributed times were used, the same means were used, but a standard deviation of one-tenth the mean values were assumed. Each of the networks as represented by a row in Fig. 10.55 was simulated 400 times to obtain the network statistics.

10.6 GERT as an Aid to Management

10.6.1 Introduction. In this section we will examine three areas: (1) modeling of production-line-type operations, (2) management decision-making, and (3) work load determination in an organization.

10.6.2 Modeling of production-line-type operations. The following problem was discussed by Pritsker and Whitehouse [35] and demonstrates the potential of GERT in this area.

A MODEL OF MANUFACTURING PROCESS

On a production line a part is manufactured at the beginning of the line. The manufacturing operation is assumed to take 4 hours. Before the finishing touches are put on the part, it is inspected, with 25% of the parts failing the inspection and requiring rework. The inspection time (including waiting for inspection) is assumed to be distributed according to the negative exponential distribution, with a mean of 1 hour. Reworking takes 3 hours, and 30% of the parts reworked fail the next inspection. The inspection of the reworked items is also distributed according to the negative exponential with mean of $\frac{1}{2}$ hour. Parts which fail this inspection are scrapped. If the part passes either of the above inspections, it is sent to the final finishing operation which takes 10 hours 60% of the time and 14 hours 40% of the time. A final inspection, which takes 1 hour, rejects 5% of the parts; these are scrapped. The GERT network for this production line is illustrated in Fig. 10.56.

$$W_{B,1}(s) = e^{4s}(0.25)(1 - s)^{-1}(e^{3s})(0.3)(1 - 2s)^{-1} + (e^{4s})(0.25)(1 - s)^{-1}(e^{3s})$$
$$\times (0.7)(1 - 2s)^{-1}(0.6e^{10s} + 0.4e^{14s})(0.05e^s) + (e^{4s})(0.75)(1 - s)^{-1}$$
$$\times (0.6e^{10s} - 0.4e^{14s})(0.05e^s)$$

$$W_{B,2}(s) = (e^{4s})((0.25)(1 - s)^{-1}(e^{3s})(0.7)(1 - 2s)^{-1} + (0.75)(1 - s)^{-1}$$
$$\times (0.6e^{10s} + 0.4e^{14s})(0.95e^s)$$

$$P_{B,2} = W_{B,2}(s)|_{s=0} = 0.87875$$

$$P_{B,1} = W_{B,1}(s)|_{s=0} = 0.12125$$

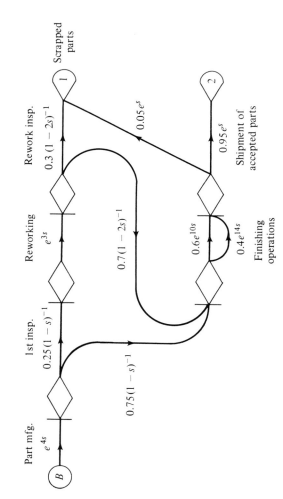

Figure 10.56. GERT representation of a production line.

$$M_{B,2}(s) = \frac{W_{B,2}(s)}{P_{B,2}}$$

$$M_{B,1}(s) = \frac{W_{B,1}(s)}{P_{B,1}}.$$

These two MGFs characterize the distribution of the time for a part to flow through the production line. In this example, transit times were not included, but they would not complicate the formulation.

It appears that GERT shows great potential in modeling production-line-type jobs. The problem discussed does not include feedback, but feedback could easily be included in the model if it was appropriate. Queueing at inspection stations in the example would turn it into an interesting GERTS application.

10.6.3 GERT as an aid to decision making. It is often necessary to compare a number of alternatives and to decide which yields the most desirable results. GERT proves to be a useful technique for these purposes.

Consider the following situation. Suppose a piece of equipment is located on a production line and produces parts continuously throughout the day. While in operation, the machine gets out of adjustment due to wear and tear. Eventually the machine will become inoperative and must be repaired. Suppose there are two stages of this reduced service. In stage 1 the machine can still do its job, but it is apparent that it will soon need repair. Stage 2 is the stage in which the machine is inoperative and must be repaired to become operative again. If the machine is repaired when it is discovered in stage 1 of deterioration, the repair task is not difficult; but if it is allowed to run until it breaks down (stage 2), then the repair takes longer. Management is thus faced with the problem of whether to take the machine out of production as soon as it reaches stage 1 or to let it run until it reaches stage 2. Suppose the deterioration of the machine is random, that is, if it is in good condition at the beginning of one of its runs, the probability is a that it will be in good condition after the run and $1 - a$ that it will be in stage 1 after the run. Similarly, if the machine is in stage 1 at the beginning of the run, the probability is b that it will remain there and $1 - b$ that it will fail (stage 2). Assume that the MGF of the time of a production run is $M_p(s)$, the MGF of the time to adjust the machine is $M_a(s)$, and the MGF of the time to repair the machine is $M_r(s)$.

Management has two doctrines to compare: (*A*) Adjust the machine as soon as it reaches stage 1 of failure, which can be represented in GERT format, as shown in Fig. 10.57, and (*B*) wait until the machine has completely failed and then repair it. The GERT representation of this is shown in Fig. 10.58.

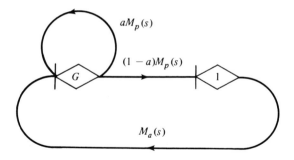

Figure 10.57. GERT representation of repair policy *A*.

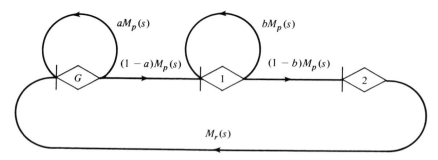

Figure 10.58. GERT representation of repair policy *B*.

The doctrines have now been represented by means of GERT. Management must now decide upon a criterion of choice. One possible choice is to select the system which has the lower probability of down-time. In doctrine *A* this will be the probability that the process is in state 1, while for doctrine *B* this will be the probability that the process is in state 2. If $E(p)$, $E(a)$, and $E(r)$ represent the expected time for a run, adjustment, and repair, these probabilities are:

Doctrine *A*:

$$p_1 = \frac{((1-a)E(a))}{((1-a)E(p)E(a) + aE(p))}$$

Doctrine *B*:

$$p_2 = \frac{((1-a)E(r))}{(((1-a)(1-b)E(p)^2E(r))/(1-bE(p))) + aE(p)}$$

There are, of course, other possible decision rules. For instance, it will cost money for the extra surveillance to determine if the process is in stage 1 of deterioration. It could be determined if the reduction down-time created by doctrine A justifies the increased cost of surveillance. Whatever the decision policy is, however, these GERT models will help management. The idea for this problem was suggested in an article by Morse [26] dealing with Markov chains.

Another form of this problem was investigated by Pritsker [34]; he called it the tolerance-type problem. This model is the same as represented in doctrine A except that total failure is also possible with probability c. This doctrine can be represented in GERT as shown in Fig. 10.59.

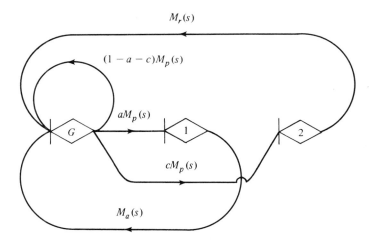

Figure 10.59. GERT representation of a tolerance system.

Decision problems are not limited to choosing between alternatives before implementing a program. Often decisions must be made as to whether to replace a present program with a proposed plan.

Consider the following case problem suggested by Stern [39] which falls under this general classification. The executive committee of the Jones Manufacturing Company is faced with the decision to replace its present advertising campaign with an extensive new plan. At present there are two other brands of the product in competition with the Jones product. Brand A claims 39% of the market, and brand B claims 33% of the market, leaving 28% for the Jones Company. The company chooses a representative section of the country to try out its new campaign. It is found that after some time in this area, 70% of the people who bought the Jones product the previous time they needed this product again purchased the Jones brand, while 20% switched to brand A and 10% to brand B. Of those who bought brand A the previous time, 70% remained with it, 10% switched to brand B, and 20% bought the Jones product. The similar figures for brand-B users were 80% who stayed with the product, while of those who switched 10% went to brand

A and 10% to the Jones product. This is a Markov process which is easily represented by GERT, as shown in Fig. 10.60.

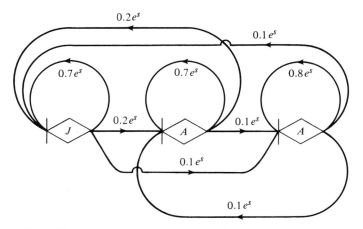

Figure 10.60. GERT representation of a marketing problem.

The Jones Company is interested in the expected share of the market after this advertising campaign has been initiated. This would be represented by the steady probability that an individual will buy the Jones product. For this example, it is 33%. The executive board must decide whether the apparent increase of 5% in sales warrants the new program. They will observe that it is probably an optimistic estimate since the other companies will probably counter with a program of their own when they observe loss of sales. But this form of analysis has given management more information than they would have had otherwise.

GERT for these two examples has proven an aid to management when choosing between alternatives. It should find many uses of this nature.

10.6.4 Distribution of work load. It is often of interest to management to establish the work load in its organization. In this section a model will be discussed which will consider both the flow of work through the organization and the duration of the time to complete a given job by a given individual. Kemeny, Schleifer, Snell, and Thompson [19] considered a Markov model similar to the one being considered here, which is actually a semi-Markov model.

Consider a small company with four employees: a president, a vice-president, and a secretary for each. Suppose that some task associated with the business is initiated and the way it passes through the company is observed. It comes to a specific employee, who may carry out the task himself or pass it to another employee. There is also a possibility that the task may be laid aside and never carried out. In the GERT model there are four nonabsorbing states corresponding to the four employees, P, VP, PS, VS, and two absorbing states, I (done) and II (laid aside). Define:

1. p_{ij} and $M_{ij}(s)$, where i and $j = P, VP, PS, VS$, to be the probability and the MGF of the time for worker i to pass the work to worker j.
2. $p_{i,\text{I}}$ and $M_{i,\text{I}}(s)$, where $i = P, VP, PS, VS$, to be the probability and the MGF of the time for worker i to complete the job.
3. $p_{i,\text{II}}$ and $M_{i,\text{II}}(s)$, where $i = P, VP, PS, VS$, to be the probability and MGF of the time for worker i to lay aside the job.

This completely defines the GERT network.

Now consider a hypothetical problem. The nature of the work is such that the president is equally likely to pass an item to his secretary or to the vice-president. Let the times for these activities be described by $M_{P,PS}(s) = M_{P,VP}(s) = e^{s+3s^2}$. The vice-president is in the habit of passing three-quarters of his work on to his secretary. Let the time for these activities be described by $M_{VP,VS}(s) = e^{s+3s^2}$. The remainder is equally likely to be done by him or laid aside. Let the times for these activities be described by $M_{VP,\text{I}}(s) = e^{3s+7s^2}$, and $M_{VP,\text{II}}(s) = e^s$. Each secretary does three-eighths of her tasks personally, lays one-eighth aside, and the rest of the items are split evenly between those items sent to the other secretary and those returned to her boss for further instruction. Let the times for these activities be described by $M_{PS,P}(s) = M_{PS,VS}(s) = M_{VS,VP}(s) = M_{VS,PS}(s) = e^{s+s^2}$, $M_{PS,\text{I}}(s) = M_{VS,\text{I}}(s) = e^{3s+7s^2}$, $M_{PS,\text{II}}(s) = M_{VS,\text{II}}(s) = e^s$. The GERT representation of this system is now easily drawn as shown in Fig. 10.61.

There is much information available from this model. To whom is it safest to give incoming work to assure that it will be acted upon? The probability that a project is acted upon, given that it is initially handed to worker i, is equal to $W_{i,\text{I}}(s)|_{s=0}$. For this model:

$$p_{P,\text{I}} = 0.7$$
$$p_{VP,\text{I}} = 0.67$$
$$p_{PS,\text{I}} = 0.73$$
$$p_{VS,\text{I}} = 0.725$$

Thus, the best strategy is to submit the work to the president's secretary if you want to have the project acted upon.

Another point of interest is to determine to whom to give the piece of work if you desire to minimize the time in the system. $W_{i,\text{I}}(s) + W_{i,\text{II}}(s)$ equals the MGF of the time in the system, given that the work is initiated with worker i. The first derivative of this expression with s equated to zero represents the expected time in the system. These expected times are 6.00,

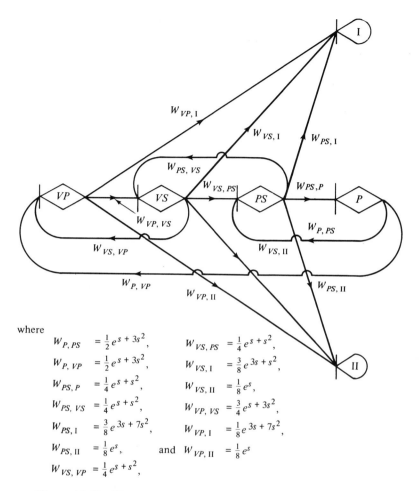

where

$$W_{P,PS} = \frac{1}{2}e^{s+3s^2},$$

$$W_{P,VP} = \frac{1}{2}e^{s+3s^2},$$

$$W_{PS,P} = \frac{1}{4}e^{s+s^2},$$

$$W_{PS,VS} = \frac{1}{4}e^{s+s^2},$$

$$W_{PS,I} = \frac{3}{8}e^{3s+7s^2},$$

$$W_{PS,II} = \frac{1}{8}e^{s},$$

$$W_{VS,VP} = \frac{1}{4}e^{s+s^2},$$

$$W_{VS,PS} = \frac{1}{4}e^{s+s^2},$$

$$W_{VS,I} = \frac{3}{8}e^{3s+s^2},$$

$$W_{VS,II} = \frac{1}{8}e^{s},$$

$$W_{VP,VS} = \frac{3}{4}e^{s+3s^2},$$

$$W_{VP,I} = \frac{1}{8}e^{3s+7s^2},$$

$$\text{and } W_{VP,II} = \frac{1}{8}e^{s}$$

Figure 10.61. GERT representation of a work load distribution.

5.06, 4.94, and 4.75 starting with the president, vice-president, president's secretary, and vice-president's secretary, respectively. The quickest exit from the system is obtained by giving the work to the vice-president's secretary.

The expected work on a given project can be found for each employee, given the person who initiated the work. From this information we determine the expected work performed by each individual, given that a particular distribution of work input into the system can be found. It might be interesting to find an input distribution of work that will balance the work in the business. This information is available from the GERT network. The concept of balancing the work load is not as important in a problem of this nature as it would be on a production line. The problem of assembly line balancing remains an extremely difficult problem to handle.

The foregoing represents another potential approach to the line-balancing problem. If the nodes in their work-load model are replaced by GERTS queue nodes, it might be possible to perform a *bottleneck* analysis for the system.

EXERCISES

1. Analyze an $M/M/2$ repairman problem with $m = 2$, $n = 2$. Assume that machines will fail at the rate of one every time period and it will take the repairman one-third of a time period to repair a machine on the average. Find:
   (a) Time to plant failure.
   (b) Duration of failure.
   (c) Busy period and idle time for the service facility.
   (d) Steady-state probabilities.
   (e) Time to regeneration given that the system does not fail.
   (f) Distribution of the number of machines failing before complete failure of the plant.
   (g) Expected number of recoveries before total failure.
   (h) Expected number of repairmen busy in a time period.

2. Consider an $M/M/1$ barbershop in which there are two chairs besides the barber chair. Men arriving when all chairs are full will go elsewhere. Men arrive at a rate of 3 per hour and the barber can service 3 men per hour on the average. Use GERT to completely analyze this problem.

3. If in Exercise 2 the barber has the opportunity of hiring another barber, how would you use GERT to analyze the merits of hiring a second barber?

4. Consider an $M/G/1$ barbershop with the same characteristics as described in Exercise 2, except that it takes exactly 20 minutes to cut hair. Use a GERT to completely analyze this problem.

5. Consider an $M/G/1$ repairman problem in which $m = 3$, $n = 0$. The machines break down at a rate of 1 per hour and it takes the repairman exactly 30 minutes to service a machine. Analyze this problem using both the "exact" and "approximate" methods of solution. Compare your results.

6. Consider the model discussed in Section 10.1.4. There is a theory that we should have storage space for one unit in front of the first server, two units in front of the second server, and three units in front of the third server. This policy is supposed to level the work load for the servicemen. Test this theory using GERTS.

7. Another theory about the recirculating conveyor is that it is insensitive to the length of the recirculation path. Test this theory using GERTS.

8. Analyze an (S, s) inventory policy where $S = 4$, $s = 1$, and $M = 1$. The probability of the number of arrivals during a week is $Pr(0) = 0.70$, $Pr(1) = (0.20)$, and $Pr(2) = 0.10$. Also, $P = \$10$, $k = \$0.10$, $N = \$5$, and $R = \$20$. Find:
   (a) Distribution of the time between receipt of orders.
   (b) Average inventory size.
   (c) Average annual cost.
   (d) The probability of stock-out before replenishment.

9. Analyze the situation described in Exercise 8 except that the review period (M) is two weeks.

10. Develop an "exact" method similar to that described in Section 10.1.3 for the queueing problem for the continuous review (S, s) policy with Poisson demands and arbitrary lead time.

11. Analyze the following periodic review (S, s) policies with Poisson demands $(\lambda = 0.2)$ and constant lead time $(T = 5)$. The review period (R) is 2.
   (a) $S = 3$, $s = 0$.
   (b) $S = 3$, $s = 2$.
   (c) $S = 4$, $s = 1$.

12. Develop an approach to the problem discussed in Exercise 11 if the lead time is equally likely to be any value between 9 and 11.

13. A manager wants an inventory reorder point such that there will be a 10% chance of a stock-out during each reorder period if the weekly demand for product is normal with mean of 50 units and a standard duration of 20. The system is a continuous review system and the probability of various lead times are $Pr(2) = 0.50$, $Pr(3) = 0.30$, and $Pr(3) = 0.20$. Use GERTS to find an acceptable policy.

14. Test the sensitivity of the assembly line problem discussed in Fig. 10.21 to the in-process inventory limitations preceding stations 1 and 2.

15. Use GERT to model and analyze the circuit in Fig. 10.62.

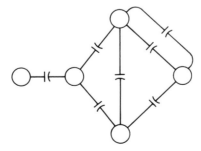

Figure 10.62. Sample electrical network.

16. Develop a GERTS approach to the analysis of circuits of the kind described in this chapter.

17. Adapt the model of the rocket recovery problem such that there are a maximum of two unsuccessful flights for each casing. Use both a GERT and GERTS approach to analyze the system.

18. Perform additional analysis on the model shown in Fig. 10.29. What additional information is available from this model?

19. Model and analyze a complex transmitter which contains two primary generators, A and B, and three secondary generators C, D, and E. The time before failure of each of the five generators is a random variable which is exponentially distributed and whose mean is 1000 hours. The transmitter works as long as there is at least (1) one primary generator and two secondary generators or (2) two primary generators operative.

20. Develop and analyze two additional replacement policies for the tube replacement problem described in Fig. 10.32.

21. Use GERTS to determine which of the following single sampling plans is best to detect a shift in quality from 0.05 to 0.10.
    (a) $n = 50$, $c = 4$.
    (b) $n = 100$, $c = 6$.
    (c) $n = 50$, $c = 1$.

22. Repeat Exercise 21 for a shift of quality level from 0.02 to 0.05.

23. Model the following double sampling plans. Compare their appropriateness for detecting shifts in quality levels from .03 to .06.
    (a) $n_1 = 50$, $n_2 = 150$, $c_1 = 1$, $c_2 = 3$, $c_3 = 5$.
    (b) $n_1 = 100$, $n_2 = 200$, $c_1 = 2$, $c_2 = 5$, $c_3 = 8$.

24. Referring to a text like Duncan's [4], find and analyze two sampling plans not discussed in this chapter using GERT and/or GERTS.

25. Discuss and give examples of how one might develop an acceptance sampling plan based upon the discussion in paragraph (b) of Section 10.4.2.

26. What information is available from a GERT and/or GERTS model of Dodge's continuous sampling plan which might be useful to a practitioner wanting to use such a plan.

27. Analyze the PERT network shown in Fig. 3.28 using GERTS.

28. Discuss and give examples of how feedback might be meaningfully included in a model such as the one shown in Fig. 10.47.

29. Attempt to use the models in Figs. 10.49, 10.50, 10.51, and 10.52 to analyze the Rand D Process.

30. Modify the brainstorming model described in this chapter to require unanimous acceptance by each participant. Analyze the model and compare your results to those shown in Fig. 10.55.

31. Modify the model in Fig. 10.56 to allow queueing of parts at the inspection nodes. Use GERTS to analyze this system.

32. Develop two management decision problems in which GERT and/or GERTS would be useful tools.

33. Analyze the tolerance-type problem discussed in Section 10.6.3 and modeled in Fig. 10.59.

34. Incorporate queue nodes into the work load system shown in Fig. 10.61 and try to perform a bottleneck analysis on this system.

REFERENCES

1. Burbridge, J. J., "An Approach to the Analysis of a Conveyor System," presented at the *38th National Meeting of O.R.S.A.* (October, 1970).

2. Derman, C. and G. J. Lieberman, "A Markovian Decision Model for a Joint Replacement and Stockout Problem." *Management Science*, Vol. 13, No. 3 (May, 1967), pp. 609–617.

3. Disney, R. L., "Some Results of Multichannel Queueing Problems with Ordered Entry—An Application of Conveyor Theory." *Journal of Industrial Engineering*, Vol. 14 (March, 1963).

4. Duncan, A. J., *Quality Control and Industrial Statistics*. Urbana, Illinois, Richard D. Irwin, Inc., 1965.

5. Elmaghraby, S. F., *The Design of Production Systems*. New York, Van Nostrand-Reinhold Publishing Corp., 1966.

6. Enlow, R. A. and A. A. B. Pritsker, *Planning R and D Projects using GERT*. NASA Report, NASA 12-2035 (June, 1969).

7. Fabens, A. J., "The Solutions of Queueing and Inventory Models by Semi-Markov Processes." *Journal of the Royal Statistical Society Series B*, Vol. 23 (1961), pp. 113–217.

8. Fabens, A. J. and S. Karlin, "A Stationary Inventory Model with Markovian Demand," in K. J. Arrow, S. Karlin, and P. Suppes, eds., *Mathematical Methods in the Social Sciences*, 1959, Stanford, Stanford University Press, 1960, Chap. 11.

9. Foster, F. G., "On Stochastic Matrices Associated with Certain Queueing Processes." *Annals of Mathematical Statistics*, Vol. 24, No. 3 (1953), pp. 355–360.

10. Fry, J. H., "Selection, Stopping Conditions, and Response Characteristics of Dodge's Continuous Sampling Plans When Incoming Quality Deteriorates to an Unacceptable Level." Master's Thesis, Lehigh University, 1966.

11. Gaver, D. P., Jr., "Renewal Theoretic Analysis of a Two Bin Inventory Control Policy." *Naval Research Logistics Quarterly*, Vol. 6 (1959), pp. 141–163.

12. Gaver, D. P., Jr., "On Base Stock Level Inventory Control." *Operations Research*, Vol. 7, No. 6 (1959), pp. 689–703.

13. Graham, P., "Profit Probability Analysis of Research and Development Expenditures." *The Journal of Industrial Engineering*, Vol. 16, No. 3 (1965), pp. 186–191.

14. Happ, W. W., "Application of Flowgraph Techniques to the Solution of Reliability Problems," in M. F. Goldberg and J. Voccaro, *Physics of Failure in Electronics.* Washington, U.S. Department of Commerce Office of Technical Services AD-434/329 (1964), pp. 375–423.

15. Howard, R. A., "Systems Analysis of Linear Models," in H. F. Scarf, D. M. Gilford, and M. W. Shelly, eds., *Multistage Inventory Models and Techniques.* Stanford, Stanford University Press (1963), Chap. 6.

16. Hsuan, E. C., "Applications of GERT to Quality Control." Master's Thesis, Lehigh University, 1969.

17. Johnson, E. L., "Optimality and Computation of (σ, S) Policies in the Multi-Item Infinite Horizon Inventory Problem." *Management Science*, No. 13, Vol. 7 (March, 1967), pp. 475–491.

18. Karlin, S. and A. Fabens, "Generalized Renewal Functions and Stationary Inventory Models." *Journal of Mathematical Analysis and Applications*, Vol. 5 (1962), pp. 471–487.

19. Kemeny, J. G., A. Schleifer, J. L. Snell, and G. L. Thompson, *Finite Mathematics with Business Applications.* Englewood Cliffs, N.J., Prentice-Hall, Inc., 1962.

20. Kendall, D. C., "Theory of Queues." *Annals of Mathematical Statistics*, Vol. 24, No. 3 (1953), pp. 338–354.

21. Lloyd, D. K. and M. Lipow, *Reliability Management Methods and Mathematics.* Englewood Cliffs, N.J., Prentice-Hall, Inc., 1962.

22. Masse, P., *Optimal Investment Decisions—Rules for Action and Criteria for Choice.* Englewood Cliffs, N.J., Prentice-Hall, Inc., 1962.

23. Moran, P. A. P., *The Theory of Storage.* London, Methuen, 1959.

24. Morse, P. M., *Queues, Inventories and Maintenance.* New York, John Wiley & Sons, Inc., 1958.

25. Morse, P. M., "Solution of a Class of Discrete-Time Inventory Problems." *Operations Research*, Vol. 1, No. 4 (January 1959), pp. 67–78.

26. Morse, P. M., "Markov Processes," in *Notes on Operations Research 1959.* Cambridge, The Technology Press, M.I.T. (1959), Chap. 3.

27. Mullen, C. H., "A Study of Short-Run Quality in Product Inspected By a Dodge Continuous Sampling Plan (CSP-1) with Emphasis on a Merged Application Employing Several Plans in Parallel." Master's Thesis, Lehigh University, 1968.

28. Naddor, E., "Markov Chains and Simulations in an Inventory System." *Journal of Industrial Engineering*, Vol. 14, No. 2 (1963), pp. 91–98.

29. Parzen, E., *Stochastic Processes.* San Francisco, Holden-Day, Inc., 1962.

30. Powell, G. E., "Development, Evaluation and Selection of a Dodge Continuous Sampling Plan When the Rectifying Operation is Not Perfect." Master's Thesis, Lehigh University, 1967.

31. Pritsker, A. A. B. and P. J. Kiviat, *Simulation with GASP II*. Englewood Cliffs, N. J., Prentice-Hall, Inc., 1969.

32. Pritsker, A. A. B., "Applications of Multichannel Queueing Results to the Analysis of Conveyor Systems." *Journal of Industrial Engineering*, Vol. 17 (January, 1966).

33. Pritsker, A. A. B. and R. A. Enlow, "Four GERT Views of Planning Rand D Projects," presented at the *Thirty-fourth National Meeting of O.R.S.A.*, November, 1968.

34. Pritsker, A. A. B., "The Optimal Control of Discrete Stochastic Processes." Unpublished Dissertation at Ohio State University, 1961,

35. Pritsker, A. A. B. and G. E. Whitehouse, "GERT: Graphical Evaluation and Review Technique, Part II, Probabilistic and Industrial Engineering Applications." *Journal of Industrial Engineering*, Vol. 17, No. 6 (June, 1966).

36. Roberts, N. H., *Mathematical Methods in Reliability Engineering*, New York, McGraw-Hill, Inc., 1964.

37. Scarf, H. E., "Analytic Techniques in Inventory," in H. E. Scarf, D. M. Gilford, and M. W. Shelly, eds., *Multistage Inventory Models and Techniques*. Stanford, Stanford University Press (1963), Chap. 7.

38. Scarf, H. E., "Stationary Operating Characteristics of an Inventory Model with Time Lag," in K. J. Arrow, S. Karlin, and H. Scarf, eds., *Studies in the Mathematical Theory of Inventory and Production*. Stanford, Stanford University Press (1958), Chap. 16.

39. Stern, M. E., *Mathematics for Management*. Englewood Cliffs, N.J., Prentice-Hall, Inc., 1963.

40. Van Slyke, R. M., "Monte Carlo Methods and the PERT Problem." *Operations Research*, Vol. 11 (September, 1963).

41. Weiss, H. K., "Estimation of Reliability Growth in a Complex System with Poisson Type Failure." *Operations Research*, Vol. 4, No. 5 (1956), pp. 532–544.

42. Whitehouse, G. E. and E. C. Hsuan, "The Application of GERT to Quality Control: A Feasibility Study." *Proceedings of the 21st National Meeting of AIIE* (1970).

43. Whitehouse, G. E., "Solution of $M/G/1$ and $GI/M/1$ Finite Queueing Problems by Graphical Means," presented at the *31st National Meeting of the Operations Society of America* (June, 1967).

44. Whitehouse, G. E., "GERT, A Useful Technique for Analyzing Reliability Problems." *Technometrics* (February, 1970).

45. Whitehouse, G. E., "The Application of Graphical Methods to Analyze Inventory Systems." *Production and Inventory Management*, 1st Quarter (1970).

46. Whitehouse, G. E., *Extensions, New Developments and Applications of GERT.* PhD. Dissertation, Arizona State University, August, 1965.

47. Zehna, P. W., "Inventory Depletion Policies," in K. J. Arrow, S. Karlin, and H. Scarf, eds., *Studies in Applied Probability and Management Science.* Stanford, Stanford University Press (1962), Chap. 15.

CASE STUDIES OF GERT APPLICATIONS

In this chapter we will see how analysts have used GERT to help them analyze four specific problem areas. We have tried to select four rather diverse areas. Each case study is independent of the others. The topics selected are:

– An analysis of a manufacturing process using the GERT approach.
– The analysis of a T-type traffic intersection using GERT.
– A model of student progress at Virginia Polytechnic Institute.
– The development of optimum stopping rules for sampling plans.

11

11.1 Case Study—An Analysis of a Manufacturing Process Using the GERT Approach*

This case is based upon the work of Settles, Pritsker, and Thompson [15, 16, 18] and deals with the application of GERT to the analysis of a manufacturing process.

A series of operations required to produce a gear at AiResearch/Phoenix is the particular manufacturing process studied. The past history of this particular gear indicated a high incidence of rework, or scrap, or both. In fact this part was singled out as the one that has caused the most problems during the period preceding this study.

In Chapter 8 we discussed some suggested steps to the analysis of systems using GERT. Settles [18] has modified the steps suggested in the chapter slightly to make them applicable to the cost domain. These steps will be used in this case study; they are:

1. Convert a qualitative description of a system or problem to a model in stochastic network form.

_____

*This section has been based on References 15, 16, and 18 and the material is used with the permission of Drs. Settles, Pritsker, and Thompson.

2. Collect the necessary data to describe the branches of the network.
3. Determine the equivalent function or functions of the network.
4. Convert the equivalent function into performance measures associated with the network. Examples of performance measures are:
   (a) the probability that a specified node is realized;
   (b) the average cost to realize the specified node;
   (c) an estimate of the standard deviation of the cost to realize the specified node;
   (d) the minimum cost observed to realize the specified node;
   (e) the maximum cost observed to realize the specified node; and,
   (f) a histogram of the costs to realize the specified node.
5. Make inferences concerning the system under study from the information obtained in item 4 above.

This case study emphasizes the network-formulation and data-collection phases of GERT.

THE MANUFACTURING PROCESS

The process begins with the release of a work order for a batch of gears and ends at final stores. The process itself can be separated into four basic groups of operations: (1) milling, (2) heat treatment, (3) grinding, and (4) inspection and final processing. The milling operations are interspersed with heat treatments and include such operations as turning and hobbing. After the milling operations are completed, the parts are subjected to an extensive heat-treatment cycle where operations correspond to stages in the cycle. It is here that the gears are hardened to the desired specifications. Finally, the gears are ground to obtain the proper finish and dimensions. Also there are some operations in this group which require additional machining and heat treatment. On completion of the grinding operations, a final inspection is performed. Acceptable gears are sent to finish stores, while those rejected are either reworked or scrapped.

In most cases, nonconforming gears are tagged when discovered and continue to flow through the shop with those gears that have no visible defects. All gears are inspected at the final inspection station; however, those which are tagged receive special attention. In many cases these discrepancies are so minor that a tagged gear can be accepted without further machining. Nevertheless, there are those parts which must be reworked at the appropriate station, or scrapped, or both. Of course there is a greater incidence of rework and scrapping among tagged gears than among untagged gears.

The final operation of the heat-treatment cycle is to inspect each gear for *white spots* caused by an improper heat treatment. If this problem occurs, the defective gears are immediately reprocessed through the heat-treatment cycle. It is assumed that only one reprocessing will be required. This is the only case in which rework is accomplished prior to determining the disposition of the gear at the final inspection station.

Although this is a brief discussion of the process, it should be sufficient for the purposes of this case study. Now the problem can be defined.

PROBLEM DEFINITION

The past history of this particular gear indicates a definite need for an analysis of the production process. Recent work order releases have encountered problems in the heat-treatment cycle and in grinding operations. In one instance, some form of rework was required on all gears in the original release.

In this process, rework and scrappage are costly items which significantly increase the unit cost of the finished product. Therefore, the production manager is interested in alternative methods to reduce the unit cost while maintaining a smooth product flow in the shop. Analyses of this type can be accomplished using the GERT approach.

DEVELOPMENT OF THE GERT NETWORK

The first step in the GERT approach is to convert a qualitative definition of the manufacturing process to a GERT network model. Using the four basic groups of operations in this process, a simplified initial network is shown in Fig. 11.1. The network represents the possible paths a single part can follow in the course of the process. The network is drawn for a part as opposed to being a description of the manufacturing operations through which parts flow, i.e., a queueing model.

The qualitative network is not complete until the difference between tagged (nonconforming) and untagged gears is resolved. Since tagged gears have a higher incidence of rework or scrap at the final inspection station, the network must reflect this difference to adequately model the process. This difference can be resolved in the network if we define an alternate set of operations which are identical to the operations being performed on the untagged parts. The final qualitative GERT network is shown in Fig. 11.2 where the alternative set of operations are designated by the 200 series nodes. Once a gear is tagged, it proceeds along the alternate path; hence, the higher incidence of rework and scrap can be reflected on the output side of node 209 in Fig. 11.2.

At this point the branches of the network are specified only in qualitative terms. In this analysis the branch parameters of interest are probability and cost; hence, data must be obtained to determine the frequency of occurrence for each branch and the cost incurred if this branch of the network is realized. This initiates the data-collection phase of the GERT approach.

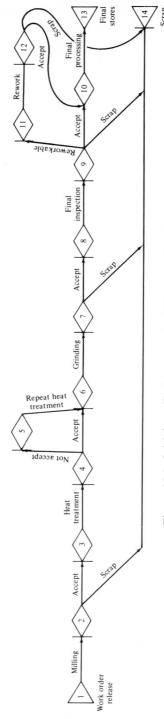

Figure 11.1. Initial qualitative network of the manufacturing process.

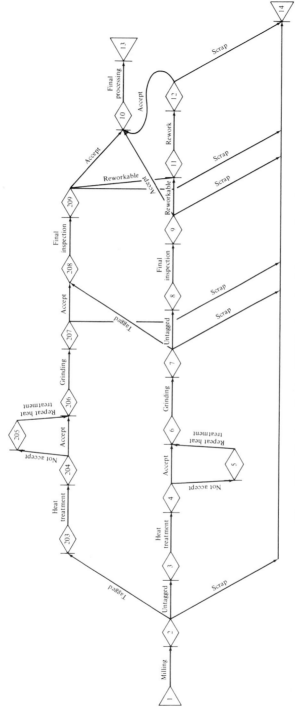

Figure 11.2. Final qualitative GERT network of the manufacturing process.

439

DATA COLLECTION

Obtaining good estimates of the branch parameters is possibly the most difficult phase of the GERT approach. The analysis is only as good as the input data; hence, the importance of accurate data should be emphasized.

In this case data were obtained from the most recent gear production history. The data base was limited to the three most recent work order releases which represented a total of 229 gears. Since production systems are subject to periodic change, older data may not truly reflect the current process.

The frequency of occurrence of each branch was relatively easy to obtain. Work orders at AiResearch/Phoenix indicate at what point in the process a gear was scrapped or designated as nonconforming (tagged). Therefore, with the aid of a production control specialist, an operation-by-operation history of discrepancies was developed. The manufacturing process was further simplified by grouping series of operations which had no discrepancies. The GERT network representation of this simplified version of the process is shown in Fig. 11.3. A description of the operations and the frequency of discrepancies can be found in Table 11.1.

The network presented in Fig. 11.3 is essentially the same as the network given in Fig. 11.2. However, there are some differences which need clarification.

The groupings of operations used in the network of Fig. 11.3 do not correspond to the four main groupings of operations discussed previously. For example, the branch joining nodes 1 and 2 of the network in Fig. 11.2 represents the basic group of milling operations. For the network in Fig. 11.3, however, this same group of milling operations corresponds to the series of branches between nodes 1 and 6. Hence, discrepancies occurred at several of the milling operations as shown in Table 11.1.

Another difference between the two networks concerns the representation preliminary in-process inspections. Since these inspections are performed directly after many operations, the inspection cost is included as part of the processing cost. At each point in the process where this occurs, the GERT network would involve a processing branch and inspection branch. Since these operations are in series, a branch representing the inspection cost is not required. Combining operations at this level reduces the size of the network.

For illustrative purposes, a detailed description of the process represented by the branch between nodes 3 and 4 of the quantitative network is given in Fig. 11.4. By combining the costs associated with processing and inspection, node 3' can be eliminated. The branches emanating from node 3'' represent the decision alternatives following inspection, and no costs are associated with any of these branches. Hence, the costs of the subsequent processing operation can be combined with the probabilities associated with the inspection decision. In other words, node 4 could replace node 3'', again

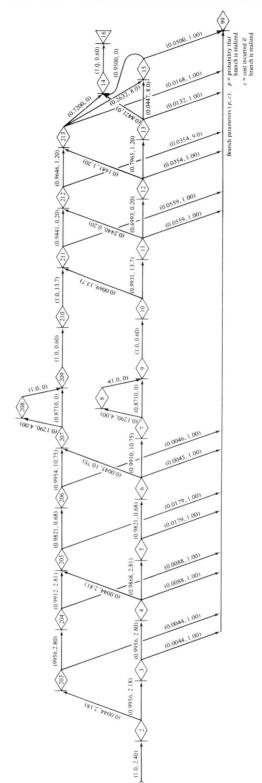

Figure 11.3. Quantitative GERT network of the manufacturing process.

441

Table 11.1. BRANCH PARAMETERS FOR A GERT NETWORK OF THE MANUFACTURING PROCESS.

Source Node	Sink Node	Operation Number	General Description of Operations	Probability		Cost ($)
				Fraction	Decimal	
1	2	10–30	Initial processing and milling	229/229	1.0000	2.40
2	3	40	Milling	228/229	.9956	2.18
2	203	40*1	Milling	1/229	.0044	2.18
3	4	50–70	Heat treatment and milling	227/228	.9956	2.80
3	99	SCRAP	Scrap after Operation 40	1/228	.0044	1.00
203	204	50–70*	Heat treatment and milling	227/228	.9956	2.80
203	99	SCRAP	Scrap after Operation 40*	1/228	.0044	1.00
4	5	80	Milling	224/227	.9868	2.81
4	205	80*	Milling	1/227	.0044	2.81
4	99	SCRAP	Scrap after Operation 70	2/227	.0088	1.00
204	205	80*	Milling	225/227	.9912	2.81
204	99	SCRAP	Scrap after Operation 70*	2/227	.0088	1.00
5	6	90	Milling	219/223	.9821	0.68
5	99	SCRAP	Scrap after Operation 80	9/223	.0179	1.00
205	206	90*	Milling	219/223	.9821	0.68
205	99	SCRAP	Scrap after Operation 80*	4/223	.0179	1.00
6	7	100–365	Milling and heat treatment	217/219	.9910	10.75
6	207	100–365*	Milling and heat treatment	1/219	.0045	10.75
6	99	SCRAP	Scrap after Operation 90	1/219	.0045	1.00
206	207	100–365*	Milling and heat treatment	218/219	.9954	10.75
206	99	SCRAP	Scrap after Operation 90*	1/219	.0046	1.00
7	9	DUMMY	Heat treatment O.K.	189/217	.8710	0.00
7	8	REWORK	Heat treatment repeat	28/217	.1290	4.00
8	9	DUMMY	Precedence relationship	28/28	1.0000	0.00
9	10	370	Milling	217/217	1.0000	0.60
207	209	DUMMY	Heat treatment O.K.	189/217	.8710	0.00
207	208	REWORK	Heat treatment repeat	28/217	.1290	4.00
208	209	DUMMY	Precedence relationship	28/28	1.0000	0.00
209	210	370*	Milling	217/217	1.0000	0.60
10	11	380–500	Milling and grinding	143/144	.9931	13.70
10	211	380–500*	Milling and grinding	1/144	.0069	13.70
210	211	380–500*	Milling and grinding	144/144	1.0000	13.70
11	12	510	100% nital etch	100/143	.6993	0.20
11	212	510*	100% nital etch	35/143	.2448	0.20
11	99	SCRAP	Scrap after Operation 500	8/143	.0559	1.00
211	212	510*	100% nital etch	135/143	.9441	0.20
211	99	SCRAP	Scrap after Operation 500*	8/143	.0559	1.00
12	13	520–550	Heat treatment and inspection	90/113	.7965	1.20
12	213	520–550*	Heat treatment and inspection	19/113	.1681	1.20
12	99	SCRAP	Scrap after Operation 510	4/113	.0354	1.00
212	213	520–550*	Heat treatment and inspection	109/113	.9646	1.20
212	99	SCRAP	Scrap after Operation 510*	4/113	.0354	1.00

Table 11.1. Continued.

Source Node	Sink Node	Operation Number	General Description of Operations	Probability Fraction	Probability Decimal	Cost ($)
13	14	DUMMY	Accepted at final inspection	64/76	.8421	0.00
13	15	REWORK	Miscellaneous rework	11/76	.1447	8.00
13	99	SCRAP	Scrap after final inspection	1/76	.0132	1.00
213	13	DUMMY	Accepted at final inspection	18/19	.7200²	0.00
213	15	REWORK	Miscellaneous rework	5/19	.2632²	8.00
213	99	SCRAP	Scrap after final inspection	—	.0168	1.00
15	14	DUMMY	Final processing	—	.9500**	0.00
15	99	SCRAP	Scrap after rework	—	.0500**	1.00
14	16	560–590	Final processing	—	1.0000	0.60

¹ Nonconforming gears are tagged and allowed to travel with acceptable gears.
* Operations performed on tagged gears.
² Adjusted to allow for scrap.
** Estimated.

reducing the number of branches required. Therefore, the detailed partial network in Fig. 11.4 has been reduced to the partial network shown in Fig. 11.5. This partial network is equivalent to the corresponding section of the quantitative network in Fig. 11.3.

The cost parameter associated with each branch of the network is often a nebulous quantity. Before proceeding further, the researcher must decide

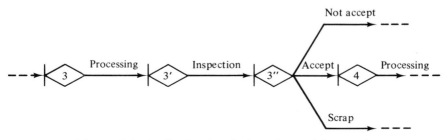

Figure 11.4. Qualitative description of a partial network.

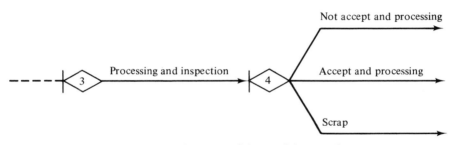

Figure 11.5. Refinement of the partial network.

the level of complexity required for the analysis of the network. This is equivalent to the level of abstraction decision in other model-building activities. If only a low-level network is required, the cost parameter can be related to the setup and processing times ignoring overhead costs. An average cost/hour can then be assigned to each operation based on actual performance or standard data. However, several operations such as heat treatment involve mainly equipment and equipment-depreciation costs. For these operations it was necessary to estimate a cost per part based on management policies. For the stated objectives of this report, costs are assumed to be deterministic. Further refinements to the network will be discussed in a subsequent section.

For this network, standard data were used to obtain the setup and processing times of most operations. An average cost/hour of $4.00 was used to relate cost to time. A thorough analysis of overhead costs is beyond the scope of this case; therefore, estimates were made for those operations which are considered as overhead items. The GERT network in Fig. 11.3 contains the costs associated with each branch in the network.

EQUIVALENT NETWORK, PERFORMANCE MEASURES, AND INFERENCES

The final three steps in the GERT approach are (1) to determine the equivalent network and the total cost distribution; (2) to convert the total cost distribution into performance measures associated with the network; and (3) to make inferences concerning the production system.

The GERT EXCLUSIVE-OR computer program discussed in Chapter 8 provides a means for obtaining the information upon which inferences can be made. The input data required by the program can be obtained directly from the GERT network given in Fig. 11.3. Output from the program includes the following information:

1. The probability and cost associated with each path in the network.
2. The probability that a part is accepted or scrapped.
3. The mean and variance of the cost incurred for accepted or scrapped parts.

Using the output from the program, the probability of an acceptable part was found to be 0.857 with a mean cost of $39.95 and a variance of 11.66. In other words, if a lot of 1000 gears was released, it is expected that 857 gears would be accepted at an average cost of $39.95. Similarly, the probability of a scrapped part is 0.143 with an expected cost of $31.37 and a variance of 158.33. Since the cost of acceptable gears does not consider scrappage, it does not reflect the true cost of producing an acceptable gear. Adjusting this cost to allow for scrappage yields a total expected cost of $44.25.

At this stage in the GERT analysis, it is possible to construct an equivalent network. Thus the GERT network in Fig. 11.3 can be reduced to the network in Fig. 11.6 (where only the expected values are displayed). Now it is desired to determine the probability mass function associated with each of the branches in Fig. 11.6.

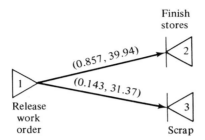

Figure 11.6. Equivalent network of the manufacturing process.

The probabilities and costs associated with each path can be used to obtain the total cost distribution for both acceptable and scrapped parts. All paths with a probability of less than 0.001 were ignored. The cost distribution for acceptable parts is shown in Fig. 11.7. The cost distribution for scrapped parts is shown in Fig. 11.8. Probability statements can now be made about the production cost of both acceptable and scrapped gears. For example, the probability that an acceptable gear will cost more than $46 is 0.024.

An improvement analysis is considered as part of the review process. However, the discussion is included in the next section because of the importance of this technique as a management tool in solving production problems.

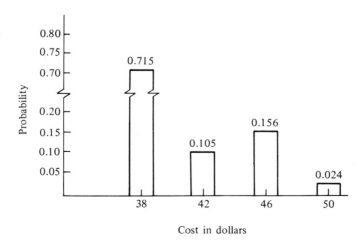

Figure 11.7. Probability mass function of cost for acceptable gears.

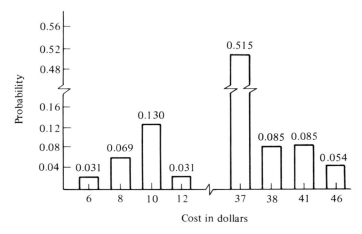

Figure 11.8. Probability mass function of cost for scrapped gears.

IMPROVEMENT ANALYSIS

An improvement analysis allows the manager to evaluate the effect of various alternatives on the total production system. Initially, each alternative must be defined and analyzed to determine the branches of the network affected. The network is then modified to reflect these changes in either network structure or network parameters. The GERT computer program can then be used to obtain the effect of the change. This process is repeated for all alternatives and a comparison can then be made to assess the worth of the proposed alternatives.

Three alternatives are considered in this research. Two of these are actual alternatives proposed by AiResearch/Phoenix. The last is a hypothetical alternative for illustrating a feature of the model.

For an initial cost of $5000, an additional quench die can be purchased for use in the heat-treatment cycle. It is estimated that this capital expenditure would increase the cost per gear by $0.50. As a result, it is anticipated that the probability of a successful heat-treatment cycle would be increased from 0.871 to 0.950.

A second alternative for improving the effectiveness of the heat-treatment cycle is to purchase a carburizing fixture for an initial cost of $2000. In this case it is estimated that the increased cost would be $0.25 per gear. For this alternative, the probability of a successful heat-treatment cycle would be increased to 0.920.

It is coincidence rather than a requirement of the model that both of these alternatives affect the same branches of the network in Fig. 11.3. An important feature of the improvement analysis is that any or all of the branches of the network can be affected by any alternative. This feature is illustrated by the third alternative.

Suppose that a heat treatment which improves the forging's machinability can be applied prior to the first machining operation. Estimates are that this improved machinability would increase the acceptance at final inspection from 0.842 to 0.942 for untagged gears and from 0.720 to 0.820 for tagged gears. It is estimated that a cost of $0.20 per gear would be incurred for the process.

The description of the alternatives can now be converted into network terminology. The modifications to the network required for each alternative are shown in Table 11.2. Hence, three new networks are created for evaluation purposes.

Table 11.2. NETWORK CHANGES FOR EACH ALTERNATIVE IN THE SENSITIVITY ANALYSIS.

Alternative	Source Node	Sink Node	Probability		Cost	
			From	To	From	To
Quench die	6	7			10.75	11.25
	6	207			10.75	11.25
	206	207			10.75	11.25
	7	9	0.871	0.950		
	7	8	0.129	0.050		
	207	209	0.871	0.950		
	207	208	0.129	0.050		
Carburizing	6	7			10.75	11.00
	6	207			10.75	11.00
	206	207			10.75	11.00
	7	9	0.871	0.920		
	7	8	0.129	0.080		
	207	209	0.871	0.920		
	207	208	0.129	0.080		
Improved machinability	1	2			2.40	2.60
	13	14	0.842	.942		
	13	15	0.144	.044		
	213	14	0.720	.820		
	213	15	0.263	.163		

It is not necessary to use the GERT computer program for evaluating the first two alternatives. A closer look at each of these alternatives reveals that only the input and output sides of nodes 7 and 207 of the network in Fig. 11.3 would be affected. These nodes are associated with untagged and tagged gears, respectively. Since the tagged and untagged sections of the network are mutually exclusive (only one will be realized in any realization of the network), an analysis of the proposed alternatives need only consider the effect in either section.

The section of the network affected by the proposed alternatives is shown in Fig. 11.9.

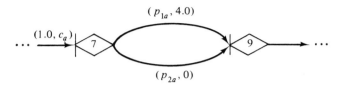

Figure 11.9. Partial network for an improvement analysis.

In this partial network, c_a represents the increased cost of the heat-treatment cycle for alternative a, p_{1a} is the probability that the gear must repeat the heat-treatment cycle at a cost of \$4, and p_{2a} is the probability that the gear can proceed directly to the subsequent operation. The equivalent W function for this partial network is:

$$W_{Ea} = p_{2a}e^{c_a s} + p_{1a}e^{(4 + c_a)s}$$

where:

$$p_{2a} = 1 - p_{1a}$$

By differentiating W_{Ea} with respect to s and setting $s = 0$, we obtain the expected increase in total cost (TC_a):

$$E[TC_a] = c_a + 4p_{1a}$$

For the original network of Fig. 11.3, $c_a = 0$ and $p_{1a} = 0.1290$, which gives a standard comparison of:

$$E[TC_a] = [0.1290][4.0] = 0.516$$

Therefore, if $E[TC_a] < 0.516$, the alternative is preferred to the original network. Now the two alternatives can be evaluated with the results shown in Table 11.3. Based on these results, both alternatives should be rejected. The expectation decision criterion is used in making these decisions.

Table 11.3. RESULTS OF IMPROVEMENT ANALYSIS FOR ALTERNATIVES 1 AND 2.

a	c_a	p_{1a}	$4p_{1a}$	$E[TC_a]$	*Decision*
1	\$0.50	0.05	0.20	\$0.70	Reject
2	0.25	0.08	0.32	0.57	Reject

A similar analysis could be performed for the third alternative; however, the calculations would be laborious. This is because two sections of the original network are affected by the alternative. Therefore, the simplest method for evaluating this alternative is to use the GERT computer program.

Using the computer output for alternative 3, the probability of an acceptable gear was found to be 0.862 with a mean cost of $39.33. The probability of a scrapped part was 0.138 with an expected cost of $31.07. Adjusting the average cost of acceptable gears to include scrappage yields $43.64 as the expected cost of an acceptable gear. This is a reduction in cost of $0.51 per gear for this alternative, over the original process given in Fig. 11.3. Therefore, alternative 3 is preferred to the original network.

This concludes the improvement analysis. The worth of GERT as a management tool in production control has been demonstrated. Perhaps the greatest benefit of the technique is that it forces the analyst to think of each alternative in terms of a network. Therefore, it is possible to gain insights into the problem perhaps unattainable by conventional means. This was particularly true of the improvement analysis for alternatives 1 and 2.

REFINEMENTS OF THE MODEL

Thus far the analysis of the manufacturing process has been based on deterministic (constant) cost parameters assigned to branches of the network. In practice, these costs are not deterministic but are random variables.

When randomness is introduced into the model, the GERT approach becomes even more useful than before. Even for a simple problem it is a laborious task to obtain system performance measures by analytical methods. By means of GERT, however, the analysis can proceed as before. Once the equivalent network is obtained, it is at least theoretically possible to obtain the cost distribution of the equivalent network.

Treating the cost parameter as a random variable requires data to form probability distributions for each branch of the network. At AiResearch/ Phoenix the necessary information was gleaned from daily tabulation runs of actual processing times charged against the operations. Costs are obtained by multiplying the processing times by an average cost/hour. In many cases, actual processing times were for a batch of gears. When this occurred an average processing time per gear was calculated. In these cases a deterministic cost parameter was assumed as equal to this average value. For those operations which were considered as overhead items, the parameter estimates were also assumed to be deterministic.

Next, the underlying probability distributions of cost for the remaining operations were determined. This is no simple task, particularly if curves must be fitted for many operations and more than one type of distribution is considered.

Three types of modifications are made to the network of Fig. 11.3.

These are:

1. Incorporating the newly obtained average costs for existing branches of the network.
2. Incorporating probability distributions for existing branches as determined by the curve-fitting program.
3. Expanding existing branches which represent groups of operations, some of which have an associated probability distribution for cost.

The revised network is shown in Fig. 11.10 where the branch parameters are represented by their respective W functions. Using Table 11.4, we can relate this network to the network presented in Fig. 11.3.

Using the GERT computer program for the modified network, the probability of an acceptable gear was found to be 0.857 with a mean cost of $39.14 and a variance of 17.68. The probability of a scrapped part is 0.143 with a mean cost of $29.89 and a variance of 192.37. Adjusting the average cost of acceptable gears to include scrappage yields $43.42 as the expected cost of an acceptable gear.

For GERT networks containing only EXCLUSIVE-OR nodes, we showed in Chapter 8 that the nth moment of the equivalent network depends only on the first n moments of the branches of the network. Thus the probability and expected costs for the deterministic and random variable cases should be the same. The expected cost differed slightly due to the revised method of computing the average costs associated with the branches. Since the branch variances are increased when the cost is considered as a random variable, the variance of the equivalent network should also increase. From the results given above this increase was 6.02 for acceptable gears and 34.04 for scrapped gears. This gives management an indication of the variations they can expect in gear production cost. If the probability density function of the equivalent network cost was obtained, probability statements regarding the percentage of time the cost/gear exceeded specified dollar amounts could be made.

11.2 Case Study—The Analysis of a T-Type Traffic Intersection Using GERT*

This case study was a direct outgrowth of the material presented in Section 10.1. It is an attempt to demonstrate and evaluate GERT's potential in modeling traffic systems. The material for this case study is based upon the works of Roth and Whitehouse [13, 21].

---

*The material in this section is adapted from Reference 13 and 24 and is used with the permission of Mr. Roth.

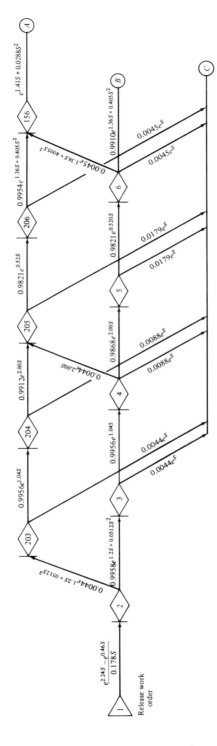

Figure 11.10. Extension and refinement of the GERT network for manufacturing process.

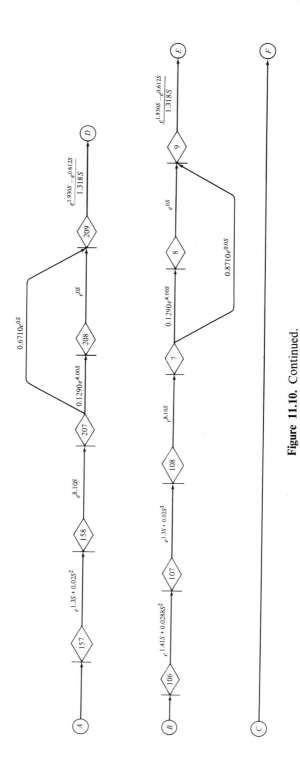

Figure 11.10. Continued.

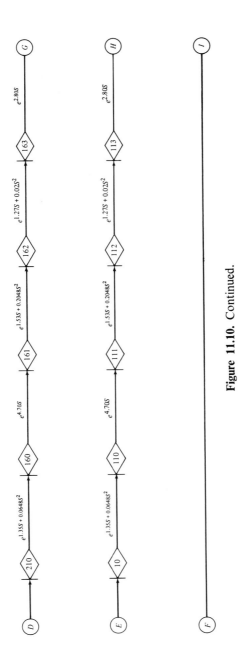

Figure 11.10. Continued.

453

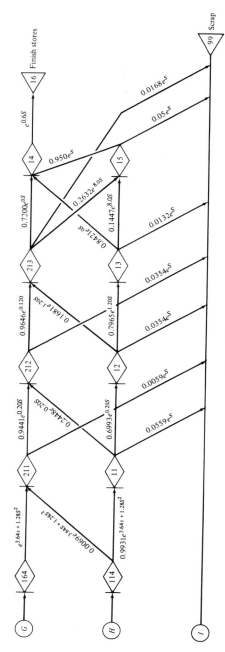

Figure 11.10. Continued.

454

Table 11.4. COST PARAMETER ESTIMATES FOR THE REVISED GERT NETWORK.

Mod. Type	Source Node	Sink Node	Operation Number	General Description of Operations	Dist.	Parameter Estimates ($)
2	1	2	10–30	Initial processing and milling	U	$a = 0.46, b = 2.24$
2	2	3	40	Milling	NO	$\mu = 1.20, \sigma = 0.32$
2	2	203	40*1	Milling	NO	$\mu = 1.20, \sigma = 0.32$
·	·	·	·	·	·	·
3	6	106	100–105	Milling	NO	$\mu = 1.36, \sigma = 0.90$
3	6	156	100–105*	Milling	NO	$\mu = 1.36, \sigma = 0.90$
—	6	99	SCRAP	Scrap after operation 90	D	$\mu = 1.00$
3	206	156	100–105*	Milling	NO	$\mu = 1.36, \sigma = 0.90$
—	203	99	SCRAP	Scrap after operation 90'	D	$\mu = 1.00$
3	106	107	110	Milling	NO	$\mu = 1.41, \sigma = 0.24$
3	156	157	110*	Milling	NO	$\mu = 1.41, \sigma = 0.24$
3	075	108	120	Milling	NO	$\mu = 1.30, \sigma = 0.20$
3	157	158	120*	Milling	NO	$\mu = 1.30, \sigma = 0.20$
3	108	7	130–365	Milling and heat treatment	D	$\mu = 8.10$
3	158	207	130–356*	Milling and heat treatment	D	$\mu = 8.10$
·	·	·	·	·	·	·
2	9	10	370	Milling	U	$a = 0.612, b = 1.93$
·	·	·	·	·	·	·
2	209	210	370*	Milling	U	$a = 0.162, b = 1.93$
3	10	110	380	Grinding	NO	$\mu = 1.35, \sigma = 0.36$
3	210	160	380*	Grinding	NO	$\mu = 1.35, \sigma = 0.35$
3	110	111	390–430	Milling and grinding	D	$\mu = 4.70$

Table 11.4. Continued.

Mod. Type	Source Node	Sink Node	Operation Number	General Description of Operations	Dist.	Parameter Estimates ($)
3	160	161	340–430*	Milling and grinding	D	$\mu = 4.70$
3	111	112	440	Grinding	NO	$\mu = 1.53,\ \sigma = 0.64$
3	161	162	440	Grinding	NO	$\mu = 1.53,\ \sigma = 0.64$
3	112	113	450	Grinding	NO	$\mu = 1.27,\ \sigma = 0.20$
3	162	163	450*	Grinding	NO	$\mu = 1.27,\ \sigma = 0.20$
3	113	114	460–480	Grinding	D	$\mu = 2.80$
3	163	164	460–480*	Grinding	D	$\mu = 2.80$
3	114	11	500	Grinding	NO	$\mu = 3.64,\ \sigma = 1.6$
3	114	211	500*	Grinding	NO	$\mu = 3.64,\ \sigma = 1.6$
3	164	211	500*	Grinding	NO	$\mu = 3.64,\ \sigma = 1.6$
.	.	.	.	.	.	.
.	.	.	.	.	.	.
.	.	.	.	.	.	.

Distribution Key
U—Uniform
NO—Normal
D—Discrete

[1] Nonconforming years are tagged and allowed to travel with acceptable gears.
*Operations performed on tagged gears.
Note: Only branches which were affected by modifications of types 2 and 3 are included in this table. Type-1 modifications are incorporated in the *w* functions of the network presented in Fig. 11.10.

11.2.1 Analysis of T-type intersections. The traffic system considered is a highway exit ramp which provides access to the secondary road through a T-type intersection. Automobiles are provided with a single lane up to a point shortly before the actual intersection. At this point, the lane widens to permit those vehicles desiring to turn right to separate themselves from the left-turning vehicles. Figure 11.11 depicts the situation. This system can

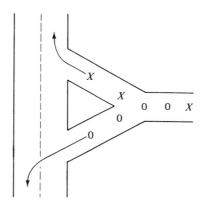

Figure 11.11. The highway exit.

be viewed as a complex queueing system. Automobiles which attain the head of a channel may be considered to be in the process of being served, with service being completed for the vehicle when it enters the secondary road. Move-up times to the service location will be assumed to be negligible since they may be accomplished as the preceding vehicle vacates the system. A service for system is equivalent to the appearance of a gap in the traffic flow. In a paper published on highway merging, Evan, Hernson, and Weiss [3] mention that for light traffic the gaps in traffic flow may be taken to be independent, identically distributed random variables with an exponential distribution function.

Completing a right turn from this system would require a process similar to the highway merge, so that service from the right lane would be exponentially distributed. For a left turn to be completed, there must be gaps appearing simultaneously in both streams of traffic. But this event, which is the joint occurrence of two independent, exponentially distributed events, is also exponentially distributed. If μ_1 is defined as the number of left-turn opportunities per unit time period and μ_2 is defined as the number of right-turn (only) opportunities per unit time period, then we have the following:

$$\mu_1 + \mu_2 = \text{number of right turns per unit time period}$$

$$\mu_1 = \text{number of left turns per unit time period}$$

To illustrate, if there is one left-hand arrival in the system, it will be served at a rate μ_1; if there is a right-hand arrival in the system, it will be served at a rate $\mu_1 + \mu_2$; but if there is one of each arrival type, the right-hand arrival may be served with a service rate μ_2 or they both may be served at the service rate μ_1.

Two cases are to be considered which provide solutions to the congestion problem by way of the specific example. In the first case, the arrival process will be assumed to be random, and in the second case no specific arrival process will be assumed. Finally the applicability of GERT will be considered.

POISSON ARRIVALS

Before proceeding, it is necessary to develop a standardized notation to describe the states of our traffic systems. In the two-digit number which describes a node, the left-hand digit indicates the number of left-hand arrivals in the system and the right-hand digit indicates the number of right-hand arrivals. A blockage resulting from a right-hand arrival is indicated by *RB*, and *LB* indicates a left blockage.

The GERT model of this system rests on the well-known result that a Poisson arrival process has exponential interarrival times. Therefore, the time transmittance encountered in leaving a node by any of its branches will have an exponential density with its parameter equal to the sum of the parameters which pertain to the node. Essentially, it is the density function of the time which will elapse before leaving the node that is determined.

To illustrate, consider state 11 where there are two people in the system, one desiring to make a right turn and one a left turn. The density function for the time elapsed before state 11 is vacated is given by:

$$f(t) = (\lambda + \mu_1 + \mu_2)e^{-(\lambda + \mu_1 + \mu_2)t}$$

While it is a certain event that the node will be vacated to at some time t, the probabilities have not been assigned to the branches. The probabilities are assigned to the branches according to the following relationship:

$$p_i = \frac{a_i}{\sum_{i=1}^{B} a_i}$$

where: p_i = probability of traversing branch,
a_i = parameter of exponential distribution for branch, and
B = number of branches exiting node.

Since the moment generating function of an exponential distribution is given by

$$M(s) = \left(1 - \frac{s}{a}\right)^{-1}$$

the W function for the branches exiting from node 11 are:

To Node	$W(s)$
00	$(\mu_1/(\lambda + \mu_1 + \mu_2)) \cdot (1 - s/(\lambda + \mu_1 + \mu_2))^{-1}$
01	$(\mu_2/(\lambda + \mu_1 + \mu_2)) \cdot (1 - s/(\lambda + \mu_1 + \mu_2))^{-1}$
LB	$(p\lambda/(\lambda + \mu_1 + \mu_2)) \cdot (1 - s/(\lambda + \mu_1 + \mu_2))^{-1}$
RB	$(q\lambda/(\lambda + \mu_1 + \mu_2)) \cdot (1 - s/(\lambda + \mu_1 + \mu_2))^{-1}$

where: p = the probability that an arrival wants to turn right, and
$q = 1 - p$ = the probability that an arrival wants to turn left.

Applying the probability relation to node 00, we have:

To Node	$W(s)$
10	$(p\lambda/\lambda) \cdot (1 - s/\lambda)^{-1}$
01	$(q\lambda/\lambda) \cdot (1 - s/\lambda)^{-1}$

The remainder of the transmittances are indicated on their appropriate branches in the GERT network of the system shown in Fig. 11.12 for the case where there is only one space available for people wanting to turn right and one for people wanting to turn left. To demonstrate more easily the information which may be obtained about the system, the expediency of a specific numerical example will be employed.

The parameters are:

$$\lambda = 2 \text{ arrivals/unit time period}$$
$$p = q = 0.50$$
$$\mu_1 = 2 \text{ services/unit time period}$$
$$\mu_2 = 2 \text{ services/unit time period}$$

Some of the information which may be obtained about the system is as follows.

PROBABILITY OF BLOCKAGE AND MOMENTS OF TIME TO BLOCKAGE FROM CLEARED SYSTEM

In this case, it is assumed that the system begins in the 00 state, and the probabilities of both left and right blockages are calculated. The system, depicted in Fig. 11.13, is the numerical equivalent of Fig. 11.12. It may be seen that the loops and paths through the system are those itemized in Table 11.5.

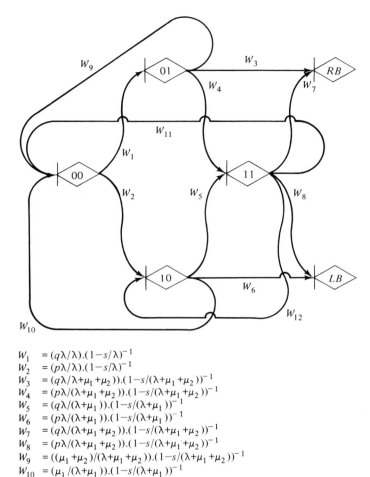

$$W_1 = (q\lambda/\lambda).(1-s/\lambda)^{-1}$$
$$W_2 = (p\lambda/\lambda).(1-s/\lambda)^{-1}$$
$$W_3 = (q\lambda/\lambda+\mu_1+\mu_2)).(1-s/(\lambda+\mu_1+\mu_2))^{-1}$$
$$W_4 = (p\lambda/(\lambda+\mu_1+\mu_2)).(1-s/(\lambda+\mu_1+\mu_2))^{-1}$$
$$W_5 = (q\lambda/(\lambda+\mu_1)).(1-s/(\lambda+\mu_1))^{-1}$$
$$W_6 = (p\lambda/(\lambda+\mu_1)).(1-s/(\lambda+\mu_1))^{-1}$$
$$W_7 = (q\lambda/(\lambda+\mu_1+\mu_2)).(1-s/(\lambda+\mu_1+\mu_2))^{-1}$$
$$W_8 = (p\lambda/(\lambda+\mu_1+\mu_2)).(1-s/(\lambda+\mu_1+\mu_2))^{-1}$$
$$W_9 = ((\mu_1+\mu_2)/(\lambda+\mu_1+\mu_2)).(1-s/(\lambda+\mu_1+\mu_2))^{-1}$$
$$W_{10} = (\mu_1/(\lambda+\mu_1)).(1-s/(\lambda+\mu_1))^{-1}$$
$$W_{11} = (\mu_1/(\lambda+\mu_1+\mu_2)).(1-s/(\lambda+\mu_1+\mu_2))^{-1}$$
$$W_{12} = (\mu_2/(\lambda+\mu_1+\mu_2)).(1-s/(\lambda+\mu_1+\mu_2))^{-1}$$

Figure 11.12. GERT representation of system with unit channel lengths.

Upon solving the network, we obtain the appropriate W functions for each of the terminal nodes.

$$W_{LB} = \frac{(48 - 14s + s^2)}{(80 - 228s + 108s^2 - 18s^3 + s^4)}$$

$$W_{RB} = \frac{(32 - 12s + s^2)}{(80 - 228s + 108s^2 - 18s^3 + s^4)}$$

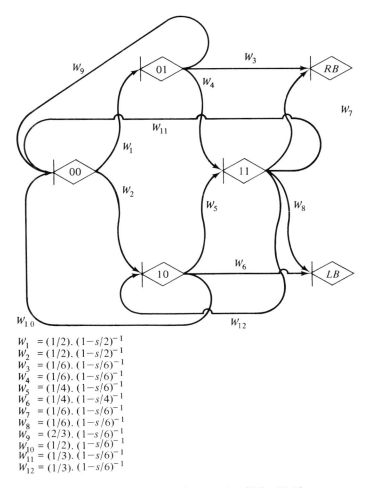

$$W_1 = (1/2) \cdot (1-s/2)^{-1}$$
$$W_2 = (1/2) \cdot (1-s/2)^{-1}$$
$$W_3 = (1/6) \cdot (1-s/6)^{-1}$$
$$W_4 = (1/6) \cdot (1-s/6)^{-1}$$
$$W_5 = (1/4) \cdot (1-s/6)^{-1}$$
$$W_6 = (1/4) \cdot (1-s/4)^{-1}$$
$$W_7 = (1/6) \cdot (1-s/6)^{-1}$$
$$W_8 = (1/6) \cdot (1-s/6)^{-1}$$
$$W_9 = (2/3) \cdot (1-s/6)^{-1}$$
$$W_{10} = (1/2) \cdot (1-s/6)^{-1}$$
$$W_{11} = (1/3) \cdot (1-s/6)^{-1}$$
$$W_{12} = (1/3) \cdot (1-s/6)^{-1}$$

Figure 11.13. Numerical example of Fig. 11.12.

Table 11.5

	Paths		Loops
To *RB*	$W_1 W_3$	First-order	$W_1 W_9$
	$W_1 W_4 W_7$		$W_1 W_4 W_{11}$
	$W_2 W_5 W_7$		$W_1 W_4 W_{12} W_{10}$
To *LB*	$W_2 W_6$		$W_2 W_5 W_{11}$
	$W_2 W_5 W_8$		$W_2 W_{10}$
	$W_1 W_4 W_8$		$W_5 W_{12}$
	$W_1 W_4 W_{12} W_6$	Second-order	$W_1 W_9 W_5 W_{12}$

The probabilities are therefore:

$$P_{LB} = W_{LB}(0) = \tfrac{48}{80} \qquad P_{RB} = W_{RB}(0) = \tfrac{32}{80}$$
$$= 0.60 \qquad\qquad\qquad = 0.40$$

Now, since:

$$M_t(s) = \frac{W(s)}{W(0)}$$

it follows that:

$$E(t) = \frac{d(M_t(s))}{ds}\bigg|_{s=0} = \frac{\dfrac{d(W(s))}{ds}\bigg|_{s=0}}{W(0)}$$

$$= \left(\frac{1}{p}\right) \cdot \frac{d(W(s))}{ds}\bigg|_{s=0}$$

Solving for the expected value of the times for each blockage, we have:

$$E_{LB}(t) = \left(\frac{1}{0.60}\right) \cdot \frac{((80)(-14) - (48)(-326))}{(80)^2}$$

$$= 2.53$$

$$E_{RB}(t) = \left(\frac{1}{0.40}\right) \cdot \frac{((80((-12) - (32)(-326))}{(80)^2}$$

$$= 2.45$$

Further moments of the distributions may be obtained by successive differentiations of the moment generating functions.

PROBABILITY OF REBLOCKAGE AND MOMENTS OF TIME TO
REBLOCKAGE, GIVEN THAT A BLOCKAGE IS CLEARED
TO A CERTAIN STATE

A most important property of the system is the duration of the time interval between successive blockages. This information may be obtained if we structure the flowgraph as in Fig. 11.14. As an example, the behavior of the system when a blockage is cleared to state 10 will be considered. Return to a 10 state implies that the event which occasioned the return was a left service, so that the transmittance to the 10 state must include the moment generating function $(1 - s/\mu_1)^{-1}$.

Solution of the flowgraph yields the respective w functions:

$$W_{LB}(s) = \frac{-2s^3 + 30s^2 - 128s + 112}{Den}$$

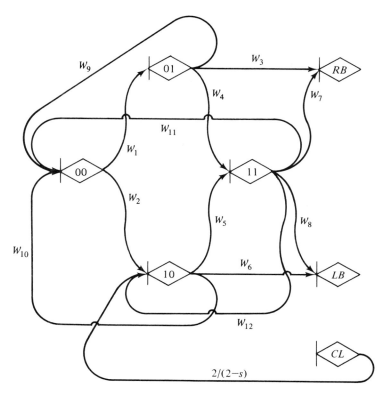

Figure 11.14. System to study reblockage.

$$W_{RB}(s) = \frac{2s^2 - 20s + 48}{Den}$$

$$Den = -s^5 + 20s^4 - 144s^3 + 442s^2 - 532s + 160$$

which indicates that the probabilities are:

$$P_{LB} = \frac{112}{160} = 0.70 \qquad P_{RB} = \frac{48}{160} = 0.30$$

and the expected time for their occurrences are:

$$E_{LB}(t) = \frac{1}{0.70} \cdot \frac{(160)(-128) - (112)(-532)}{(160)^2} = 2.19$$

$$E_{RB}(t) = \frac{1}{0.30} \cdot \left(\frac{(160)(-20) - (48)(-532)}{(160)^2} \right.$$

The information gained in this section for the 10 state may also be compiled for the 11 and 01 states.

TIME FOR RETURN TO IDLE, GIVEN THAT NO BLOCKAGE HAS OCCURRED

This requires that the flowgraph for the system be redrawn as in Fig. 11.15. Since all loops which returned the system to the 00 state are now

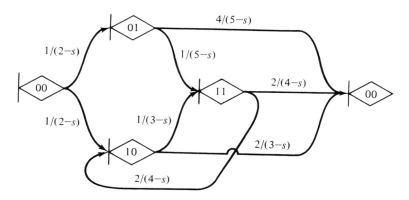

Figure 11.15. System to study idle time given no blockage.

considered paths to the idle state, the new network is far less complex. The equivalent W function between 00 and 00' is given by:

$$W_E(s) = \frac{6s^2 - 50s + 1000}{s^4 - 14s^3 + 69s^2 - 140s + 100}$$

so that the expected time for the recurrence of the idle state is:

$$E(t) = \frac{(100)(-50) - (100)(-140)}{(100)^2} = 0.90$$

Calculations of the recurrence times may also be obtained for the other states in the system if we reconstruct the graph as an open system.

NUMBER OF TIMES THE SYSTEM VISITS THE IDLE STATE BEFORE REBLOCKAGE, GIVEN THAT A BLOCKAGE IS CLEARED TO A CERTAIN STATE

Counters may be introduced into the flowgraph to register the number of times a given node is visited. By introducing an e^c tag on all transmittances entering the 00 state, an equivalent transmittance for the flowgraph may be obtained which is a function of both s and c. Since e^c is the moment generating function of a constant of value 1, the expected number of times the 00 state is occupied before reblockage (of either type) occurs, given a blockage is cleared to state 11 by a right service is found as follows:

$$M(c) = M(s, c)|_{s=0}$$
$$= \frac{480 - 160e^c}{1056 - 736e^c}$$

which gives the expected value as:

$$E_{(count)} = \frac{d(M(c))}{dc}\Big|_{c=0}$$
$$= \left[(1056 - 736e^c)(-160e^c) - \frac{(480 - 160e^c)(-736e^c)}{(1056 - 736e^c)}\right]\Big|_{c=0}$$
$$= 1.80$$

There are many more ways in which counters and W generating functions may be employed in studying the system. They will be left to the ingenuity of the reader.

11.2.2 Generalized arrivals. Roth [13] has studied in detail the effect of generalized arrivals on the T-type intersection. This section of the case study reports Roth's general approach. In Section 10.1, consideration was given to the imbedded Markov chain approach of Kendall [6]. In the source paper, Kendall considers a $GI/M/s$ queueing system which he solves by using the concept of *arrival epochs*. While it might have been desirable to obtain a solution for the system based on *departure epochs* so that service distributions other than exponential might have been imposed on the serving facility, the presence of two servicing rates makes such a formulation difficult to conceive. Fortunately, for the example under discussion, the exponential distribution is a good assumption as previously noted.

The concept of arrival epochs examines the system at the instant immediately following the event of an arrival to the system. Transition probabilities from a given state are developed as a direct result of the interrelationship between the exponential and Poisson distributions. Successive states of the system may be incremented only in unit steps, whereas no such restriction is placed upon decrementing the number in the system. It follows that as the number in the system increases, the number of nodes reachable from that state increases. It also follows that if the "snapshots" of the system are only taken immediately following an arrival, the imbedded process will never attain an idle, or 00 state.

Consider now the example with which we have been working modified to have channels containing two waiting spaces. Suppose that the system is in state 01. Examining the system at time t^+, where t is the time of an arrival, we would find the system in one of four possible states—01, 10, 11, or 02. The first two may be attained only if a service occurred during the interarrival interval; the latter two, only if no service occurred. Since the service times are exponential, the probability of the state of the system being either 11 or 02 is:

$$P\{2 \text{ units in system}\} = P\{\text{no service before next arrival}\}$$

$$= \int_0^\infty [(\mu_1 + \mu_2)t]^0 \cdot \frac{e^{-(\mu_1+\mu_2)t}}{0!} \, dF_A(t)$$

where:

$$dF_A(t) = \text{distribution of interarrival times}$$

If one service occurred in the interval between arrivals, that would be sufficient to guarantee that the system would be in either state 10 or 01 at time t^+. It is obvious that gaps in traffic do not occur only when there is a car waiting to turn. Therefore, the probability that there will be one unit in the system at time t is:

$$P_1 = P\{1 \text{ unit in system}\}$$
$$= P\{1, 2, \ldots, m, \ldots \text{ services before next arrival}\}$$
$$= 1 - \int_0^\infty [(\mu_1 + \mu_2)t]^0 \cdot \frac{e^{-(\mu_1+\mu_2)t}}{0!} \, dF_A(t)$$

Now if the arrivals generated from the generalized interarrival time distribution desire to turn left with probability p and right with probability $(1 - p) = q$, then the preceding results may be rewritten as:

$$P_{11} = P\{\text{system goes to state } 11\}$$
$$= P\{\text{left arrival}\} \cdot P\{\text{no right service before arrival}\}$$
$$= p \cdot \int_0^\infty [(\mu_1 + \mu_2)t]^0 \cdot e^{-(\mu_1+\mu_2)t} \, dF_A(t)$$

$$P_{02} = P\{\text{system goes to state } 02\}$$
$$= q \cdot \int_0^\infty [(\mu_1 + \mu_2)t]^0 \cdot e^{-(\mu_1+\mu_2)t} \, dF_A(t)$$

$$P_{10} = P\{\text{system goes to state } 10\}$$
$$= p \cdot (1 - P_{11} - P_{02})$$

$$P_{01} = P\{\text{system goes to state } 01\}$$
$$= q \cdot (1 - P_{11} - P_{02})$$

This same sort of straightforward analysis may not, however, be applied to states with more than one unit in them at the beginning of the arrival epoch. Suppose now that the system is in state 22 at time t and the inter-arrival period is t_a. The states which the system may assume at time $(t + t_a)^+$ are RB, LE, 22, 12, 21, 11, 20, and 01. Which of these states the system attains is a result not only of the number of services in the interarrival period, but also of the type of services. For example, the state 21 may be reached as the result of a combination of one left service, no right service, and a left arrival; or as the result of no left service, right services of number greater than or equal to two, and a right arrival. Symbolically:

$$P_{21} = P \{\text{system goes to state 21}\}$$

$$= p \cdot \int_0^\infty (e^{-\mu_1 t}) \cdot \left(\frac{\mu_1 t}{1!}\right) \cdot (e^{-\mu_2 t}) \, dF_A(t)$$

$$+ q \int_0^\infty (e^{-\mu_1 t}) \cdot (1 - e^{-\mu_2 t} - e^{-\mu_2 t}) \cdot \frac{(\mu_2 t)}{1!} \, dF_A(t)$$

Following arguments similar to those just presented, Roth was able to model the generalized input model. From his model, he was able to obtain all of the information available from the Poisson arrival case.

11.2.3 The GERTS approach.

We managed to obtain interesting results by applying GERT to this complex system. There are, however, some short-comings to the system that we modeled:

1. A major advantage of GERT is considered to be the graphical display of the system. The advantage of GERT as a visual aid for this system is severely reduced because of the density of the arcs.
2. The number of nodes which represent the two waiting channel systems is given by the relation:

$$n = (w + 1)^2 + 2$$

where: n = number of nodes in GERT network and
$\quad\quad\ w$ = number of waiting spaces per channel.

This indicates that, for very large systems (say, greater than five waiting spaces per channel), a GERT solution becomes impractical. Merely to itemize the transmittances of the system would become an extremely laborious task.

A GERTS approach to this problem, however, will eliminate these objections.

Consider the GERTS model shown in Fig. 11.16. Node 2 represents

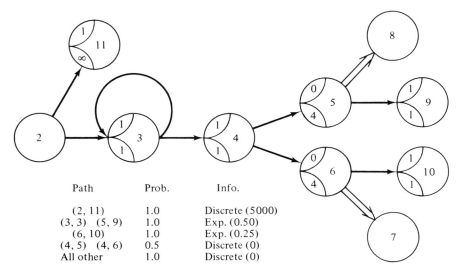

Path	Prob.	Info.
(2, 11)	1.0	Discrete (5000)
(3, 3) (5, 9)	1.0	Exp. (0.50)
(6, 10)	1.0	Exp. (0.25)
(4, 5) (4, 6)	0.5	Discrete (0)
All other	1.0	Discrete (0)

Figure 11.16. GERTS model of the T-type intersection.

the start of the network and node 11 represents the terminal node of the study. The path (2, 11) which takes 5000 time units regulates the length of the simulation. The self-loop on node 3 generates the arrival of cars to the system. The interarrival time is distributed with an exponential distribution whose mean is 0.5. Node 3 is also defined as a mark node to allow the collection of total time in the system statistics. The branching from node 4 represents the likelihood of cars desiring to make right and left turns. In this case there is an equal probability of turning right or left. Queue nodes 5 and 6 represent the queueing of cars desiring to make right and left turns. The time between mergings is distributed according to an exponential distribution and is represented by paths $(5 \rightarrow 9; 6 \rightarrow 10)$. For those desiring to make a right turn, the mean time is 0.25, while it is 0.5 for those wishing to make a left turn. If we want to investigate a design where there are five positions available for those desiring to make each type of turn, then the maximum queue length on nodes 5 and 6 will be set to 4. The person waiting to turn will represent the fifth position. Queue node 5 overflows to node 8 representing left-hand turn blockages, and queue node 6 overflows to node 7 representing right-hand turn blockages. Statistics on the time between realizations of nodes 8 and 9 are collected to determine the time between blockages. Total time in the system is collected at nodes 9 and 10 as the time for a car to move there from node 3. The statistics for the GERTS run can be summarized as follows:

Node	Mean	Standard Deviation	No. of Obs.	Min.	Max.	Node Type
7	1351.2	2099.1	3	137.3	3775.0	B
8	85.1	127.8	58	0.0	632.3	B
9	0.8987	1.0214	4825	0.0002	12.6011	I
10	0.3249	0.3475	4939	0.0000	4.0024	I

	5	6
Average Number in Queue	0.3833	0.0793
Average Busy Time	0.4846	0.2416

The output tells us that there are three right-hand turn blockages and 58 left-hand turn blockages during the simulation. The mean time intervals between blockages are 1351.2 and 85.1, respectively. The average time in the system for someone wanting to turn right is 0.3249, and it is 0.8987 for someone wanting to turn left. Realizing that the average number in the system is equal to the average number in queue plus the average busy time for this problem, we see that the average number of cars desiring to turn left at any time is 0.8679 and there are 0.3209 cars on the average wanting to turn right.

By varying various parameters of the GERTS model in Fig. 11.16, the design engineer can test various intersection configurations to determine their performance.

11.3 Case Study—A Model of Student Progress at Virginia Polytechnical Institute*

This case study attempts to show the applicability of GERTS to the modeling of a socioeconomic environment. This case will be devoted to a study of matriculation of students through V.P.I. The case was taken from a study by Burgess [2] who also studied the registration procedure at V.P.I. using GERTS.

11.3.1 GERTS model of the university. A GERTS model will be developed that graphically shows the movement of flow of students into, through, and out of the University. This network model is formulated in order to obtain an estimate of the probability that a student will end up in a certain state, given that he applied for admission to the University after graduating from high school. Estimates of the number of students in each terminal state will

---

*The material in this section is adapted from Reference 2 and is used with the permission of Mr. Burgess.

be established based on the number of high school graduates who apply for admission to the University. In addition, the average time to reach each state is also estimated.

The student progress model will be considered from two levels. Student progress model A will consider an entire entering freshman class, and student progress model B will consider only freshman students in the engineering curriculum.

STUDENT PROGRESS MODEL A

If we assume that the end objective of the student is a terminal degree, then a simple GERT network would be as shown in Fig. 11.17. Each branch

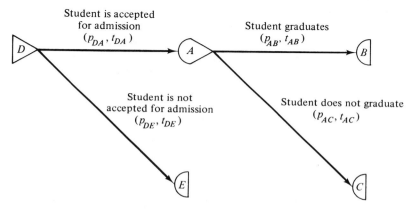

Figure 11.17. Student progress model A-1.

in Fig. 11.17 represents a possible activity that may take place, given that a student applies for admission to the University. For this network it is assumed that if a student is accepted for admission he will enroll in the University. Associated with each branch or activity of the network is (1) the probability, p, that the branch is taken and (2) the time, t, that is required to traverse the branch that is taken. For example, if we assume 50% of college students graduate and it takes exactly four years, then $p_{AB} = 0.50$ and $t_{AB} = 4$ years. The time needed to graduate, t_{AB}, might more accurately be represented by a random variable since one study has revealed that undergraduates earn a bachelor's degree anywhere from two and one-half years to ten years after freshman entrance [14].

To further describe student progress model A we may add enrollment, registration, and academic work activities to the network as shown in Fig. 11.18. Nodes labeled C, B, and E represent end or sink nodes which designate final states for the students. These nodes signify, respectively, (1) students

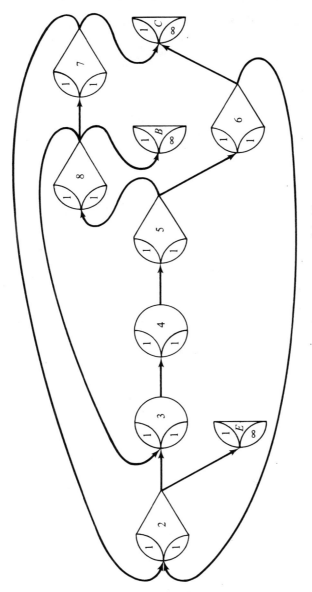

Figure 11.18. Student progress model *A*-2.

Table 11.6. ACTIVITY DESCRIPTION FOR STUDENT PROGRESS MODEL *A*-2

Start Node	End Node	Activity Description
2	3	Student is accepted for admission.
2	E	Student is not accepted for admission.
3	4	Registration.
4	5	Academic work.
5	6	Student is dismissed for failing to meet QCA requirements.
5	8	Student meets QCA requirements.
6	C	Student does not graduate.
6	2	Reapply for admission.
7	C	Student does not graduate.
7	2	Reapply for admission.
8	B	Student graduates.
8	3	Register for next quarter.
8	7	Student leaves school.

who do not graduate, (2) students who graduate, and (3) students who are not accepted for admission. Table 11.6 describes the network activities.

This model further illustrates the characteristics of GERT networks. To explain the network, consider node 5 which occurs after the academic work activities. This node has a probabilistic output side indicating that the student may or may not meet quality credit average (QCA) requirements. Feedback into the system is also incorporated in the model. For example, a student may not meet the QCA requirements (branch 5 to 6) and later reapply for admission (6 to 2). Continuing students (8 to 3) may be thought of as endogenously generated inputs to the system. Students also often leave school for various other reasons (8 to 7), and some come back to school (7 to 2) after a time delay which may be represented by a random variable.

Another student progress model, modified somewhat from Fig. 11.18, is shown in Fig. 11.19. The progress of a student from the time of application to the University is shown. For this network, assumptions were made in accordance with Werdelin [19], who hypothesized that the number of students in the system depends on three factors:

> (1) Only a certain part of the students remain in the system from one year to the next, while others drop out without having completed school or leave after graduation; (2) of those who remain in school, only some are promoted to the next grade, while others repeat their previous one; and (3) each year a number of students enter the school system.

Of the enrollment patterns that may occur after admission, that is, continuing students and returning students, only the former case is represented in the network diagram. The latter case could be considered, but for

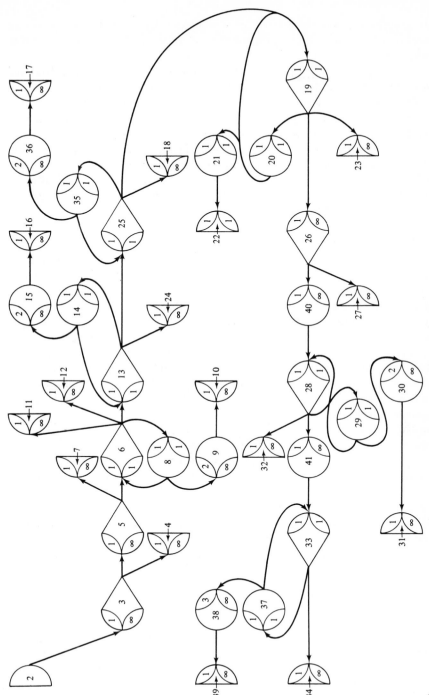

Figure 11.19. Student progress model *A*-3.

this analysis the activity is left out as suggested by Werdelin [19] who has found that for nearly all cases the number of students who return to school is negligible.

The network in Fig. 11.19 has a few characteristics that need explanation. All nodes that have no output branches are sink nodes, e.g., node number 16. The realization of any sink node represents one student who terminates at that node. Table 11.7 gives a complete verbal description of each activity and the values that were assigned to the probabilistic nodes.

This network also offers the capability of an information channel. For example, if the branch from node 8 to node 9 (representing the repeat of the freshman year by the student) is traversed two times, the activity from node 9 to node 10 is taken, which implies that the student is expelled because he has already repeated his freshman year one time before.

STUDENT PROGRESS MODEL *B*

Figure 11.20 shows a GERTS network model of the progress or flow of entering freshmen in the engineering curriculum. The model is formulated in order to predict the possible states in which a student may terminate, given that he entered the University as a freshman in engineering (activity 2 to 3). The average time that was required to reach each possible state in the system will also be predicted by the GERT simulation program.

Values for the probabilistic nodes were adapted from Koenig [7] and others who present student transition tables which show the probability of a student moving from one state to another. For example, given that a student is a freshman in engineering, the probability is that he will be a freshman in engineering, natural science, business, or other curriculum during the next year. Many other researchers have investigated the possible transitions students may make within a system of this type [1, 4, 9, 11, 14, 19, and 20].

Each node in Fig. 11.20 which does not have activities emanating from its output side is a terminal state or sink node. For example, in terms of the network in Fig. 11.20, assume the student is at node number 4, which represents the beginning of the sophomore year. Table 11.8 gives the state and the probability of being in that state, given that the student had achieved node number 4. Other possible paths that each student may take are easily followed in the figure. Table 11.9 gives a verbal description of the network activities.

11.3.2 Analysis of the GERTS models of the university. Since the main purpose of this study is to formulate educational network models and show the utility of GERTS as an analysis tool, general data-collection procedures were not employed. The values assigned to the probabilistic nodes in the network were established through conversation with the personnel in the

Table 11.7. ACTIVITY DESCRIPTION FOR STUDENT PROGRESS *A*-3

Start Node	End Node	Param. No.	Dist. Type	Count Type	Prob.		Activity Description and Number
2	3	3	1	0	1.000	0	Apply for admission
3	5	3	1	0	0.650	0	Registrar accept student
3	4	3	1	0	0.350	0	Registrar reject student
5	6	3	1	0	0.650	0	Student accept admission
5	7	3	1	0	0.350	0	Student reject admission
6	13	1	1	0	0.810	0	Complete freshman year
6	11	1	1	0	0.070	0	Grades—flunk out
6	12	1	1	0	0.070	0	Drop out
6	8	1	1	0	0.050	0	Repeat freshman year
8	6	3	1	0	1.000	0	Feedback loop
8	9	3	1	0	1.000	0	Information flow
9	10	3	1	0	1.000	0	Freshman repeat—flunk
13	25	1	1	0	0.800	0	Complete sophomore year
13	14	1	1	0	0.100	0	Repeat sophomore year
13	24	1	1	0	0.100	0	Leave after sophomore year
14	13	3	1	0	1.000	0	Feedback loop
14	15	3	1	0	1.000	0	Information flow
15	16	3	1	0	1.000	0	Sophomore repeat—flunk
19	26	1	1	0	0.700	0	Graduate with BS
19	20	1	1	0	0.200	0	Repeat senior year
19	23	1	1	0	0.100	0	Leave during senior year
20	19	1	1	0	1.000	0	Feedback loop
20	21	3	1	0	1.000	0	Information flow
21	22	3	1	0	1.000	0	Senior repeat—flunk
25	19	1	1	0	0.750	0	Finish junior year
25	35	1	1	0	0.150	0	Repeat junior year—1 quarter
25	18	1	1	0	0.100	0	Leave after junior year
26	27	3	1	0	0.930	0	Employment or military
26	40	3	1	0	0.070	0	Begin on MS degree
28	32	1	1	0	0.650	0	Finish MS
28	29	1	1	0	0.290	0	Second year on MS
28	41	1	1	0	0.060	0	Begin on PhD work
29	28	3	1	0	1.000	0	Feedback loop
29	30	3	1	0	1.000	0	Information flow
30	31	3	1	0	1.000	0	Not finish MS in two years
33	37	1	1	0	0.700	0	Start next year on PhD
33	34	1	1	0	0.300	0	Finish PhD
35	25	3	1	0	1.000	0	Feedback loop
35	36	3	1	0	1.000	0	Information flow
36	17	3	1	0	1.000	0	Junior repeat—flunk
37	33	3	1	0	1.000	0	Feedback loop
37	38	3	1	0	1.000	0	Information flow
38	39	3	1	0	1.000	0	Not finish PhD in 3 years
40	28	3	1	0	1.000	0	
41	33	3	1	0	1.000	0	

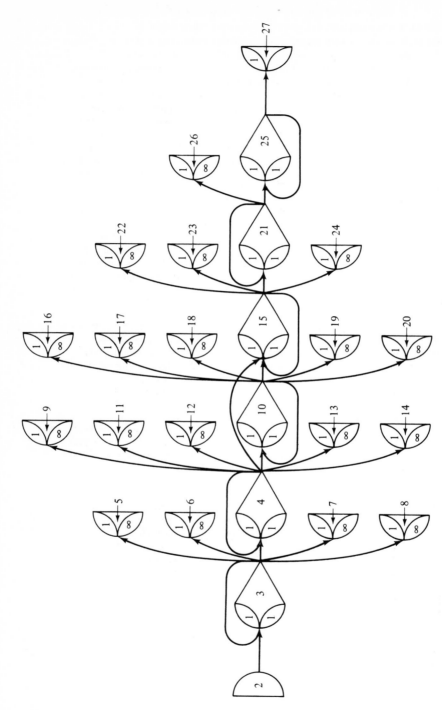

Figure 11.20. Student progress model *B*.

476

Table 11.8. ENGINEERING STUDENT TRANSITIONS FROM NODE 4

Probability	Node Number	Description
0.19	9	Leave school
0.05	4	Repeat sophomore year in engineering
0.48	10	Begin junior year in engineering
0.01	15	Begin senior year in engineering
0.03	11	Begin junior year in natural science
0.04	12	Begin junior year in business
0.04	13	Begin sophomore year in some other curriculum
0.16	14	Begin junior year in some other curriculum

Table 11.9. ACTIVITY DESCRIPTION FOR STUDENT PROGRESS MODEL *B*

Start Node	End Node	Param. No.	Dist. Type	Count Type	Prob.		Activity Description and Number
2	3	2	1	0	1.000	0	Enter freshman engineer
3	4	1	1	0	0.550	0	Complete freshman year
3	8	1	1	0	0.220	0	Leave school—first year
3	7	1	1	0	0.160	0	Transfer—other sophomore
3	3	1	1	0	0.030	0	Repeat freshman year
3	5	1	1	0	0.030	0	Transfer—business sophomore
3	6	1	1	0	0.010	0	Transfer—other freshman
4	10	1	1	0	0.480	0	Complete sophomore year
4	9	1	1	0	0.190	0	Leave school—second year
4	14	1	1	0	0.160	0	Transfer—other junior
4	4	1	1	0	0.050	0	Repeat sophomore year
4	12	1	1	0	0.040	0	Transfer—business junior
4	13	1	1	0	0.040	0	Transfer—other sophomore
4	11	1	1	0	0.030	0	Transfer-natural science junior
4	15	1	1	0	0.010	0	Skip junior year
10	15	1	1	0	0.760	0	Complete junior year
10	20	1	1	0	0.080	0	Leave school—third year
10	19	1	1	0	0.070	0	Transfer senior
10	10	1	1	0	0.050	0	Repeat junior year—1 quarter
10	16	1	1	0	0.020	0	Transfer—natural science senior
10	17	1	1	0	0.010	0	Transfer—business senior
10	18	1	1	0	0.010	0	Transfer—other junior
15	24	1	1	0	0.550	0	Graduate—BS in Engineering
15	15	1	1	0	0.320	0	Repeat senior year—1 quarter
15	21	1	1	0	0.070	0	Begin work on MS degree
15	22	1	1	0	0.050	0	Begin MS degree in B. Ad.
15	23	1	1	0	0.010	0	Transfer—other senior
21	26	1	1	0	0.620	0	Graduate—MS in Engineering
21	21	1	1	0	0.310	0	Second year on MS degree
21	25	1	1	0	0.070	0	Begin work on PhD
25	25	1	1	0	0.710	0	Start next year work—PhD
25	27	1	1	0	0.290	0	Graduate—PhD in Engineering

Registrar's Office, Admissions Office, College of Engineering Office, and the Industrial Engineering Department at Virginia Polytechnical Institute. Data also were adapted from studies made by Zinter [23], Brown and Savage [1], Suslow [17], Oliver [12], Max [10], and Koenig et al. [8].

STUDENT PROGRESS MODEL *A*

Consider first the analysis of student progress model *A*. Table 11.7 contained a verbal description of the network under the section entitled "Activity Description." The numbers listed under the "Param. No." column refer to the time parameter that is associated with a specified branch of the network. For example, the branch from node 6 to node 8 utilizes parameter set 2 with a distribution type 1. Since the type-1 distribution specifies a constant time value and the first value in parameter set 2 is 1.0, the activity will require 1.0 time unit to be completed.

Table 11.10 gives a description of each node and the corresponding attributes of that node. For example, node number 4 requires that any one activity incident to node 4 be realized before the node is realized. The number of times that activities incident to node 4 must be realized in order to realize the node a second time is equal to 9999. (The number 9999 is used in the computer to represent infinity.) The output type is deterministic (*D*), which implies that all activities emanating from node 4 will be scheduled when the node is realized. Also, node 4 is a statistics node with calculations based on all (*A*) realizations of the node rather than only the first (*F*) realization of the node.

The final results of the simulation (Table 11.11) give the performance measures that have been discussed previously.

To explain further the results of the simulation of student progress model *A*, consider node number 27 which signifies the student who graduates from the University and begins employment or military service. Table 11.11 shows that node number 27 was realized 20.89% of the time based on 5615 simulations of the network. In other words, of the total number of students who apply to the University in a given year, approximately 20% will graduate and begin employment after having been at the University for an average of 13.74 quarters. Final results for the remaining statistics nodes may be easily understood if we refer to Table 11.11 which describes the activities that terminate at the sink nodes.

The results of two other simulations of this model are shown in Tables 11.12 and 11.13. Table 11.12 shows the effect on the system of lowering the University's admission standards. This is done by changing the probability for branch 3 to 5 from 0.65 to 0.75 and branch 3 to 4 from 0.35 to 0.25. In addition, the probability of a student dropping out at the end of each year of the undergraduate levels was assumed to increase by 2% as a result of

Table 11.10. NODE DESCRIPTION FOR STUDENT PROGRESS MODEL *A*

Node	Number Releases	Number Releases for Repeat	Output Type	Removal of Event	Statistics Descriptor
2	1	9999	D		
3	1	9999	P		
4	1	9999	D		A
5	1	9999	P		
6	1	1	P		
7	1	9999	D		A
8	1	1	D		
9	2	9999	D		
10	1	9999	D		A
11	1	9999	D		A
12	1	9999	D		A
13	1	1	P		
14	1	1	D		
15	2	9999	D		
16	1	9999	D		A
17	1	9999	D		A
18	1	9999	D		A
19	1	1	P		
20	1	1	D		
21	2	9999	D		
22	1	9999	D		A
23	1	9999	D		A
24	1	9999	D		A
25	1	1	P		
26	1	9999	P		
27	1	9999	D		A
28	1	1	P		
29	1	1	D		
30	2	9999	D		
31	1	9999	D		A
32	1	9999	D		A
33	1	1	P		
34	1	9999	D		A
35	1	1	D		
36	2	9999	D		
37	1	1	D		
38	3	1	D		
39	1	9999	D		A
40	1	9999	D		A
41	1	9999	D		A

Table 11.11. FINAL RESULTS FOR STUDENT PROGRESS MODEL A (5615 SIMULATIONS)

Node	Prob./Count	Mean	Std. Dev.	Min.	Max.	Node Type
4	0.3507	0.0	0.0	0.0	0.0	A
7	0.2237	0.0	0.0	0.0	0.0	A
10	0.0011	6.0000	0.0	6.0000	6.0000	A
11	0.0285	3.2250	0.7927	3.0000	6.0000	A
12	0.0349	3.1684	0.6922	3.0000	6.0000	A
16	0.0036	9.0000	0.0	9.0000	9.0000	A
17	0.0068	12.4737	1.1086	12.0000	15.0000	A
18	0.0410	9.7565	1.3930	9.0000	15.0000	A
22	0.0094	18.7924	1.4592	18.0000	24.0000	A
23	0.0385	14.1528	3.0331	12.0000	24.0000	A
24	0.0383	6.4744	1.1349	6.0000	12.0000	A
27	0.2089	13.7417	2.6690	12.0000	24.0000	A
31	0.0014	18.7500	1.3887	18.0000	21.0000	A
32	0.0121	17.0294	2.9064	15.0000	27.0000	A
34	0.0007	22.5000	3.0000	18.0000	24.0000	A
39	0.0005	25.0000	1.7320	24.0000	27.0000	A
41	0.0012	17.5714	2.6992	15.0000	21.0000	A
40	0.0148	13.5181	2.7469	12.0000	24.0000	A

Table 11.12. RESULTS OF STUDENT PROGRESS MODEL A WITH ALTERED ADMISSION STANDARDS (5615 SIMULATIONS)

Node	Prob./Count	Mean	Std. Dev.	Min.	Max.	Node Type
4	0.2497	0.0	0.0	0.0	0.0	A
7	0.2639	0.0	0.0	0.0	0.0	A
10	0.0016	6.0000	0.0	6.0000	6.0000	A
11	0.0443	3.2169	0.7785	3.0000	6.0000	A
12	0.0431	3.1612	0.6778	3.0000	6.0000	A
16	0.0028	9.5625	1.2093	9.0000	12.0000	A
17	0.0075	12.2143	0.7820	12.0000	15.0000	A
18	0.0465	9.6322	1.3078	9.0000	18.0000	A
22	0.0110	18.8226	1.5526	18.0000	24.0000	A
23	0.0422	13.6962	2.5925	12.0000	21.0000	A
24	0.0504	6.4028	1.0553	6.0000	12.0000	A
27	0.2196	13.8987	2.7358	12.0000	24.0000	A
31	0.0014	18.7500	1.3887	18.0000	21.0000	A
32	0.0148	17.1687	2.7842	15.0000	24.0000	A
34	0.0007	21.7500	2.8723	18.0000	24.0000	A
39	0.0004	24.0000	0.0	24.0000	24.0000	A
41	0.0011	16.5000	2.5100	15.0000	21.0000	A
40	0.0173	13.4227	2.4530	12.0000	21.0000	A

Table 11.13. STUDENT PROGRESS MODEL *A* WITH ALTERED STUDENT ACCEPTANCE
PROBABILITY (5615 SIMULATIONS)

Node	Prob./Count	Mean	Std. Dev.	Min.	Max.	Node Type
4	0.3539	0.0	0.0	0.0	0.0	*A*
7	0.1671	0.0	0.0	0.0	0.0	*A*
10	0.0012	6.0000	0.0	6.0000	6.0000	*A*
11	0.0310	3.1552	0.6663	3.0000	6.0000	*A*
12	0.0365	3.1610	0.6777	3.0000	6.0000	*A*
16	0.0041	9.1304	0.6255	9.0000	12.0000	*A*
17	0.0075	12.4286	1.0625	12.0000	15.0000	*A*
18	0.0461	9.7992	1.4542	9.0000	15.0000	*A*
22	0.0107	18.8000	1.4474	18.0000	24.0000	*A*
23	0.0420	13.9703	2.9368	12.0000	24.0000	*A*
24	0.0426	6.4393	1.0979	6.0000	12.0000	*A*
27	0.2399	13.7572	2.6658	12.0000	24.0000	*A*
31	0.0012	18.4286	1.1339	18.0000	21.0000	*A*
32	0.0153	17.0930	2.8393	15.0000	27.0000	*A*
34	0.0005	24.0000	0.0	24.0000	24.0000	*A*
39	0.0004	24.0000	0.0	24.0000	24.0000	*A*
41	0.0009	18.0000	3.0000	15.0000	21.0000	*A*
40	0.0175	13.5000	2.6642	12.0000	24.0000	*A*

the change in the admission standards. For example, node 27 in Table 11.12 shows an increase over node 27 in Table 11.11 of approximately 1% in the number of students who graduate and begin employment.

Table 11.13 gives the effects of a 10% increase in the percentage of students who decide to accept their admission to V.P.I. This was accomplished by changing the branching probabilities for node 5. The probability for students who accept their admissions (node 5 and 6) was increased to 0.75, and the probability for those who refuse their admissions (node 5 to node 7) was decreased to 0.25. By comparing node 27 in Table 11.11 and Table 11.13, we see an increase of more than 3% in the percentage of students who graduate and begin employment as a result of the increase in student acceptance of admission.

Simulations of the type just presented have the power to allow educational planners to estimate and anticipate the effects of social and economic change in the area of enrollment projection.

STUDENT PROGRESS MODEL *B*

The GERTS program also was used to simulate the possible transitions a beginning freshman engineering student may make during the course of study at the University. The possible transitions were shown graphically in Fig. 11.20 and were described in Table 11.9. The results of the simulation

(Table 11.14) are presented in the same format as the student progress network *A*. For example, of the total number of students who began their freshman year in engineering, 0.4% received a Doctor of Philosophy degree in
engineering (node 27) after having spent an average of 25.5 quarters at the
University. Some students were able to accomplish this in only 18 quarters,
whereas some required as much as 36 quarters.

Table 11.14. RESULTS OF STUDENT PROGRESS MODEL *B* (1000 SIMULATIONS)

Node	Prob./Count	Mean	Std. Dev.	Min.	Max.	Node Type
5	0.0290	3.2069	0.7736	3.0000	6.0000	*A*
6	0.0100	3.0000	0.0	3.0000	3.0000	*A*
7	0.1490	3.1208	0.5918	3.0000	6.0000	*A*
8	0.2480	3.1694	0.6938	3.0000	6.0000	*A*
9	0.1100	6.4091	1.0342	6.0000	9.0000	*A*
11	0.0210	6.1429	0.6547	6.0000	9.0000	*A*
12	0.0300	6.1000	0.5477	6.0000	9.0000	*A*
13	0.0250	6.3600	0.9950	6.0000	9.0000	*A*
14	0.1120	6.2946	0.8968	6.0000	9.0000	*A*
16	0.0040	9.0000	0.0	9.0000	9.0000	*A*
17	0.0030	10.0000	1.7320	9.0000	12.0000	*A*
18	0.0050	9.0000	0.0	9.0000	9.0000	*A*
19	0.0180	9.8333	1.3827	9.0000	12.0000	*A*
20	0.0230	9.3913	1.0331	9.0000	12.0000	*A*
22	0.0180	13.1667	2.0934	9.0000	18.0000	*A*
23	0.0040	12.7500	1.5000	12.0000	15.0000	*A*
24	0.0650	13.4364	2.7834	9.0000	27.0000	*A*
26	0.0220	18.5454	4.4048	12.0000	33.0000	*A*
27	0.0040	25.5000	7.5498	18.0000	36.0000	*A*

One other simulation was made to show the effect of changes in the
branching probabilities for student progress model *B*. To show the effect on
the total system of an increase in the percentage of students who continue
in engineering after their freshman year, an increase of 10% was assumed for
freshmen who complete their freshman year (branch 3 to 4) and a corresponding decrease of 5% was used for freshmen who transfer to another
curriculum (branch 3 to 7) and freshmen who withdraw from school (branch
3 to 8). For example, node 24 in Table 11.15 (which represents students
who graduate in engineering) shows approximately a 31% increase over
the previous simulation in Table 11.14.

The effect of various other changes in the parameters associated with
the network are easily simulated using GERTS. Hence, the utility of using
GERTS to manipulate the variables in a model of a system for the purpose of
understanding, experimenting with, and predicting the behavior of the system
is seen.

Table 11.15. RESULTS OF STUDENT PROGRESS MODEL *B* WITH MODIFIED PROBABILITY
FOR FRESHMEN

Node	Prob./Count	Mean	Std. Dev.	Min.	Max.	Node Type
5	0.0300	3.2000	0.7611	3.0000	6.0000	*A*
6	0.0100	3.0000	0.0	3.0000	3.0000	*A*
7	0.1090	3.0826	0.4931	3.0000	6.0000	*A*
8	0.1960	3.1684	0.6922	3.0000	6.0000	*A*
9	0.1370	6.3723	0.9927	6.0000	9.0000	*A*
11	0.0250	6.1200	0.6000	6.0000	9.0000	*A*
12	0.0300	6.0000	0.0	6.0000	6.0000	*A*
13	0.0290	6.5172	1.1533	6.0000	9.0000	*A*
14	0.1170	6.2308	0.8028	6.0000	9.0000	*A*
16	0.0040	9.0000	0.0	9.0000	9.0000	*A*
17	0.0030	9.0000	0.0	9.0000	9.0000	*A*
18	0.0050	9.6000	1.3416	9.0000	12.0000	*A*
19	0.0200	9.9000	1.4105	9.0000	12.0000	*A*
20	0.0300	9.3000	0.9154	9.0000	12.0000	*A*
22	0.0220	13.2273	1.9984	9.0000	18.0000	*A*
23	0.0040	12.0000	0.0	12.0000	12.0000	*A*
24	0.2010	13.6269	2.9538	9.0000	27.0000	*A*
26	0.0240	18.6250	4.3320	15.0000	33.0000	*A*
27	0.0040	25.5000	7.5498	18.0000	36.0000	*A*

11.4 Case Study—The Development of Optimum Stopping Rules for Sampling Plans*

A number of GERT models of sampling were developed in Section 10.4. Now the effective use of one of these models will be discussed. Two fundamental questions arise in using Dodge's continuous sampling plans. These are: (1) how to select an optimum plan to begin with and (2) when to stop the inspection process if incoming quality deteriorates to an unacceptable level. Fry [5] proposed a stopping rule and a cost model for CSP-1 which will now be discussed.

A stopping rule permits stopping of the manufacturing process for analysis and repair when the incoming quality deteriorates to an unacceptable level. A cost model can be developed from which the optimal inspection plan can be selected under given input conditions. When the probability of a unit being defective changes from p_0 (normal process) to p_1 (out-of-control condition), it is said that the quality has shifted. A GERT model for the

---

*The material in this section was adapted from Reference 5 and is used with the permission of Mr. Fry.

detailing state is formulated as shown in Fig. 11.21. Let $q_0 = 1 - p_0$. From Mason's rule, we obtain:

$$W_{0,i}(c) = \frac{q_0^i}{1 - p_0 e^c(1 + q_0 + q_0^2 + \cdots + q^{i-1})}$$

$$P_{0,i} = W_{0,i}(c)\big|_{c=0} = 1$$

$$M_{0,i}(c) = W_{0,i}(c) = \frac{q_0^i}{1 - p_0 e^c(1 + q_0 + q_0^2 + \cdots + q_0^{i-1})} = \frac{q_0^i}{1 - e^c(1 - q_0^i)}$$

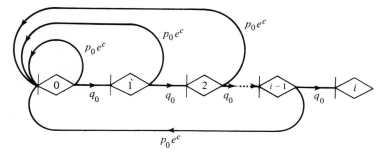

Figure 11.21. GERT model of the CSP-1 sampling plan.

The moment generating function can be expanded further to a power series:

$$M_{0,i}(c) = q_0^i + q_0^i(1 - q_0^i)e^c + q_0^i(1 - q_0^i)^2 e^{2c} + \cdots$$

$$P(x) = q_0^i(1 - q_0^i)^x = \text{the probability of } x \text{ defects in clearing}$$
$$\text{a sequence of } i \text{ units}$$

Let:

$N_c = $ critical number of defects which, if found while we attempt to clear i units, will cause the process to be stopped and investigated

Thus:

$$N_c = \min_{x=0,1,2,\ldots} \sum_{j=0}^{x} q_0^i(1 - q_0^i)^j$$

This stopping rule can be interpreted as: The process will be stopped as soon as the total number of defects found equals or exceeds N_c. Tables of N_c for various p_0's and α's (where α is the confidence level associated with stopping the process) can be found in Fry's paper [5]. Once the shift in quality occurs, the process is on a course to be stopped. The stopping cycle can be depicted as shown in Fig. 11.22.

The expected number of units passed under the sampling procedure in

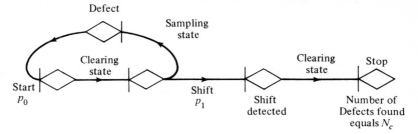

Figure 11.22. GERT model of a CSP-1 sampling plan.

the p_1 state before a defect is:

$$E(N_s) = \frac{1}{f p_1}$$

as shown earlier.

The stopping process in the detailing state can be modeled in a GERT network as shown in Fig. 11.23. Let $q_1 = 1 - p_1$. From Mason's rule, we obtain:

$$W_{0,N_c}(s) = \left(\frac{p_1 e^s}{1 - q_1 e^s}\right)^{N_c}$$

$$\because P_{0,N_c} = W_{0,N_c}(s)|_{s=0} = 1$$

$$\therefore M_{0,N_c}(s) = W_{0,N_c}(s) = \left(\frac{p_1 e^s}{1 - q_1 e^s}\right)^{N_c}$$

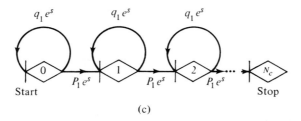

(c)

Figure 11.23. GERT model for stopping rules in CSP-1.

The expected number of units inspected until the process is stopped can be obtained from:

$$E(N_c) = \frac{d}{ds} M_{0,N_c}(s)|_{s=0}$$

$$= \frac{d}{ds}\left(\frac{p_1 e^s}{1 - q e^s}\right)^{N_c}\Bigg|_{s=0}$$

$$= \frac{N_c}{p_1^2}$$

Thus, the total expected number of units produced in the Z_1 state is:

$$E(N_{Z_1}) = \frac{1}{f p_1} + \frac{N_c}{p_1^2}$$

Let N be the total number of units produced in a stopping cycle, and let β be the probability of entering the Z_1 state. Then:

$$E(N_{Z_1}) = \beta N$$

$$\beta N = \frac{1}{f p_1} + \frac{N_c}{p_1^2}$$

or

$$N = \left(\frac{1}{\beta} \frac{1}{f p_1} + \frac{N_c}{p_1^2} \right)$$

The expected number of units produced in the Z_0 state can be found from:

$$E(N_{Z_0}) = (1 - \beta)N = \frac{1 - \beta}{\beta} \left(\frac{1}{f p_1} + \frac{N_c}{p_1^2} \right)$$

Thus, the expected number of units produced in a stopping cycle has been derived in both states Z_0 and Z_1.

The expected number of units inspected in the Z_1 state can be derived from:

$$E(N_{P_1(\text{inspect})}) = f\left(\frac{1}{f p_1} \right) + \frac{N_c}{p_1^2} = \frac{1}{p_1} + \frac{N_c}{p_1^2}$$

And then the expected number of units uninspected in the Z_1 state can be obtained from:

$$E(N_{Z_1(\text{uninspect})}) = E(N_{Z_1}) - E(N_{Z_1(\text{inspect})})$$

$$= \frac{1}{f p_1} + \frac{N_c}{p_1^2} - \frac{1}{p_1} - \frac{N_c}{p_1^2}$$

$$= \frac{1 - f}{f p_1}$$

These expected values are used to set up a cost model from which the optimal inspection plan can be selected under given input conditions. Let:

$C_1 = $ the unit cost if inspecting an item

$C_2 = $ the penalty cost of letting a defect slip through

C_3 = the cost of erroneously stopping the process

F_{Z_0} = the total fraction of units inspected when process is in control

Fry (5) formulated the following relation:

$$F_{Z_0} = \frac{f}{f + q_0^i(1 - f)}$$

Defining the expected total cost of a stopping cycle as *ETC*, we then have:

ETC = cost of inspecting units in Z_0 state

    + cost of uninspected defective units in Z_0 state

    + cost of inspecting units in Z_1 state

    + cost of uninspected defective units in Z_1 state

    + cost of erroneously stopping job in Z_0 state

$$= C_1\left(\frac{f}{f + q_0^i(1 - f)}\right)\left(\frac{1 - \beta}{\beta}\right)\left(\frac{1}{fp_1} + \frac{N_c}{p_1^2}\right)$$

$$+ C_2\left(\frac{q_0^i(1 - f)}{f + q_0^i(1 - f)}\right)p_0\left(\frac{1 - \beta}{\beta}\right)\left(\frac{1}{fp_1} + \frac{N_c}{p_1^2}\right)$$

$$+ C_1\left(\frac{1}{p_1} + \frac{N_c}{p_1^2}\right) + C_2\left(\frac{1 - f}{f}\right) + C_3(1 - \alpha)(1 - \beta)$$

This equation represents the cost model for a stopping cycle. It is by minimization of this cost that the optional plan for a given set of conditions can be determined.

REFERENCES

1. Brown, B. W., Jr. and I. R. Savage, "Methodological Studies in Educational Attendance Prediction." University of Minnesota, Department of Statistics, 1960.

2. Burgess, R. R., *GERT Models of the University*, Master's Thesis, Virginia Polytechnic Institute, June, 1970.

3. Evan, D. H., R. Hernson, and G. H. Weiss, "The Highway Merging and Queueing Problem." *Operations Research*, Vol. 12, No. 6 (1964), pp. 832–857.

4. "Ford Foundation Project in University Administration." Progress Report, March, 1969, Research Project No. 69-2.

5. Fry, J. H., "Selection, Stopping Conditions, and Response Characteristics of Dodge's Continuous Sampling Plans When Incoming Quality Deteriorates to an Unacceptable Level." Master's Thesis, Lehigh University, 1966.

6. Kendall, M. G., "Theory of Queues." *Annals of Mathematical Statistics*, Vol. 24, No. 3 (1953).

7. Koenig, H. E., "Systems Models and Their Application in Management Planning and Resource Allocation in Institutions of Higher Education." East Lansing, Michigan, November, 1969.

8. Koenig, H. E., M. G. Keeney, and R. Zemach, "A Systems Model for Management Planning, and Resource Allocation in Institutions of Higher Education." Final Report, Project C-518, National Science Foundation, Washington, D. C., 1968.

9. Lombaers, H. J. M., ed., *Project Planning by Network Analysis*. Amsterdam, North-Holland Publishing Co., 1969, pp. 147–153.

10. Max, Pearl, "How Many Graduate?" *College and University*, Vol. 45, No. 1 (Fall, 1969).

11. Mood, A. M., "Operations Analysis of American Education." Washington, D.C., U. S. Office of Education, September, 1966.

12. Oliver, R. M., "Models for Predicting Gross Enrollments at the University of California." Research Report No. 68-3 (August, 1968).

13. Roth, C. A., *A Complex Queueing Network As An Application of GERT*, Master's Thesis, Lehigh University, August, 1968.

14. Schure, A., "Educational Escalation Through Systems Analysis." Project ULTRA at New York Institute of Technology, *Audiovisual Instruction* (May, 1965), pp. 371–377.

15. Settles, F. S., "GERT Network Models of Production Economics." Unpublished PhD. Dissertation, Arizona State University, June, 1969.

16. Settles, F. S., "GERT Network Models of Production Economics," a paper presented at the National AIIE Convention, Houston, Texas, May, 1968.

17. Suslow, S., "Student Enrollment Predictions." University of California, Berkeley, California, 1968.

18. Thompson, W. J., F. S. Settles, and A. A. B. Pritsker, "An Analysis of a Manufacturing Process Using the GERT Approach." Research Report of NASA Contract NASA-12-2035, Arizona State University, June, 1969.

19. Werdelin, I., "A School Enrollment Model." *Educational and Psychological Interactions*, No. 12 (1966).

20. Werdelin, I., "Statistics for Educational Planning and Administration V: Planning a School System." *Educational and Psychological Interactions*, No. 26 (July, 1967).

21. Whitehouse, G. E. and C. A. Roth, "The Analysis of a T-Type Traffic Intersection Using GERT." presented at the *35th National Meeting of the Operations Research Society of America* (June, 1969).

22. Winters, William K., *Dynamod II in a Time Sharing Environment*, National Center for Education Statistics, U. S. Office of Education, Technical Note Number 45, October 23, 1967.

23. Zinter, Judith R., "Dynamod II Transition Probabilities for Student-Teacher Population's Growth Model." Technical Note Number 39, Washington, D.C., U. S. Office of Education, September 18, 1967.

FLOWGRAPH COMPUTER PROGRAM

This appendix contains the FORTRAN coding for a computer program which can be used to solve open flowgraphs. The use of this program is described in Chapter 6. The program was written by Mr. Donald McIlvain and is used with his permission.

APPENDIX

```
C       PROGRAM TO SOLVE FLOWGRAPHS
C
C       D R MCILVAIN     OCTOBER, 1970
C
        DIMENSION NS(99),NSP(40),A(99,99)
C
C       A(I,J) IS TRANSMITTANCE FROM I TO J
C       NS(I) IS ONE IF I IS A SOURCE NODE, ZERO OTHERWISE
C
        KEND=0
C
C       READ TITLE AND PRINT HEADING, ZERO MATRIX
C
     1  READ(5,100,END=95) NSP
        WRITE(6,200) NSP
        DO 2 I=1,99
        NS(I)=0
        DO 2 J=1,99
     2  A(I,J)=0.0
C
C       READ SOURCE NODES
C
        READ(5,101,END=95) NSP
        KSW=0
```

```
        DO 3 J=1,40
        I=NSP(J)
        IF(I.GT.0) NS(I)=1
      3 CONTINUE
C
C       READ DATA (END OF FILE OR BLANK CARD IS END OF PROBLEM)
C
     10 READ(5,102,END=20) I,J,T
        WRITE(6,201) I,J,T
        IF(I) 13,11,12
     11 IF(J) 13,21,13
     12 IF(J) 13,13,15
     13 WRITE(6,202)
     14 KSW=KSW+1
        GO TO 10
     15 A(I,J)=A(I,J)+T
        GO TO 10
C
C       IDENTIFY THE SOURCE NODES
C
     20 KEND=1
     21 IF(KSW.EQ.0) GO TO 23
     22 WRITE(6,204) KSW
        GO TO 90
     23 DO 28 J=1,99
        IF(NS(J).EQ.0) GO TO 28
        WRITE(6,206) J
     28 CONTINUE
C
C       REMOVE SELF LOOPS
C
        DO 49 J=1,99
        IF(A(J,J)) 41,49,41
     41 T=1.0-A(J,J)
        IF(T) 45,43,45
     43 WRITE(6,203) J
        KSW=KSW+1
        GO TO 49
     45 A(J,J)=0.0
        DO 46 I=1,99
     46 A(I,J)=A(I,J)/T
     49 CONTINUE
        IF(KSW.NE.0) GO TO 22
C
C       PROCESS TRANSMITTANCES FROM SINKS
C
        DO 69 I=1,99
        IF(NS(I).NE.0) GO TO 69
        DO 65 J=1,99
        T=A(I,J)
        IF(T) 61,65,61
```

```
    61  A(I,J)=0.0
        DO 62 L=1,99
    62  A(L,J)=A(L,J)+T*A(L,I)
    65  CONTINUE
    69  CONTINUE
C
C       PRINT OUT RESULTS
C
        WRITE(6,210)
        DO 89 I=1,99
        DO 89 J=1,99
        IF(A(I,J)) 82,89,82
    82  WRITE(6,211) I,J,A(I,J)
    89  CONTINUE
        WRITE(6,212)
C
C       IF END OF DECK QUIT, IF NOT GO BACK
C
    90  IF(KEND.EQ.O) GO TO 1
    95  WRITE(6,220)
        CALL EXIT
C
C       FORMAT STATEMENTS
C
   100  FORMAT(40A2)
   101  FORMAT(40I2)
   102  FORMAT(2I2,F10.0)
   200  FORMAT(' 1SOLUTION OF FLOWGRAPH'/'0',40A2/ '0INPUT DATA'/
      1  ' FROM  TO     TRANSMITTANCE')
   201  FORMAT(2I4,F16.4)
   202  FORMAT(10X,'PREVIOUS ENTRY INVALID')
   203  FORMAT(10X,'UNITY SELF LOOP ON NODE', 13)
   204  FORMAT('0',I6,' ERRORS — PROBLEM REJECTED')
   206  FORMAT(I8,' SPECIFIED AS A SOURCE')
   210  FORMAT(/'ORESULTING EFFECTIVE TRANSMITTANCES'/
      1  ' FROM  TO     TRANSMITTANCE')
   211  FORMAT(2I4,F16.8)
   212  FORMAT(' END OF PROBLEM')
   220  FORMAT('OEND OF DATA — END OF PROGRAM')
        END
```

INDEX

496

499